AF785282

A Theme Issue Honoring Professor Peter Proksch's 70th Birthday: Bioactive Compounds from the Ocean

A Theme Issue Honoring Professor Peter Proksch's 70th Birthday: Bioactive Compounds from the Ocean

Editors

Bin-Gui Wang
Haofu Dai

MDPI

Basel • Beijing • Wuhan • Barcelona • Belgrade • Novi Sad • Cluj • Manchester

Editors
Bin-Gui Wang
Chinese Academy of Sciences
Qingdao
China

Haofu Dai
Chinese Academy of Tropical
Agricultural Sciences
Haikou
China

Editorial Office
MDPI AG
Grosspeteranlage 5
4052 Basel, Switzerland

This is a reprint of articles from the Special Issue published online in the open access journal *Marine Drugs* (ISSN 1660-3397) (available at: https://www.mdpi.com/journal/marinedrugs/special_issues/QF0H3Z44JS).

For citation purposes, cite each article independently as indicated on the article page online and as indicated below:

Lastname, A.A.; Lastname, B.B. Article Title. *Journal Name* **Year**, *Volume Number*, Page Range.

ISBN 978-3-7258-1777-1 (Hbk)
ISBN 978-3-7258-1778-8 (PDF)
doi.org/10.3390/books978-3-7258-1778-8

Contents

About the Editors

Bin-Gui Wang

Bin-Gui Wang is currently a professor of marine natural products at the Institute of Oceanology, Chinese Academy of Sciences. He received his B.Sc. degree from Lanzhou University in 1986 and obtained several years of experience in natural antioxidants study at the Xi'an Oils and Fats Institute. In 1997, he completed his Ph.D. in Organic Chemistry at Lanzhou University and worked as a postdoctoral fellow at the Kunming Institute of Botany of the Chinese Academy of Sciences (1997–1999). After his studies, he worked as a DAAD fellow at Heinrich-Heine-Universität Düsseldorf in Germany (2000–2002, with Prof. Peter Proksch). His research involves studies of bioactive natural compounds from marine organisms such as marine algae, endophytic fungi, and deep-sea-sourced microbes. He has published approximately 250 research papers, reviews, and book chapters and is the co-inventor of 26 patents. He is a member of the Chinese Pharmaceutical Association and serves on the editorial boards of *Biochemical Systematics and Ecology*, *Chemistry & Biodiversity*, *Marine Drugs*, and *Marine Life Science & Technology*.

Haofu Dai

Hao-Fu Dai is currently a professor of natural products chemistry at the Institute of Tropical Bioscience and Biotechnology, Chinese Academy of Tropical Agricultural Sciences. He received his B.Sc. degree from Jiangxi Normal University in 1996 and then completed his Ph.D. at the Kunming Institute of Botany, Chinese Academy of Sciences (1996–2001). After his studies, he worked as a DAAD fellow at Heinrich-Heine-Universität Düsseldorf in Germany (2001–2003, with Prof. Peter Proksch). His research involves studies of bioactive natural compounds from tropical bio-resources such as tropical medicinal plants and marine endophytic fungi. He has published approximately 300 research papers, reviews, and book chapters and is the co-inventor of 50 patents. He is a member of the Chinese Pharmaceutical Association.

marine drugs

MDPI

Article

Metabolomics-Guided Discovery of New Dimeric Xanthones from Co-Cultures of Mangrove Endophytic Fungi *Phomopsis asparagi* DHS-48 and *Phomopsis* sp. DHS-11

Jingwan Wu, Dandan Chen, Qing Li, Ting Feng and Jing Xu *

Collaborative Innovation Center of Ecological Civilization, School of Chemistry and Chemical Engineering, Hainan University, Haikou 570228, China; 20081700110009@hainan.edu.cn (J.W.); 21220856000083@hainanu.edu.cn (D.C.); 22210710000026@hainan.edu.cn (Q.L.); 20081700210004@hainanu.edu.cn (T.F.)

* Correspondence: happyjing3@hainanu.edu.cn; Tel.: +86-898-6627-9226

Abstract: The co-culture strategy, which mimics natural ecology by constructing an artificial microbial community, is a useful tool for the activation of biosynthetic gene clusters (BGCs) to generate new metabolites, as well as to increase the yield of respective target metabolites. As part of our project aiming at the discovery of structurally novel and biologically active natural products from mangrove endophytic fungi, we selected the co-culture of a strain of *Phomopsis asparagi* DHS-48 with another *Phomopsis* genus fungus DHS-11, both endophyted in mangrove *Rhizophora mangle* considering the impart of the taxonomic criteria and ecological data. The competition interaction of the two strains was investigated through morphology observation and scanning electron microscopy (SEM), and it was found that the mycelia of the DHS-48 and DHS-11 compacted and tangled with each other with an interwoven pattern in the co-culture system. A new approach that integrates HPLC chromatogram, ^{1}HNMR spectroscopy, UPLC-MS-PCA, and molecular networking enabled the targeted isolation of the induced metabolites, including three new dimeric xanthones phomoxanthones L-N (**1–3**), along with six known analogs (**4–9**). Their planar structures were elucidated by an analysis of their HRMS, MS/MS, and NMR spectroscopic data and the absolute configurations based on ECD calculations. These metabolites showed broad cytotoxic activity against the cancer cells assessed, of which compounds **7–9** displayed significant cytotoxicity towards human liver cells HepG-2 with IC_{50} values ranging from 4.83 μM to 12.06 μM. Compounds **1–6** exhibited weak immunosuppressive activity against the proliferation of ConA-induced (T-cell) and LPS-induced (B-cell) murine splenic lymphocytes. Therefore, combining co-cultivation with a metabolomics-guided strategy as a discovery tool will be implemented as a systematic strategy for the quick discovery of target bioactive compounds.

Keywords: mangrove endophytic fungi; *Phomopsis asparagi*; *Phomopsis* sp.; co-culture; metabolomics; molecular network; xanthone dimers

Citation: Wu, J.; Chen, D.; Li, Q.; Feng, T.; Xu, J. Metabolomics-Guided Discovery of New Dimeric Xanthones from Co-Cultures of Mangrove Endophytic Fungi *Phomopsis asparagi* DHS-48 and *Phomopsis* sp. DHS-11. *Mar. Drugs* **2024**, 22, 102. https://doi.org/10.3390/md22030102

Academic Editors: Bin-Gui Wang and Haofu Dai

Received: 6 February 2024
Revised: 20 February 2024
Accepted: 21 February 2024
Published: 23 February 2024

1. Introduction

Mangrove-derived microorganisms in intertidal microenvironments are tolerant to numerous stresses, including highly saline or brackish water, high solar irradiation, and tidal gradients. The special ecological niche induced certain buds being evolved to encode unique biosynthetic genes that have the potential to generate unique bioactive metabolites [1,2], which increasingly attracted the attention of both pharmaceutical and natural product chemists [3,4]. A therapeutic molecule from the mangrove environment is now available in the market and exemplified by salinosporamide A (Marizomib®), which is a fermentation product of mangrove-derived *Salinispora tropica* CNB-392 [5]. Salinosporamide A is a highly potent inhibitor of the 20S proteasome that was granted an orphan drugs designation by the European Medicines Agency (EMA) for the treatment of multiple myeloma [6]. It may be expected that further promising candidates obtained from mangrove-derived microorganisms will follow in the future.

Chemical investigations of mangrove-derived microorganisms, especially endophytic fungi, have shown a sharp increase in recent years, and more than 250 endophytic fungi strains produced 79% (1090) by the mangrove-derived fungi originated 1300 new compounds [7]. Nevertheless, genomic sequencing techniques began to reveal that most mangrove endophytic fungi possess significantly more biosynthetic gene clusters (BGCs) than the number of compounds they produce under standard laboratory culture conditions [8–11]. Several approaches have been developed to aid in the activation of these dormant BGCs by employing modern biological or chemical techniques, such as genomics, transcriptomics, proteomics, and metabolomics [12–14]. In addition to the techniques that demand prior knowledge of the genome sequences of the studied microorganisms, several genome sequence-independent tools have been developed [14]. One of these approaches adopted to enhance the expression of silent BGCs in fungal cultures is co-cultivation, which is based on the premise that two or more microorganisms growing within a confined environment respond to environmental cues, which trigger the activation of BGCs to produce often new, bioactive secondary metabolites that cannot be otherwise detected in the corresponding monocultures [15,16]. There are several insights to be gained from co-culture experiments with mangrove endophytic fungi, as the activated secondary metabolites encompass a broad range of structural diversity, such as alkaloids [17,18], isocoumarins [19], xanthones [20], phenols [21,22], and peptides [23,24]. These successes indicate that co-cultivation is a rational and effective approach to induce the production of new metabolites, as well as to increase the yields of respective target metabolites with pharmacological potential. However, studies examining the co-culturing the pairing endophytic fungi within the same mangrove host plant to mimic the co-exist occurring interactions in naturally ecological situation are limited and cannot confirm a conclusion.

Metabolomics is the comprehensive analysis of small molecule metabolites in a biological system to reflect a phenotype response to chemical and biological stimuli, providing insight into the biological functions [25]. Many spectroscopic (NMR, MS, MS–MS) and chromatographic (HPLC, GC, GC–MS, LC–MS, TLC, etc.) methods are widely used for metabolomic analysis [26]. The use of HPLC chromatograms followed by LC−MS-based principal component analysis (PCA) has been a useful tool to distinguish whether metabolomic changes engendered in the co-culture and monoculture samples are based on their LC–MS profiles [27]. Nonetheless, reliable dereplication is often very challenging when solely using MS-1 data and available databases, especially at m/z values <600 [28]. ^{1}H NMR-based approaches have shown to be very effective in discriminating between alternative candidate structures in the dereplication process. It can provide comprehensive characteristic fingerprints since the co-cultivation can greatly affect the fingerprints of the extracts, based on structural features of specific chemotypes that can be easily determined from inspection of the chemical shift in the ^{1}H NMR spectrum [29]. Meanwhile, molecular networking is the main analysis tool to create structured networking based on the observation that structurally related molecules share similar MS/MS fragmentation patterns. It has emerged as an effective technique for dereplication and offers information about the chemical diversity in selected extracts before isolation [30,31]. Particularly, it is available on the Global Natural Products Social Molecular Networking "http://gnps.ucsd.edu (accessed on 4 February 2024)" open-access web-based platform [32]. Indeed, a variety of studies have been conducted by employing metabolomics analyses coupled with molecular networking to screen the chemical diversity of fungal extracts and rapidly target the isolation of novel natural products from complex mixtures in co-cultivation [27,33–35].

In continuing the search for structurally novel and biologically active natural products from mangrove endophytic fungi [36–45], a strain of *Phomopsis asparagi* DHS-48 isolated from the fresh root of *Rhizophora mangle* was found to be particularly productive with regard to the accumulation of a series of immunosuppressive chromones and cytochalasins [46–48]. With the goal of tapping the metabolic potential of this titled fungal strain, we initiated the co-culture fermentation process of well-characterized DHS-48 with other fungal strains. Considering the impart of taxonomic criteria and ecological data (sharing an ecological niche), another *Phomopsis* genus fungus DHS-11 endophyted in *R. mangle* was selected for pairing, which is known as a producer of isocoumarin and pyrone derivatives [49]. The modifications of the morphological features of mycelia by the co-culture on a small-scale on solid potato dextrose agar medium (PDA) plates were investigated through scanning electron microscopy (SEM). The EtOAc extraction of the upscaled fermentation of the co-culture on a rice solid-substrate medium were subjected to metabolomics analysis of the mono- and co-cultures to observe specific induction of unexpressed pathways, such as HPLC chromatogram, ^{1}HNMR spectroscopy, UPLC-MS-PCA, and molecular networking. A follow up metabolomics-guided isolation (Scheme 1) led to the discovery of three new dimeric xanthones phomoxanthones L-N (**1–3**), along with six known analogs (**4–9**). Herein, we report on the co-culture morphological features and in-depth metabolomics analyses, followed by the isolation and structure elucidation of the induced dimeric xanthones (**1–9**) (Figure 1), as well as their cytotoxic and immunosuppressive activities.

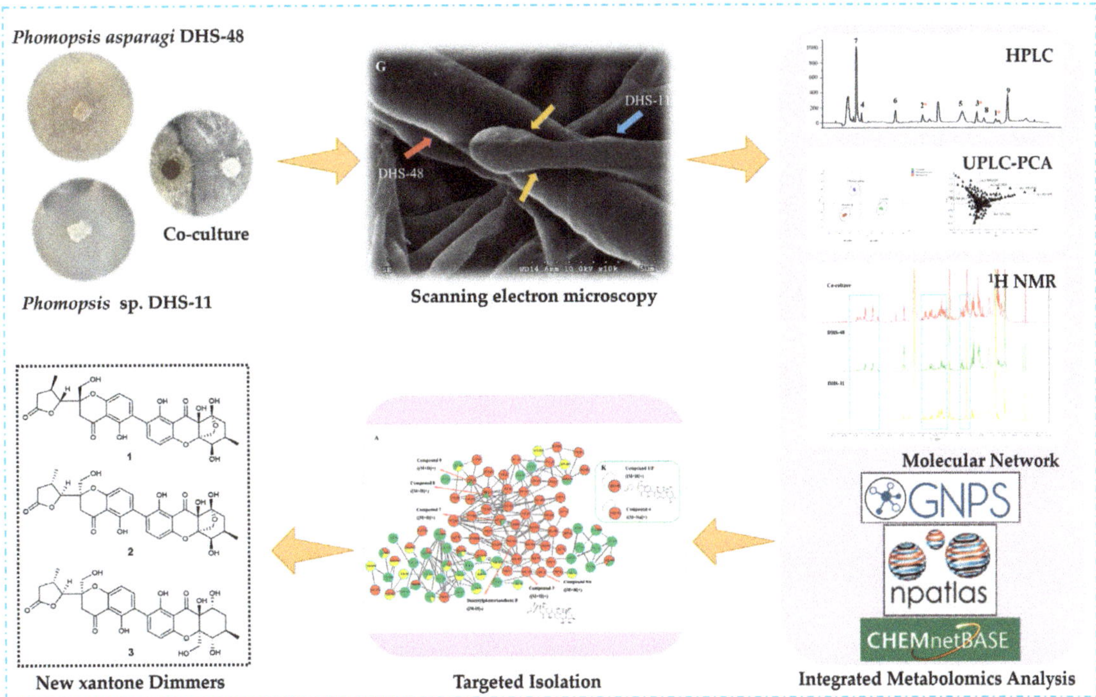

Scheme 1. Workflow of targeted isolation of new dimeric xanthones from co-cultures of mangrove endophytic fungi *Phomopsis asparagi* DHS-48 and *Phomopsis* sp. DHS-11 based on integrated metabolomics-guided discovery.

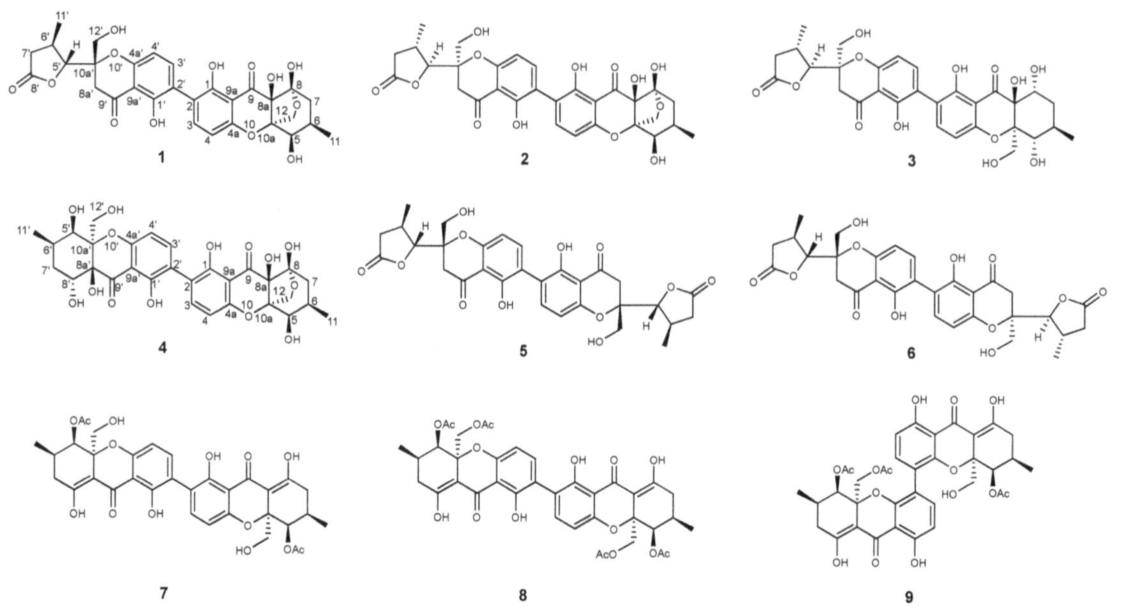

Figure 1. Structures of the isolated compounds **1–9**.

2. Results and Discussions

2.1. Morphology of Co-Culture Systems

Co-cultivation can induce morphological changes and colony growth, leading to an interaction with the competing colony that permits the morphology of the species involved in the interactions to be monitored with the unaided eye [50]. Solid media was preferred over liquid cultures for the screening of induction phenomena in fungal co-cultures [51]. On the PDA medium, the Phomopsis asparagi DHS-48 colony appeared yellow–brown in the middle, but a concentric annular circle appeared on the reverse side of the medium and reached a diameter of 60 mm on day 7 (Figure 2A). When *Phomopsis* sp. DHS-11 was inoculated on the PDA medium at 28 °C, initially, it was cottony and white, then turned to yellowish, pale filaments around the agar plug. The fungal colony grew to 90 mm in diameter and covered the whole medium on day 7 (Figure 2B). When the DHS-48 and DHS-11 strains were co-cultured together on the PDA medium for 7 days, a dark black precipitate was observed in the confrontation fronts of the two fungal colonies in the middle, morphologically different from the axenic cultures (Figure 2C). It was revealed that the DHS-48 strain could coexist well with DHS-11, and the growth rate of the DHS-11 was higher than that of the DHS-48, with a larger DHS-11 colony on the PDA medium. The SEM provided keen insight on the mycelial microstructure, as well as that of the mat without considering the composition; however, it was able to determine the hyphae and porosity dimensions [52]. The SEM observation showed that the monocultures of DHS-48 had cylindrical to obclavate hyphae with slight surface wrinkles with a diameter of 31.87 ± 0.88 μm (Figure 2D). The diameter of the DHS-11 mycelia was about 11.78 ± 1.08 μm and was similar microscopically to DHS-48 (Figure 2E). The mycelia of the DHS-48 and DHS-11 compacted and tangled with each other with an interwoven pattern in the co-culture system (Figure 2F,G). The distinguishable competitive interactions between DHS-48 and DHS-11 denote significant metabolic induction phenomena are likely to occur, indicating the exploitation potential of stimulated new secondary metabolites that are not produced under standard monoculture conditions.

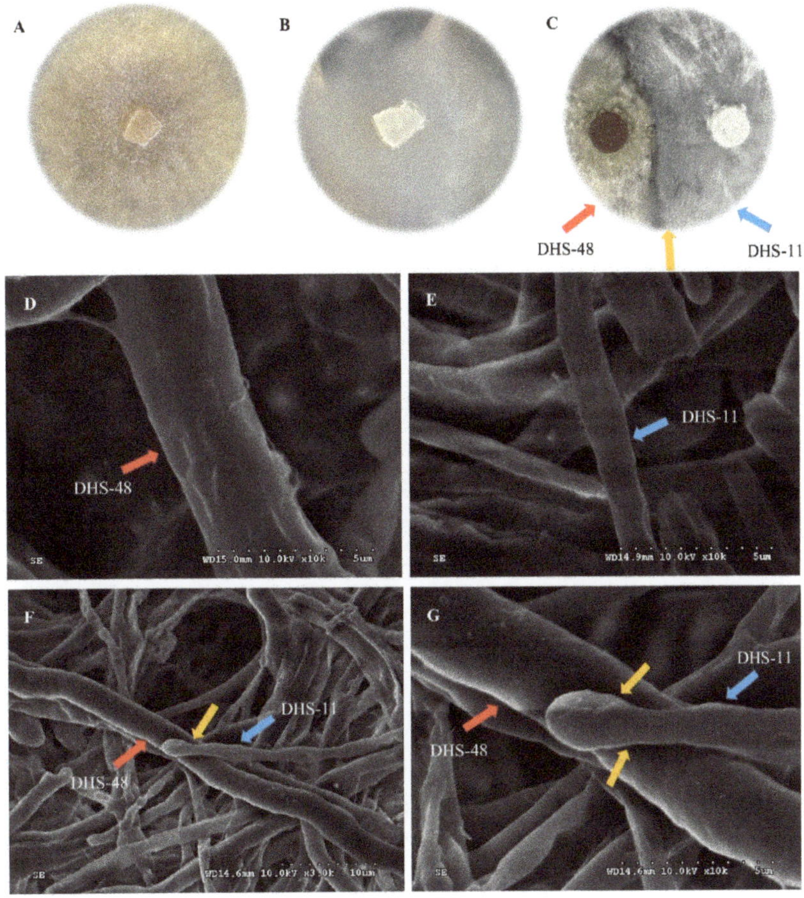

Figure 2. Mycelia morphology observation of *Phomopsis asparagi* DHS-48 and *Phomopsis* sp. DHS-11 in PDA medium's co-culture. Colony morphology of (**A**) DHS-48, (**B**) DHS-11, (**C**) DHS-48, and DHS-11 in co-culture. Scanning electron micrographs of (**D**) DHS-48, (**E**) DHS-11, (**F**) DHS-48, and DHS-11 in co-culture (**F**). Enlarged areas of (**G**). The red arrow points to strain DHS-48, the blue arrow points to strain DHS-11, and the yellow arrow points to the junction of the two species.

2.2. Metabolomics Analysis of Large-Scale Fermentation of Co-Cultures

An HPLC chromatogram with UV detection (HPLC-UV) of the crude EtOAc extract of a 30-day solid rice medium whole co-culture of DHS-48 and DHS-11 showed unexpected and significantly suppressed peaks and newly induced compounds **1–6** (Figure 3A), which were absent in the monocultures (Figure 3B,C). In addition, productions of the known metabolites **7–9** were substantially upregulated compared with their respective monocultures. Previous co-cultivation studies have provided evidence for the induction and suppression of biosynthetic pathways [53]. It is interesting to note that co-cultivation involving the suppression of BGCs or upregulation of known, functional metabolites mediates the competitive interaction by the other. The common characteristic UV absorption maxima around 203, 225, and 347 nm (UV data of compound **7**) showed a coincidence to related literature values [54], indicating that **1–9** should be dimeric xantone derivatives. Continuously, these differences were also supported by the fact that the ^{1}H NMR metabolic profile (Figure S1) of the EtOAc extracts showed several significant enhanced hydrogen resonances at 6.5–7.7 ppm (aromatic protons), 3.5–4.5 ppm (alkoxy protons), and 2.5–2.75 ppm (methine/methylene protons) compared with the control groups.

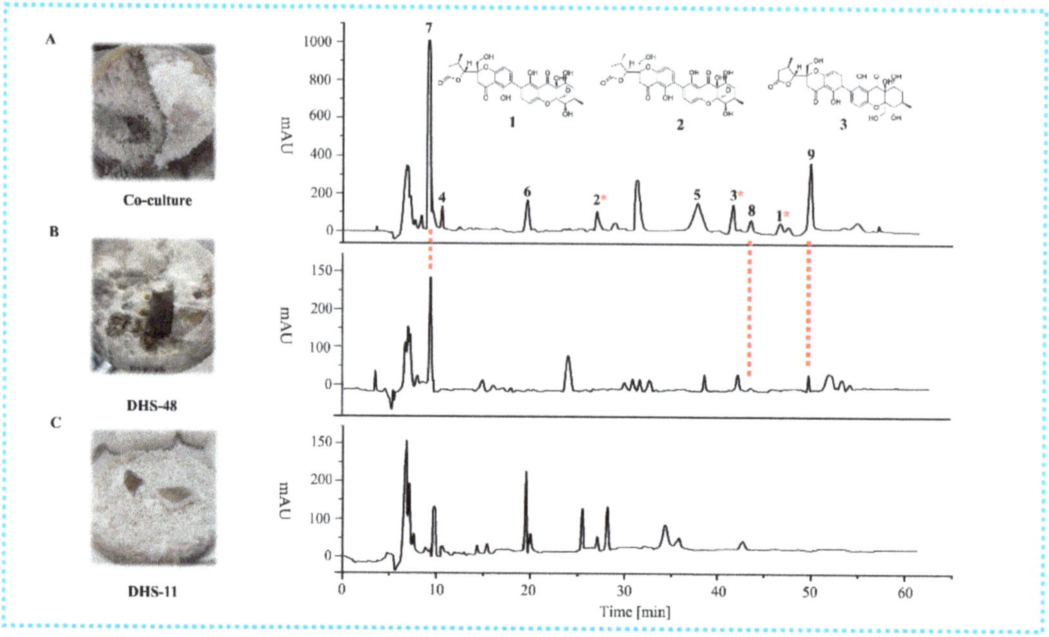

Figure 3. HPLC chromatograms of the EtOAc extracts deriving from (**A**) the whole co-culture of DHS-48 and DHS-11 and the monocultures of (**B**) DHS-48 and (**C**) DHS-11. * Compounds **1–3** in (**A**) represent the new compounds stimulated by the co-culture.

A combination of ultra-high-performance liquid chromatography (UHPLC) and high-resolution mass spectrometry (HRMS) was employed to analyze the metabolomes. An unsupervised principal component analysis (PCA) was conducted to compare the metabolic features and identify statistically significant differences. All the co-cultures and monocultures had five independent biological replicates, ensuring the reliability of the results [55]. The examination of the scores plot revealed clustering of the samples in three groups, indicating that the co-culture fingerprints did not overlap with the two corresponding monoculture clusters (Figure S2A,B). The separation meant that the datasets contained information that allowed for the discrimination of the chemical composition of the co-cultures from that of the monocultures, implying that microbial interactions modulated the biosynthetic pathways for the production of secondary metabolites.

In order to further investigate the metabolome and to obtain more information on the chemical diversity induced by co-culture, UPLC-ESI-MS/MS-based molecular networking was generated with the crude extracts of the fungal co-culture and the corresponding monocultures through the online Global Natural Products Social Molecular Networking (GNPS) platform. The generated molecular network (MN) (Figure 4A) consisted of 936 nodes in total, out of which 205 nodes were organized into 33 molecular clusters containing at least 2 nodes. As expected, co-cultivation significantly increased the chemical space of the fungi, indicated by the increased size of several molecular clusters with co-culture-induced derivatives (red-only nodes) of compounds (green or yellow nodes) produced in the axenic monocultures, such as clusters 1, 4, 6, 7, 12–14, 25, 27, and 30. Moreover, we observed seven clusters (8, 9, 15, 16, 24, 26, and 31) harboring only red nodes that were exclusively induced in the co-cultures. Unfortunately, no ion belonging to the latter clusters match to known compounds in the reference databases or could be purified in sufficient quantity to allow chemical identification. It is worth mentioning that the number of nodes in the network did not correspond exactly to the number of metabolites, as different adducts or charges of the same compounds could generate different nodes [56,57]. Additionally, it should be noted that a definite identification of any known compounds by MS/MS alone is not possible, as it is not possible to deduct the absolute configuration, and the potential presence of isomers cannot ultimately be ruled out [56]. Subsequent isolation efforts would be deemed worthwhile when the analysis of the GNPS clusters in question suggests the presence of potentially unknown compounds.

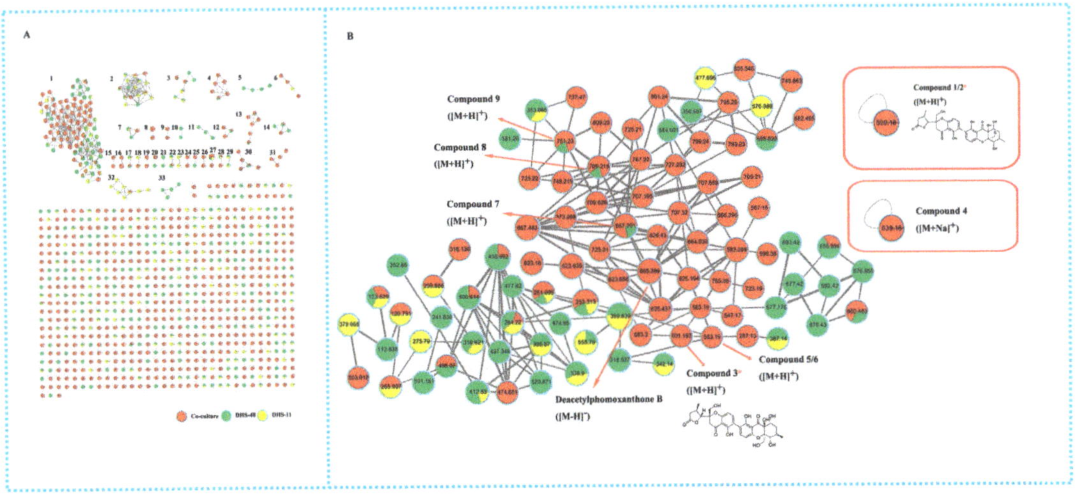

Figure 4. Molecular network (MN) of DHS-48 (green), DHS-11 (yellow), and their co-culture (red). (**A**) All 33 clusters containing at least two nodes are numbered. Thickness of the edges between nodes indicates the degree of similarity between their respective MS/MS spectra. (**B**) Enlarged cluster corresponding to xantone dimers (**1–9**) generated by GNPS, of which compounds **3**, **5–9** are annotated in cluster 1, while compounds **1**, **2**, and **4** were displayed as singletons in the MN. * represent the new compounds **1–3** stimulated by the co-culture.

An in-depth analysis of the MN of the crude extract of the co-culture extracts and respective monocultures (Figure 4B) allowed us to track the distribution of nodes according to xantone dimers down to cluster 1 and two singletons. The biggest cluster 1, of which 92 nodes comprising several m/z values were found in a mixed origin of both co-culture and monocultures, was annotated to a xantone dimer class of compounds. The first node m/z 667.2021 $[M+H]^+$ was determined to be the most abundant constituent dicerandrol A (**7**), previously described as a secondary metabolite isolated from the monoculture of DHS-48 [58], and dramatically up-regulated the amounts in the co-culture. It was purified and identified with the help of the MS/MS cleavage pattern (Figure S3A), as well as a comparison with previously reported NMR data. The molecular network quickly allowed us to establish the structural relationships of dicerandrol A and two additional nodes at m/z 583.1811 $[M+H]^+$ (compounds **5**/**6**) and m/z 709.2129 $[M+H]^+$ (compound **9**) within this cluster. They were purified and identified as the known phomopsis-H76 A (**5**), diaporthochromone B (**6**), and 12-O-deacetyl-phomoxanthone A (**9**) (Figure S3B,C). Node m/z at 751.2232 $[M+H]^+$ (compound **8**) was found to cluster as an adjacent node with m/z 709.2129, which is 42 Da larger than that of m/z 709.2129 and suggested the presence of an extra acetyl group. A comparison of its fragment ions (Figure S3D) with a manual search in the database confirmed the annotation of dicerandrol C (**8**). Another node detected at m/z 601.1916 $[M+H]^+$ (compound **3**) directly linked to node m/z 583.1811 has a mass difference of 18 Da, suggesting high structural similarity and a putative H_2O increment. The predicted molecular formulae $C_{30}H_{32}O_{13}$ of this node returned no hits with a dimeric xantone scaffold, hence it is putatively new. We tentatively proposed its structure based on the high MS/MS spectral similarity score (>0.65) as implemented in the GNPS-MN platform

(Figure S3D). Follow-up targeted purification of m/z 601.1916 $[M+H]^+$ (Figure S4A) allowed the assignment of the structure as new phomoxanthone N (**3**), confirming the cluster annotation. Other putative annotation was based on manual dereplication, including deacetylphomoxanthone B (m/z 665.389 $[M-H]^-$) and an isomer of curtisian C (m/z 709.21 $[M-H]^-$). In addition, several nodes clustered with the discriminatory nodes belonging to cluster 1 were m/z 623.635, m/z 626.43, m/z 664.939, m/z 725.21, and m/z 767.22, which could not be purified in sufficient quantity to allow chemical identification. These nodes identified by GNPS provided no hits, which strongly suggests the presence of additional analogs in this chemotype.

Other nodes, such as (singleton) at an average m/z of 599.1747 (compounds **1**/**2**) and m/z of 617.1865 (compound **4**), that had no matches in the GNPS database were attributed to dimeric xanthone derivatives. Compounds **1**, **2**, and **4** did not cluster with dimeric xanthone in cluster 1 because the product ions displayed a similarity score (cosine score) of <0.65. Subsequently, purification of the m/z 599.1759 and metabolites led to a pair of configurational xantone dimers named here Phomoxanthones L (**1**) and M (**2**), and m/z phomoxanthone D (**4**). In addition, we tabulated the structures of some nodes within clusters 1–10 annotated from molecular networking in Table S1.

Thereafter, from a careful metabolomics analysis (PCA, HPLC chromatogram, ^{1}HNMR spectroscopy, and molecular networking) of the co-cultivation, the targeted dimeric xantone derivatives including three new induced compounds were prioritized for isolation and chemical characterization.

2.3. Structure Elucidation of New Compounds

Phomoxanthone L (**1**) was isolated as a light-yellow amorphous powder, and its molecular formula was established as $C_{30}H_{30}O_{13}$ based on HRESIMS data (m/z 599.1755 $[M+H]^+$, calcd for $C_{30}H_{31}O_{13}$ 599.1765), indicating 16 indices of unsaturation. The ^{1}H and ^{13}C nuclear magnetic resonance (NMR) data (Table 1) of **1** in association with distortionless enhancement by polarization transfer (DEPT) and heteronuclear single quantum coherence (HSQC) spectrum showed the presence of a series of characteristic signals for two secondary methyls [δ_H 1.12, (d, J = 6.8 Hz), δ_C 14.7, q, 11-CH_3; 1.09, (d, J = 7.1 Hz), δ_C 19.6, q, 11′-CH_3], two methylenes [δ_H 2.09 (d, J = 13.8 Hz), δ_H 1.69 (dd, J = 13.8, 5.8 Hz), δ_C 37.1, t, CH_2-7; δ_H 2.17 (dd, J = 17.7, 9.0 Hz), δ_H 1.95 (dd, J = 17.7, 9.2 Hz), δ_C 35.6, t, CH_2-7′], two methines (δ_H 2.33, m, δ_C 29.2, d, CH-6; δ_H 2.61, m, δ_C 28.7, d, CH-6′), two oxygenated methylenes [δ_H 3.89 (d, J = 9.7 Hz), δ_H 3.83 (d, J = 9.7 Hz), δ_C 64.6, t, CH_2-12; δ_H 3.82 (d, J = 11.6 Hz), δ_H 3.71 (d, J = 11.6 Hz), δ_C 61.0, t, CH_2-12′], two oxygenated methines [δ_H 4.11 (d, J = 4.1 Hz), δ_C 76.4, d, CH-5; δ_H 4.30 (d, J = 3.4 Hz), δ_C 87.4, d, CH-5′], four adjacent olefinic methines [δ_H 7.47 (d, J = 8.5 Hz), δ_C 141.8, d, CH-3; δ_H 6.62 (d, J = 8.5 Hz), δ_C 107.4, d, CH-4; δ_H 7.36 (d, J = 8.5 Hz), δ_C 139.6, d, CH-3′; δ_H 6.52 (d, J = 8.5 Hz), δ_C 108.6, d, CH-4′], and 15 quaternary carbons (including two conjugated carbonyl at δ_C 199.76 (C-9) and δ_C 197.94 (C-9′)), and one ester carbonyl at δ_C 177.3 (C-8′)]. These spectroscopic data suggest that **1** was a heterodimer, with one monomeric moiety similar to a monomeric unit of **4** and the other identical to that of **5**, which were also isolated from the culture medium. The heteronuclear multiple bond correlation (HMBC) correlations (Figure 5) from H-3/C-2′ and H-3′/C-2 deduced the connection of two monomeric moieties should be connected via 2,2′- linkage and, consequently, permitted the construction of the planar structure of **1**. The relative configuration of **1** was established by analyzing the NOESY spectrum on the basis of the monomeric moieties. The NOE cross-peaks (Figure 6) between H-5/H-6, H-5/H_2-12, H-6/H_2-12, and H-6/H_a-7 revealed their co-facial relationship and were arbitrarily assigned as α-orientations, whereas those between H_3-11 and H_b-7 indicated that these protons have β orientations in the **4** moiety. The NOE interactions of H-5′/H_3-11′/H_b-7′ and H-5′/H_2-12′ suggested that they were positioned on the same face, and the correlations between H-6′ and H_a-7′ revealed that they were positioned on the other face, which was further confirmed by the similar coupling constants of these protons in **5**. Since compound **1** was co-isolated with **4**–**6** in our study, it was expected also to contain a monomer with

5*R*, 6*R*, 8*R*, 8a*S*, and 10a*R* absolute configuration on the basis of biogenetic considerations. Accordingly, four diastereomeric starting structures with (5*R*, 6*R*, 8*R*, 8a*S*, 10a*R*, 5′*R*, 6′*R*, 10a′*S*; 5*S*, 6*S*, 8*S*, 8a*R*, 10a*S*, 5′*S*, 6′*S*, 10a′*R*; 5*R*, 6*R*, 8*R*, 8a*S*, 10a*R*, 5′*S*, 6′*S*, 10a′*R*; 5*S*, 6*S*, 8*S*, 8a*R*, 10a*S*, 5′*R*, 6′*R*, 10a′*S*) configuration of **1** and its diastereomer **2** were subjected to calculation of the electronic circular dichroism (ECD) spectrum at the B3LYP/6-31 + g (d, p) level to compare with the experimental circular dichroism (CD) spectrum (Figure 7). Therefore, the absolute configuration of **1** was assigned to 5*R*, 6*R*, 8*R*, 8a*S*, 10a*R*, 5′*R*, 6′*R*, and10a′*S*, and the trivial name phomoxanthone L was ascribed.

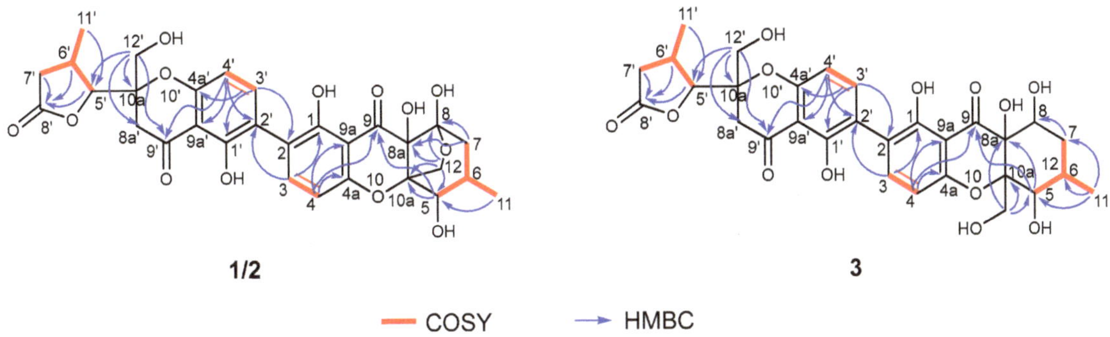

Figure 5. Key COSY and HMBC correlations of compounds **1–3**.

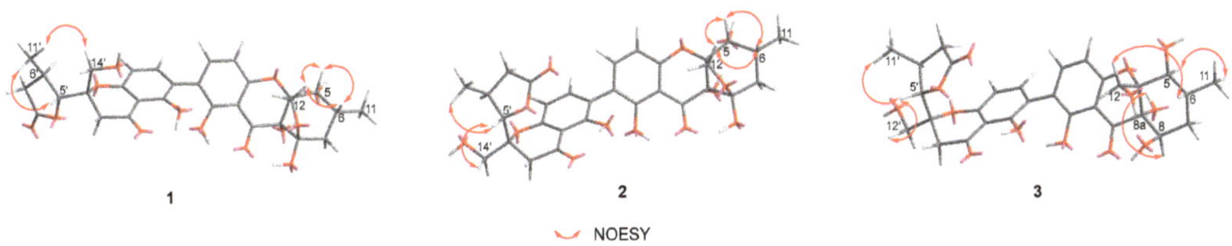

Figure 6. Key NOESY correlations of compounds **1–3**.

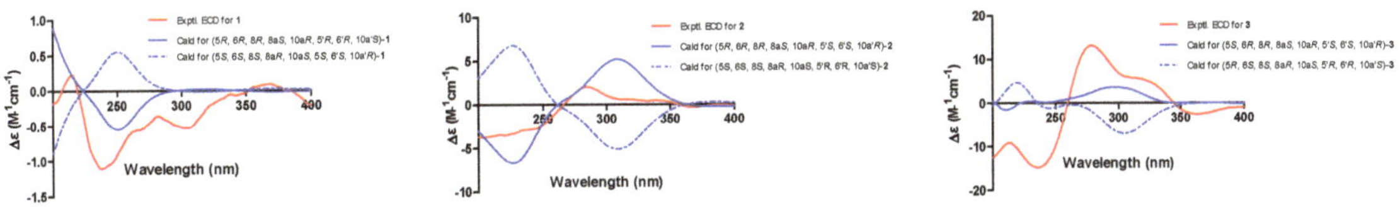

Figure 7. Experimental and calculated electronic circular dichroism (ECD) spectra of **1–3**.

Phomoxanthone M (**2**) possesses the same molecular formula of $C_{30}H_{30}O_{13}$ as **1**. Its NMR data (Table 1) and ^{1}H–^{1}H COSY, HMBC (Figure 5), and NOESY (Figure 6) correlations were very similar, and the planar structure was elucidated to be the same as that of **1**. The differences between the ^{1}H and ^{13}C NMR chemical shifts of CH-5′ [δ_H 4.43 (d, *J* = 4.0 Hz), δ_C 88.6, d, CH-5′], CH-6′ (δ_H 2.86, m, δ_C 31.0, d), CH_2-7′(δ_H 2.88, m, δ_H 2.28, m, δ_C 37.3, t) and 11′-CH_3 [1.23, (d, *J* = 5.9 Hz), δ_C 20.8, q] were larger than those between the other portions in **2** and those of **1**, thus suggesting that **2** might be a diastereomer of **1** at the β-methyl-γ-lactone ring. The absolute stereochemistry of phomoxanthone M (**2**) was ultimately established by means of comparing the experimental and calculated ECD spectra using TDDFT (Figure 7), and were determined as 5*R*, 6*R*, 8*R*, 8a*S*, 10a*R*, 5′*S*, 6′*S* and 10a′*R*. Thus, the structure of **2** was determined, and it was named phomoxanthone M.

Table 1. [1]H (400 MHz) and ^{13}C (100 MHz) NMR data of **1** and **2** in CD_3OD and **3** in DMSO.

Position	**1**		**2**		**3**	
	δ_C Type	δ_H (*J* in Hz)	δ_C Type	δ_H (*J* in Hz)	δ_C Type	δ_H (*J* in Hz)
1	159.9, C		161.5, C		158.9, C	
2	118.4, C		119.0, C		116.1, C	
3	141.8, CH	7.47, d, 8.5	143.5, CH	7.55, d, 8.4	139.6, CH	7.39, d, 8.5
4	107.4, CH	6.62, d, 8.5	108.4, CH	6.59, d, 8.4	107.9, CH	6.55, d, 8.5
4a	158.2, C		159.4, C		158.2, C	
5	76.4, C	4.11, d, 4.1	77.8, C	4.10, d, 3.8	72.3, C	4.16, s
6	29.2, C	2.33, m	30.5, C	2.30, m	27.4, C	2.11, m
7	37.1, CH_2	H_a 2.09, d, 13.8 H_b 1.69, dd, 13.8, 5.8	38.5, CH_2	H_a 2.10, t, 13.5 H_b 1.68, dd, 13.5, 5.8	31.2, CH_2	H_a 1.87, t, 10.9 H_b 1.41, d, 10.9
8	106.5, C		107.9, C		66.4, CH	4.23, s
8a	71.8, C		73.0, C		74.8, C	
9	199.76, C		199.27, C		194.9, C	
9a	108.6, C		107.0, C		106.85, C	
10a	81.9, C		83.2, C		84.4, C	
11	14.7, CH_3	1.12, d, 6.8	16.1, CH_3	1.12, d, 6.8	17.9, CH_3	0.98, d, 6.5
12	64.6, CH_2	H_a 3.89, d, 9.7 H_b 3.83, d, 9.7	66.0, CH_2	H_a 3.95, d, 9.6 H_b 3.78, d, 9.6	58.3, CH_2	H_a 4.12, d, 13.2 H_b 3.57, d, 13.2
1′	161.1, C		159.97, C		158.0, C	
2′	115.5, C		117.8, C		116.3, C	
3′	139.6, CH	7.36, d, 8.5	141.7, CH	7.45, d, 8.4	140.0, CH	7.40, d, 8.5
4′	108.6, CH	6.52, d, 8.5	108.0, CH	6.49, d, 8.4	106.67, CH	6.47, d, 8.5
4a′	156.2, C		160.4, C		158.8, C	
5′	84.7, CH	4.30, d, 3.4	88.6, CH	4.43, d, 4.0	86.6, CH	4.36, d, 4.2
6′	28.7, CH	2.61, m	31.0, CH	2.86, m	29.2, CH	2.80, m
7′	35.6, CH_2	H_a 2.17, dd, 17.7, 9.0 H_b 1.95, dd, 17.7, 9.2	37.3, CH_2	Ha 2.88, m Hb 2.28, m	35.9, CH_2	Ha 2.77, m Hb 2.23, m
8′	177.3, C		178.7, C		176.1, C	
8a′	37.3, CH_2	H_a 3.21, m H_b 2.92, m	37.8, CH_2	2.88, m	37.2, CH_2	H_a, 3.15, d, 17.1 H_b 2.91, d, 17.1
9′	197.94, C		198.46, C		197.5, C	
9a′	107.0, C		108.5, C		106.92, C	
10a′	83.6, C		84.9, C		84.0, C	
11′	19.6, CH_3	1.09, d, 7.1	20.8, CH_3	1.23, d, 5.9	20.1, CH_3	1.13, d, 6.3
12′	61.0, CH_2	H_a 3.82, d, 11.6 H_b 3.71, d, 11.6	63.1, CH_2	3.82, d, 2.1	62.3, CH_2	3.67, d, 4.1

The molecular formula of phomoxanthone N (**3**) was determined to be $C_{30}H_{32}O_{13}$ by the positive HRESIMS ion at m/z 601.1916 $[M+H]^+$ (calcd for $C_{30}H_{33}O_{13}$ 601.1921), implying 15 indices of unsaturation. A detailed analysis of the1D and 2D NMR spectral data of **3** revealed its feature of the asymmetric tetrahydroxanthone-chromanone heterodimer, and the planar structures of the two segments were the same as those of chromanone monomer of **2** and phaseolorin D, independently. In the same way, on the basis of the H-C long-range correlations (Figure 5) of H-3/C-2′ and H-3′/C-2 and the NMR values for the two linked benzene rings, it was illustrated that the connection of two monomeric moieties has the same 2,2′-linkage as **1** and **2**. Key NOESY correlations (Figure 5) of H-3/C-2′ and H-3′/C-2, as well as the NMR values for the two linked benzene rings illustrated the connection of two monomeric moieties has the same 2,2′-linkage as **1** and **2**. Key NOESY correlations (Figure 6) of H-5/H_3-11, H_3-11/H-8, and H-8/8a-OH suggesting that these protons are co-facial and have β-orientations, correlations between H-6/H_2-12, and the absence of correlations between H-8/H-6 and 8a-OH/H_2-12 indicate the α-orientations of protons H-6 and H_2-12. In addition, the almost identical NMR data and NOESY correlations as described for **2** confirmed the absolute configuration of the chromanone unit in **3**. ECD calculations (Figure 7) were further employed to determine the absolute configuration. As shown, the 5*S*, 6*R*, 8*R*, 8a*S*, 10a*R*, 5′*S*, 6′*S*, and 10a′*R* configuration was defined based on the calculated ECD spectrum in good accordance with the experimental curve.

2.4. Biological Evaluation of Isolated Compounds

All the isolated metabolites were evaluated for their cytotoxic and immunosuppressive activities according to the previously described methods [46–48]. Compounds **1–9** showed broad cytotoxic activity (Table 2). Amongst these, compounds **7–9** exhibited significant cytotoxicity towards human liver cells HepG-2 (IC_{50} values of 4.83 ± 0.22, 13.99 ± 1.13, and 12.06 ± 0.55 μM) and cervical cancer cells Hela (IC_{50} values of 18.96 ± 0.88, 23.42 ± 2.55, and 20.36 ± 1.99 μM). The results indicate that the tetrahydroxanthone moieties and hydroxyl groups attached at C-12 and C-12′ were the key functional architectures contributing to the cancer cells' proliferation inhibitory effect. Compounds **1–6** exhibited weak immunosuppressive activity against the proliferation of ConA-induced (T-cell) and LPS-induced (B-Cell) murine splenic lymphocytes (Table 3).

Table 2. Cytotoxicity of compounds **1–9**.

Compound	IC_{50} (μM) [a]	
	HepG2	**Hela**
1	53.72 ± 1.22	69.53 ± 2.25
2	50.25 ± 1.08	67.66 ± 1.89
3	67.32 ± 0.88	87.32 ± 0.98
4	45.69 ± 0.69	41.25 ± 0.26
5	48.75 ± 0.85	48.11 ± 1.00
6	48.10 ± 0.26	47.22 ± 0.34
7	4.83 ± 0.22	18.96 ± 0.88
8	13.99 ± 1.13	23.42 ± 2.55
9	12.06 ± 0.55	20.36 ± 1.99
Adriamycin [b]	\	0.88 ± 0.71
Fluorouracil [c]	179.03 ± 25.82	\

[a] Data are presented as mean ± SD from three separate experiments. [b] Hela cell positive control. [c] Hepg2 cell positive control.

Table 3. Immunosuppressive activity of compounds **1–9**.

Compound	IC_{50} (μM) [a]	
	ConA-Induced T-Cell Proliferation	**LPS-Induced B-Cell Proliferation**
1	55.53 ± 0.93	89.27 ± 2.25
2	60.25 ± 1.58	87.66 ± 2.76
3	75.75 ± 1.78	102.65 ± 1.38
4	44.84 ± 1.26	77.76 ± 1.47
5	55.71 ± 1.85	119.84 ± 1.12
6	57.39 ± 1.24	87.72 ± 0.44
7–9	-	-
cyclosporin A [b]	4.39 ± 0.02	25.11 ± 0.43

[a] Data are presented as mean ± SD from three separate experiments. [b] Positive control. '-' stands for no inhibitory effect at 200 μM.

3. Materials and Methods

3.1. General Procedures

The optical rotations were acquired using the ATR-W2 HHW5 digital Abbe refractometer (Shanghai Physico-optical Instrument Factory, Shanghai, China). The UV spectra were obtained using a Shimadzu UV-2600 PC spectrophotometer (Shimadzu Corporation, Tokyo, Japan), while the ECD spectra were measured on a JASCO J-715 spectra polarimeter (Japan Spectroscopic, Tokyo, Japan). All the LC/MS data were collected by an LCMS-IT-TOF instrument (Shimadzu Corporation, Tokyo, Japan) with an ESI source. The ^{1}H, ^{13}C, and

2D NMR spectra were acquired on a Bruker AV 400 NMR spectrometer using TMS as an internal standard. TLC and column chromatography (CC) were executed on silica gel (200–400 mesh, Qingdao Marine Chemical Inc., Qingdao, China) or a Sephadex-LH-20 (18−110 μm, Merck, Darmstadt, Germany), respectively. The HPLC analysis was measured on Wasters e2695 (Waters Corporation, Milford, MA, USA) using a C18 column (Waters, 5 μm, 10 × 150 mm). Semi-preparative HPLC was obtained on an Agilent Technologies 1200 LC with a C18 column (Agilent Technologies 10 mm × 250 mm). High-speed centrifugation was performed on a TGL-16B Anting centrifugal machine (Anting Scientific instrument Factory, Shanghai, China). The constant temperature water bath was in HH-2 thermostat water baths (Hervey Biotechnology Corporation, Jinan, China). The purity of the isolated compounds was determined via high-performance liquid chromatography (HPLC), which was performed on an Agilent 1200 instrument and a reverse-phase column (4.6 × 150 mm, 5 μm). The UV wavelength for detection was 210 nm. All the crude extracts and compounds were eluted with a flow rate of 0.8 mL·min-1 over a 50 min gradient (solvents: A, H_2O; B, MeOH), as follows: 0–5 min, 25% B; 5–15 min, 25–30% B; 15–30 min, 30–55% B; 30–40 min, 55–75% B; 40–50 min, 70–90% B; 50–60 min, 90–100% (Figure S47).

3.2. Fungal Material

The endophytic fungi Phomopsis asparagi and *Phomopsis* sp. were isolated from the fresh root of the mangrove plant Rhizophora mangle collected in Dong Zhai Gang-Mangrove Garden on Hainan Island, China, in October 2015. The fungi were identified as Phomopsis asparagi (strain no.DHS-48) and *Phomopsis* sp. (strain no.DHS-11) by ITS gene sequence (GenBank Accession No.MT126606 and No. OR801625). Two voucher strains were deposited at one of the authors' laboratories (J.X.).

3.3. Preparation of Phomopsis asparagi, Phomopsis sp., Co-Cultivation, and Morphological Observation

The two fungi, *Phomopsis asparagi* and *Phomopsis* sp., were cultured independently on PDA for 7 days at 28 °C. For co-cultivation, two 1 cm^2 pieces of agar from each fungus were placed 5 cm distance from each other on a new agar plate (9 cm in diameter). The co-cultures were incubated at 28 °C for 14 days. In parallel, mono-cultures of each strain used for co-cultivation were prepared and cultivated in the same conditions for comparison. After 14 days, the agar blocks with mycelium of Phomopsis asparagi, *Phomopsis* sp., and their co-culture were cut and mixed with 2.5% glutaraldehyde solution at 4 °C for 4 h and meticulously rinsed in phosphate buffered saline solution (PBS) three times for 15 min each. The segments were dehydrated three times in graded ethanol in a series (50%, 70%, 80%, 90%, and 100%) over a period of time for 30 min each time, then lyophilized, fixed on radio with conductive gel, sprayed with gold for 30 s with an IB-3 ion coating apparatus, and, finally, observed by the Hitachi S-3000N scanning electron microscope.

3.4. Sample Preparation of Phomopsis asparagi, Phomopsis sp., Co-Culture, and Large-Scale Fermentation and Extracts

The two fungi were independently cultivated on PDA at 28 °C for 14 days. After that, the two fungi colonies were simultaneously inoculated into an autoclaved rice solid-substrate medium in Erlenmeyer flasks (130 × 1 L); each contained 100 g of rice and 100 mL of 0.3% saline water and were fermented at 28 °C for 30 days. At the same time, *Phomopsis asparagi* and *Phomopsis* sp. were separately cultured under the same culture conditions in 20 Erlenmeyer flasks.

Following the fermentation process, a random selection of 5 bottles was made from the 130 co-cultured fermentation mixes, and similarly, 5 bottles were randomly chosen from the 20 bottles of Phomopsis asparagi and *Phomopsis* sp. monocultured fermentation mixes. The co-cultured and monocultured fermentation mixes that were selected were extracted three times with EtOAc, and the filtrate was then distilled under reduced pressure to obtain the crude extracts. The operations outlined above are to ensure that all the co-cultures and monocultures have five independent biological replicates, ensuring the reliability and of the results. The dried extracts were re-dissolved in an appropriate amount of UPLC/MS-grade methanol (MeOH) and pipetted into a pre-weighed 1.5 mL-amber glass vial through a 13 mm syringe filter with a 0.22 mm PTFE membrane to prepare a 1 mg·mL^{-1} solution for future use. The crude extracts were analyzed using HPLC, UPLC-MS/MS, and ^{1}H NMR. The remaining 125 bottles of the co-cultured fermentation mixes were similarly extracted three times with EtOAc to obtain 30 g of crude extract for subsequent isolation.

3.5. UPLC-ESI-MS/MS Analysis

The combination of the Japanese Shimazu liquid phase system and mass spectrometry system was used to accomplish the UPLC-ESI-MS/MS analysis. To keep the instrument in the best possible condition before analysis, the chromatographic column and instrument system have to be cleaned with the appropriate solvents. The samples were injected and chromatographically separated on a Waters ACQUITY UPLC BEH CI8 column (10 mm × 210 mm, 1.7 μm) at a temperature of 25 °C with an injection volume of 5 μL and the PDA detector set at 190–400 nm. The mobile phase adopted a binary elution system, with the A phase containing water and the B phase containing methanol. The elution procedure is 0–5 min, 25% B; 5–15 min, 25–30% B; 15–30 min, 30–55% B; 30–40 min, 55–75% B; 40–50 min, 70–90% B; 50–60 min, 90–100%, with a flow rate of 0.3 mL·m^{-1}.

The ESI ion source parameters were set as follows: sensitivity mode under positive and negative ions; ionization voltage, 2.00 kV; cone hole voltage, 40 V; ion source voltage compensation, 80 V; ion source temperature, 100 °C; temperature, 250 °C; cone hole gas flow rate, 50 L/h; flow rate, 600 Lh; the and cone-shaped gas were high-purity nitrogen. The scanning ion range was m/z 100–2000 Da, and the energy setting was 2 V for low energy and 40–80 V for high energy; the scanning time was 0.10 s. An MS survey scan was performed in the range of 100–1500 Da, while MS/MS scanned over a mass range of 50–1500 Da at the same time. As controls, the solvent (MeOH) and non-inoculated medium were injected under the same conditions. The data were collected and analyzed using the LCMSsoultion Ver. 3 software (Shimazu, Japanese).

3.6. Data Processing, Molecular Networking, Dereplication, and Multivariate Data Analysis

After the collection was completed, MSConvert was used to convert the data into mzXML format and construct a molecular network in GNPS (Global Natural Products Social Molecular Network "http://gnps.ucsd.edu (accessed on 4 February 2024)" [59]. The MS/MS molecular network was constructed using the classic online workflow (METABOLOMICS-SNETS-V2) in GNPS. The parameters were set to at least four matching peaks, a minimum cosine similarity score of 0.65, a parent mass tolerance of 1.0 Da, and a fragment ion tolerance of 0.3 Da. The spectra in the network were then searched against GNPS spectral libraries to annotate and identify metabolites through the database search of the MS/MS spectra. The MN work on GNPS can be found at http://gnps.ucsd.edu/ProteoSAFe/status.jsp?task=4ae8b81709dc4f3899d56a9e2f185f10, accessed on 4 February 2024. The resulting MN was visualized using Cytoscape 2.8.3 for composition and displayed with 'directed' style. The thickness of the edges between the nodes is adjusted by the cosine value, and the color of the nodes is determined based on the source of the spectral file. To simplify the analysis of the network, background nodes originating from the cultivation medium rice and solvents (MeOH) were removed from the MN [60]. Node colors were mapped based on the source of the spectra files, as Phomopsis asparagi are green, *Phomopsis* sp. are yellow, and the nodes of the co-culture are red. To improve the dereplication, a manual annotation was conducted in LCM solution Ver. 3 software by searching the microbial natural products database, such as the Dictionary of Natural Products (DNP) "http://dnp.chemnetbase.com (accessed on 5 February 2024)", the Natural Products Atlas "www.npatlasorg (accessed on 5 February 2024)", and the NPASS "http://bidd.group/NPASS (accessed on 5 February 2024)". Based on the predicted accurate mass value, the error range is within the target MW ± 1 Da, UV features, MS/MS spectrum, and taxonomic unit information (mainly within the *Phomopsis* sp. And, if necessary, extended to the fungal kingdom) [27,56,61–63]. All the data have been uploaded to the MassIVE Database of the GNPS Web site "https://gnps.ucsd.edu/ProteoSAFe/static/gnps-splash2.jsp (accessed on 15 February 2024)" and are publicly available through access number MSV000094093. Dereplication was also achieved by isolation and characterization by NMR and HR-ESIMS.

The HR-MS/MS data were analyzed using Mzmine 2.5.3. Markers between 150 and 1500 Da were collected with an intensity threshold of 10k counts and the retention time and mass windows of 0.2 min and 0.1 Da, respectively. The noise level was set to 3.0×10^3, and the raw data were deisotoped. A statistical analysis of the data was conducted using SIMCA (version 14.0, Umetrics, Umea, Sweden) [28]. With the parameters applied for the collection of markers, approximately 300 separate markers for each survey were identified in total. The markers from the extracts were compared through PCA.

3.7. Isolation of Compounds

After the completion of fermentation, the co-cultured fermentation mixes were extracted three times with EtOAc, and the filtrate was then distilled under reduced pressure to obtain 30 g of crude extract. Using stepped gradient elution with CH_2Cl_2-MeOH (0–100%) on silica gel column chromatography (CC), the crude extracts were separated into nine fractions (Fr. 1–Fr. 9). The fraction Fr. 3 was subjected to open silica gel CC using gradient elution with CH_2Cl_2-MeOH-100:0–1:1, *v*/*v* to obtain 6 fractions (Fr. 3.1–Fr. 3.6). Fr. 3.3 was chromatographed on a Sephadex LH-20 CC by eluting with MeOH to yield compounds **7** (20 mg) and **9** (500 mg). Fr. 4 was subjected to open silica gel CC using gradient elution with CH_2Cl_2-MeOH-100:0–1:2, *v*/*v* to obtain 7 fractions (Fr. 4.1–Fr. 4.7). The subfraction Fr. 4.3 was applied to ODS CC with a gradient elution of MeOH/H_2O mixtures (*v*/*v*, 1:4, 3:7, 2:3, 1:1, 3:2, 7:3, 4:1, 0:1) and obtained five subfractions (Fr. 4.3.1–Fr. 4.3.5). Then, Fr. 4.3.4 was purified by semi-preparative reversed-phase HPLC using MeOH-H_2O (60:40, *v*/*v*, 2 mL·m^{-1}, UV λ_{max} 210 nm) to afford compounds **6** (10 mg), **1** (5 mg), and **5** (7 mg), and Fr. 4.3.5 was purified by semi-preparative reversed-phase HPLC using MeOH-H_2O (60:40, *v*/*v*, 2 mL·m^{-1}, UV λ_{max} 210 nm) to afford compound **2** (10 mg) and compound **8** (10 mg). Fr. 7 was subjected to open silica gel CC using gradient elution with CH_2Cl_2-MeOH-100:2–1:2, *v*/*v* to obtain 5 fractions (Fr. 7.1–Fr. 7.5). Fr. 7.2 was chromatographed on a Sephadex LH-20 CC by eluting with MeOH to yield three fractions (Fr. 7.2.1–Fr. 7.2.3). Fr. 7.2.3 was purified by semi-preparative reversed-phase HPLC using MeOH-H_2O (60:40, *v*/*v*, 2 mL·m^{-1}, UV λ_{max} 210 nm) to afford compound **3** (6 mg). Fr. 8 was applied to ODS CC with gradient elution of MeOH/H_2O mixtures (*v*/*v*, 1:4, 3:7, 2:3, 1:1, 3:2, 7:3, 4:1, 0:1) and obtained five subfractions (Fr. 8.1–Fr. 8.5). Fr. 8.2 was chromatographed on a Sephadex LH-20 CC by eluting with MeOH to yield three fractions (Fr. 8.2.1–Fr. 8.2.3). Fr. 8.3 was purified by HPLC (MeOH/H_2O, 70:30 and 60:40, *v*/*v*; 2 mL·m^{-1}, UV λ_{max} 210 nm) to yield compound **4** (8 mg).

Phomoxanthone L (**1**): yellow amorphous powder (MeOH); $[\alpha]^{20}_D$ −20 (c 0.0001, MeOH); UV (MeOH) λ_{max} 205, 251, 361 nm (the absorptions due to aromatic rings); ^1H and ^{13}C NMR data, see Table 1; HRESIMS *m*/*z* 597.1613 $[M - H]^-$ (calcd for $C_{30}H_{29}O_{13}$ 597.1614).

Phomoxanthone M (**2**): yellow amorphous powder (MeOH); $[\alpha]^{20}_D$ −10 (c 0.0001, MeOH); UV (MeOH) λ_{max} 207, 254, 364 nm (the absorptions due to aromatic rings); ^1H and ^{13}C NMR data, see Table 1; HRESIMS *m*/*z* 599.1759 $[M+H]^+$ (calcd for $C_{30}H_{31}O_{13}$ 599.1759).

Phomoxanthone N (**3**): yellow amorphous powder (MeOH); $[\alpha]^{20}_D$ +10 (c 0.0001, MeOH); UV (MeOH) λ_{max} 212, 255, 366 nm (the absorptions due to aromatic rings); ^1H and ^{13}C NMR data, see Table 1; HRESIMS *m*/*z* 601.1916 $[M+H]^+$ (calcd for $C_{30}H_{33}O_{13}$ 601.1916).

3.8. Theory and Calculation Details

Detailed Monte Carlo conformational analyses were performed utilizing Spartan's 14 software and the Merck molecular force field (MMFF). Conformers exceeding a Boltzmann population of 0.4% were selected for electronic circular dichroism (ECD) calculations as presented in Tables S2–S7. Subsequently, these conformers underwent initial optimization at the B3LYP/6-31G(d) level in the gas phase, complemented by the polarizable conductor calculation model based on the polarizable continuum model (PCM). The stable conformations identified at the B3LYP/6-31G(d) level were then used in magnetic shielding constants. The theoretical calculation of ECD was conducted in MeOH using the time-dependent density functional theory (TD-DFT) at the B3LYP/6-31 + g (d, p) level for all the conformers of compounds **1**, **2**, and **3**. The ECD spectra were generated with the aid of the SpecDis 1.6 program (University of Würzburg, Würzburg, Germany) and GraphPad Prism 5 (University of California, San Diego, CA, USA) through the conversion of dipole-length rotational strengths into band shapes modeled by Gaussian functions with a standard deviation of 0.3 eV.

3.9. Cytotoxicity Assay

The liver cancer cell line, HepG2, and the cervical cancer cell line, Hela, were obtained from the Type Culture Collection of the Chinese Academy of Sciences in Shanghai, China. The cells were cultivated using RPMI-1640 culture medium. To assess cytotoxicity against HepG2 and HeLa cells, the 3-(4,5-dimethylthiazol-2-yl)-2,5-diphenyltetrazolium bromide (MTT) method, sourced from Sigma-Aldrich in St. Louis, Missouri, USA, was employed as described previously [47]. Furthermore, adriamycin (from Shanghai Macklin Biochemical Co., Ltd., with a purity of 99.8%) (Shanghai, China) and 5-fluorouracil (5-FU) (from Beijing Solarbio Science and Technology Co., Ltd., with a purity of 99.8%) (Beijing, China), both served as positive controls.

3.10. Splenocyte Proliferation Assay

Spleen cells were collected from BALB/c mice under aseptic conditions, plated in a 96-well plate at a concentration of 1×10^7 cells/mL per well, and activated by Con A (5 μg/mL) or LPS (10 μg/mL) in the presence of various concentrations of the compounds or cyclosporine A (CsA) at 37 °C and 5% CO_2 for 48 h. Then, 20 μL CCK-8 was added to each well 4 h before the end of the incubation. The absorbance at OD_{450} was measured on an ELISA reader, and the IC_{50} value was calculated from the correlation curve between the compound concentration and the OD_{450}.

3.11. Statistical Analysis

All the cell data are presented as the mean standard deviation of the means (S.D.), and a one-way analysis of variance (ANOVA) was used to evaluate the statistical significance of the differences between the groups by GraphPad Prism.

4. Conclusions

Mangrove endophytic fungi are considered one of the most promising sources of novel biologically active compounds. In the current study, co-cultivation of mangrove *Rhizophora mangle* endophytic fungi *Phomopsis asparagi* DHS-48 and *Phomopsis* genus fungus DHS-11 led to global changes of their metabolic profiles. Further analysis using the metabolomic approach integrated HPLC chromatogram, ^{1}HNMR spectroscopy, UPLC-MS-PCA, and molecular networking, indicating the presence of a series of induced dimeric xantones, resulting in the targeted isolation and structure elucidation of three new phomoxanthones L-N (**1**–**3**) and six known analogs (**4**–**9**). Meanwhile, compounds **7**–**9** significantly suppressed the proliferation of human liver cells HepG-2. Our study highlights that combining co-cultivation with a metabolomics-guided strategy as a discovery tool will be implemented as a systematic strategy for the quick discovery of target bioactive compounds.

Supplementary Materials: The following supporting information can be downloaded at: https://www.mdpi.com/article/10.3390/md22030102/s1, Figure S1: ^{1}H-NMR of EtOAc extracts of *Phomopsis asparagi* DHS-48, *Phomopsis* sp. DHS-11 and their Co-culture measured in CD_3OD. Figure S2: PCA of metabolomics data of co-culture and their corresponding mono-cultures. Figure S3: The MS/MS spectrum and possible fragmentation patterns of compound **5**–**9** and deacetylphomoxanthone B. Figure S4: The MS/MS spectrum and possible fragmentation patterns of compound **1**–**4**. Figure S5: ^{1}H-NMR of (**1**). Figure S6: ^{13}C-NMR of (**1**). Figure S7: DEPT of (**1**). Figure S8: ^{1}H-^{1}H COSY of (**1**).

Figure S9: HSQC of (**1**). Figure S10: HMBC of (**1**). Figure S11: NOSEY of (**1**). Figure S12: HR-ESI-MS of (**1**). Figure S13: ^{1}H-NMR of (**2**). Figure S14: ^{13}C-NMR of (**2**). Figure S15: DEPT of (**2**). Figure S16: ^{1}H-^{1}H COSY of (**2**). Figure S17: HSQC of (**2**). Figure S18: HMBC of (**2**). Figure S19: NOSEY of (**2**). Figure S20: HR-ESI-MS of (**2**). Figure S21: ^{1}H-NMR of (**3**). Figure S22: ^{13}C-NMR of (**3**). Figure S23: DEPT of (**3**). Figure S24: ^{1}H-^{1}H COSY of (**3**). Figure S25: HSQC of (**3**). Figure S26: HMBC of (**3**). Figure S27: NOSEY of (**3**). Figure S28: HR-ESI-MS of (**3**). Figure S29: ^{1}H-NMR of (**4**). Figure S30: ^{13}C-NMR of (**4**). Figure S31: HR-ESI-MS of (**4**). Figure S32: ^{1}H-NMR of (**5**). Figure S33: ^{13}C-NMR of (**5**). Figure S34: HR-ESI-MS of (**5**). Figure S35: ^{1}H-NMR of (**6**). Figure S36: ^{13}C-NMR of (**6**). Figure S37: HR-ESI-MS of (**6**). Figure S38: ^{1}H-NMR of (**7**). Figure S39: ^{13}C-NMR of (**7**). Figure S40: HR-ESI-MS of (**7**). Figure S41: ^{1}H-NMR of (**8**). Figure S42: ^{13}C-NMR of (**8**). Figure S43: HR-ESI-MS of (**8**). Figure S44: ^{1}H-NMR of (**9**). Figure S45: ^{13}C-NMR of (**9**). Figure S46: HR-ESI-MS of (**9**). Figure S47: HPLC spectrum for the purity of tested compounds. Table S1: Putative annotation of metabolites produced in the non-cultures of *Phomopsis asparagi* DHS-48 and *Phomopsis* sp. DHS-11, and their co-culture. Table S2: Gibbs free energies[a] and equilibrium populations[b] of low-energy conformers of phomoxanthone L (**1**). Table S3: Cartesian coordinates for the low-energy reoptimized MMFF conformers of phomoxanthone L (**1**) at B3LYP/6-31G(d,p) level of theory in gas. Table S4: Gibbs free energies[a] and equilibrium populations[b] of low-energy conformers of phomoxanthone M (**2**). Table S5: Cartesian coordinates for the low-energy reoptimized MMFF conformers of phomoxanthone M (**2**) at B3LYP/6-31G(d,p) level of theory in gas. Table S6: Gibbs free energies[a] and equilibrium populations[b] of low-energy conformers of phomoxanthone N (**3**). Table S7: Cartesian coordinates for the low-energy reoptimized MMFF conformers of phomoxanthone N (**3**) at B3LYP/6-31G(d,p) level of theory in gas.

Author Contributions: J.X. designed and supervised this research, structured the elucidation, and wrote the draft and final revision of the manuscript. J.W. performed the isolation. D.C. and Q.L. carried out the biological evaluation. T.F. assisted in the SEM. The final revision of the manuscript was revised by all the authors. All authors have read and agreed to the published version of the manuscript.

Funding: This work was co-financed by grants of the National Natural Science Foundation of China (No. 82160675/81973229), the Key Research Program of Hainan Province (ZDYF2021SHFZ108), the Collaborative Innovation Center Foundation of Hainan University (XTCX2022STB01), and the Guangdong Key Laboratory of Marine Materia Medica Open Fund (LMM2021-4). They are gratefully acknowledged.

Institutional Review Board Statement: Not applicable.

Data Availability Statement: The original data presented in the study are included in the article/Supplementary Material; further inquiries can be directed to the corresponding author.

Conflicts of Interest: The authors declare no conflicts of interest.

References

1. Wu, M.J.; Xu, B.; Guo, Y.W. Unusual Secondary Metabolites from the Mangrove Ecosystems: Structures, Bioactivities, Chemical, and Bio-Syntheses. *Mar. Drugs* **2022**, *20*, 535. [CrossRef] [PubMed]
2. Lin, W.; Li, G.; Xu, J. Bio-Active Products from Mangrove Ecosystems. *Mar. Drugs* **2023**, *21*, 239. [CrossRef] [PubMed]
3. Xu, J. Biomolecules Produced by Mangrove-Associated Microbes. *Curr. Med. Chem.* **2011**, *18*, 5224–5266. [CrossRef]
4. Li, K.; Chen, S.; Pang, X.; Cai, J.; Zhang, X.; Liu, Y.; Zhu, Y.; Zhou, X. Natural Products from Mangrove Sediments-Derived Microbes: Structural Diversity, Bioactivities, Biosynthesis, and Total Synthesis. *Eur. J. Med. Chem.* **2022**, *230*, 114117. [CrossRef] [PubMed]
5. Williams, P.G.; Buchanan, G.O.; Feling, R.H.; Kauffman, C.A.; Jensen, P.R.; Fenical, W. New Cytotoxic Salinosporamides from the Marine Actinomycete *Salinispora tropica*. *J. Org. Chem.* **2005**, *70*, 6196–6203. [CrossRef]
6. Buckingham, L. EU/3/18/2119. Available online: https://www.ema.europa.eu/en/medicines/human/orphan-designations/eu3182119 (accessed on 8 May 2020).
7. Chen, S.; Cai, R.; Liu, Z.; Cui, H.; She, Z. Secondary Metabolites from Mangrove-Associated Fungi: Source, Chemistry and Bioactivities. *Nat. Prod. Rep.* **2022**, *39*, 560–595. [CrossRef]
8. Challis, G.L. Genome Mining for Novel Natural Product Discovery. *J. Med. Chem.* **2008**, *51*, 2618–2628. [CrossRef]
9. Nicault, M.; Zaiter, A.; Dumarcay, S.; Chaimbault, P.; Gelhaye, E.; Leblond, P.; Bontemps, C. Elicitation of Antimicrobial Active Compounds by *Streptomyces*-Fungus Co-Cultures. *Microorganisms* **2021**, *9*, 178. [CrossRef]

10. Valayil, J.M. Activation of Microbial Silent Gene Clusters: Genomics Driven Drug Discovery Approaches. *Biochem. Anal. Biochem.* **2016**, *5*, 276–280. [CrossRef]
11. Kumar, A.; Sørensen, J.L.; Hansen, F.T.; Arvas, M.; Syed, M.F.; Hassan, L.; Benz, J.P.; Record, E.; Henrissat, B.; Pöggeler, S.; et al. Genome Sequencing and Analyses of Two Marine Fungi from the North Sea Unraveled a Plethora of Novel Biosynthetic Gene Clusters. *Sci. Rep.* **2018**, *8*, 10187. [CrossRef]
12. Gaudêncio, S.P.; Bayram, E.; Lukić Bilela, L.; Cueto, M.; Díaz-Marrero, A.R.; Haznedaroglu, B.Z.; Jimenez, C.; Mandalakis, M.; Pereira, F.; Reyes, F.; et al. Advanced Methods for Natural Products Discovery: Bioactivity Screening, Dereplication, Metabolomics Profiling, Genomic Sequencing, Databases and Informatic Tools, and Structure Elucidation. *Mar. Drugs* **2023**, *21*, 308. [CrossRef] [PubMed]
13. Schroeckh, V.; Scherlach, K.; Nützmann, H.-W.; Shelest, E.; Schmidt-Heck, W.; Schuemann, J.; Martin, K.; Hertweck, C.; Brakhage, A.A. Intimate Bacterial–Fungal Interaction Triggers Biosynthesis of Archetypal Polyketides in *Aspergillus nidulans*. *Proc. Natl. Acad. Sci. USA* **2009**, *106*, 14558–14563. [CrossRef] [PubMed]
14. Bertrand, S.; Bohni, N.; Schnee, S.; Schumpp, O.; Gindro, K.; Wolfender, J.-L. Metabolite Induction via Microorganism Co-Culture: A Potential Way to Enhance Chemical Diversity for Drug Discovery. *Biotechnol. Adv.* **2014**, *32*, 1180–1204. [CrossRef] [PubMed]
15. Knowles, S.L.; Raja, H.A.; Roberts, C.D.; Oberlies, N.H. Fungal–Fungal Co-Culture: A Primer for Generating Chemical Diversity. *Nat. Prod. Rep.* **2022**, *39*, 1557–1573. [CrossRef] [PubMed]
16. Caudal, F.; Tapissier-Bontemps, N.; Edrada-Ebel, R.A. Impact of Co-Culture on the Metabolism of Marine Microorganisms. *Mar. Drugs* **2022**, *20*, 153. [CrossRef] [PubMed]
17. Zhu, F.; Chen, G.; Wu, J.; Pan, J. Structure Revision and Cytotoxic Activity of Marinamide and Its Methyl Ester, Novel Alkaloids Produced by Co-Cultures of Two Marine-Derived Mangrove Endophytic Fungi. *Nat. Prod. Res.* **2013**, *27*, 1960–1964. [CrossRef] [PubMed]
18. Ding, W.; Lu, Y.; Feng, Z.; Luo, S.; Li, C. A New Nonadride Derivative from the Co-Culture Broth of Two Mangrove Fungi. *Chem. Nat. Compd.* **2017**, *53*, 691–693. [CrossRef]
19. Li, C.; Zhang, J.; Zhong, J.; He, B.; Li, R.; Liang, Y.; Lin, Y.; Zhou, S. Isolation and identication of the metabolites from the mixed fermentation broth of two mangrove endophytic fungi. *J. South China Agric. Univ.* **2011**, *32*, 117–119. [CrossRef]
20. Li, C.; Zhang, J.; Shao, C.; Ding, W.; She, Z.; Lin, Y. A new xanthone derivative from the co-culture broth of two marine fungi (strain no. E33 and K38). *Chem. Nat. Compd.* **2011**, *47*, 382–384. [CrossRef]
21. Shao, C.L.; Guo, Z.Y.; Xia, X.K.; Liu, Y.; Huang, Z.-J.; She, Z.-G.; Lin, Y.-C.; Zhou, S.-N. Five Nitro-Phenyl Compounds from the South China Sea Mangrove Fungus. *J. Asian Nat. Prod. Res.* **2007**, *9*, 643–648. [CrossRef]
22. Shao, C.; Hu, G.; Zhang, X.; Wang, C.; She, Z.; Lin, Y. A New Protocatechuic Acid Derivative from the Mangrove Endophytic Fungus B60. *Acta Sci. Nat. Univ. Sunyatseni* **2008**, *5*, 133–134.
23. Huang, H.; She, Z.; Lin, Y.; Vrijmoed, L.L.P.; Lin, W. Cyclic Peptides from an Endophytic Fungus Obtained from a Mangrove Leaf (*Kandelia candel*). *J. Nat. Prod.* **2007**, *70*, 1696–1699. [CrossRef]
24. Li, C.; Cox, D.; Huang, S.; Ding, W. Two New Cyclopeptides from the Co-Culture Broth of Two Marine Mangrove Fungi and Their Antifungal Activity. *Pharmacogn. Mag.* **2014**, *10*, 41. [CrossRef] [PubMed]
25. Covington, B.C.; McLean, J.A.; Bachmann, B.O. Comparative Mass Spectrometry-Based Metabolomics Strategies for the Investigation of Microbial Secondary Metabolites. *Nat. Prod. Rep.* **2017**, *34*, 6–24. [CrossRef] [PubMed]
26. Yuliana, N.D.; Jahangir, M.; Verpoorte, R.; Choi, Y.H. Metabolomics for the Rapid Dereplication of Bioactive Compounds from Natural Sources. *Phytochem. Rev.* **2013**, *12*, 293–304. [CrossRef]
27. Oppong-Danquah, E.; Parrot, D.; Blümel, M.; Labes, A.; Tasdemir, D. Molecular Networking-Based Metabolome and Bioactivity Analyses of Marine-Adapted Fungi Co-Cultivated with Phytopathogens. *Front. Microbiol.* **2018**, *9*, 2072. [CrossRef] [PubMed]
28. Xu, X.-Y.; Shen, X.-T.; Yuan, X.-J.; Zhou, Y.-M.; Fan, H.; Zhu, L.-P.; Du, F.-Y.; Sadilek, M.; Yang, J.; Qiao, B.; et al. Metabolomics Investigation of an Association of Induced Features and Corresponding Fungus during the Co-Culture of *Trametes Versicolor* and *Ganoderma applanatum*. *Front. Microbiol.* **2018**, *8*, 2647. [CrossRef]
29. Stierle, A.A.; Stierle, D.B.; Decato, D.; Priestley, N.D.; Alverson, J.B.; Hoody, J.; McGrath, K.; Klepacki, D. The Berkeleylactones, Antibiotic Macrolides from Fungal Coculture. *J. Nat. Prod.* **2017**, *80*, 1150–1160. [CrossRef] [PubMed]
30. Quinn, R.A.; Nothias, L.-F.; Vining, O.; Meehan, M.; Esquenazi, E.; Dorrestein, P.C. Molecular Networking as a Drug Discovery, Drug Metabolism, and Precision Medicine Strategy. *Trends Pharmacol. Sci.* **2017**, *38*, 143–154. [CrossRef] [PubMed]
31. Watrous, J.; Roach, P.; Alexandrov, T.; Heath, B.S.; Yang, J.Y.; Kersten, R.D.; Van Der Voort, M.; Pogliano, K.; Gross, H.; Raaijmakers, J.M.; et al. Mass Spectral Molecular Networking of Living Microbial Colonies. *Proc. Natl. Acad. Sci. USA* **2012**, *109*, E1743–E1752. [CrossRef]
32. Wang, M.; Carver, J.J.; Phelan, V.V.; Sanchez, L.M.; Garg, N.; Peng, Y.; Nguyen, D.D.; Watrous, J.; Kapono, C.A.; Luzzatto-Knaan, T.; et al. Sharing and Community Curation of Mass Spectrometry Data with Global Natural Products Social Molecular Networking. *Nat. Biotechnol.* **2016**, *34*, 828–837. [CrossRef] [PubMed]
33. Um, S.; Seibel, E.; Schalk, F.; Balluff, S.; Beemelmanns, C. Targeted Isolation of Saalfelduracin B–D from *Amycolatopsis saalfeldensis* Using LC-MS/MS-Based Molecular Networking. *J. Nat. Prod.* **2021**, *84*, 1002–1011. [CrossRef] [PubMed]
34. Vallet, M.; Vanbellingen, Q.P.; Fu, T.; Le Caer, J.-P.; Della-Negra, S.; Touboul, D.; Duncan, K.R.; Nay, B.; Brunelle, A.; Prado, S. An Integrative Approach to Decipher the Chemical Antagonism between the Competing Endophytes *Paraconiothyrium variabile* and *Bacillus subtilis*. *J. Nat. Prod.* **2017**, *80*, 2863–2873. [CrossRef] [PubMed]

35. Chagas, F.O.; Caraballo-Rodríguez, A.M.; Dorrestein, P.C.; Pupo, M.T. Expanding the Chemical Repertoire of the Endophyte *Streptomyces albospinus* RLe7 Reveals Amphotericin B as an Inducer of a Fungal Phenotype. *J. Nat. Prod.* **2017**, *80*, 1302–1309. [CrossRef]
36. Xu, Z.; Xiong, B.; Xu, J. Chemical Investigation of Secondary Metabolites Produced by Mangrove Endophytic Fungus *Phyllosticta capitalensis*. *Nat. Prod. Res.* **2021**, *35*, 1561–1565. [CrossRef]
37. Hemberger, Y.; Xu, J.; Wray, V.; Proksch, P.; Wu, J.; Bringmann, G. Pestalotiopens A and B: Stereochemically Challenging Flexible Sesquiterpene-Cyclopaldic Acid Hybrids from *Pestalotiopsis* sp. *Chem. Eur. J.* **2013**, *19*, 15556–15564. [CrossRef]
38. Deng, Q.; Li, G.; Sun, M.; Yang, X.; Xu, J. A New Antimicrobial Sesquiterpene Isolated from Endophytic Fungus *Cytospora* sp. from the Chinese Mangrove Plant *Ceriops tagal*. *Nat. Prod. Res.* **2020**, *34*, 1404–1408. [CrossRef]
39. Sun, M.Y.; Zhou, D.D.; Wu, J.W.; Zhou, J.; Xu, J. Sdy-1 Executes Antitumor Activity in HepG2 and HeLa Cancer Cells by Inhibiting the Wnt/β-Catenin Signaling Pathway. *Mar. Drugs* **2022**, *20*, 125. [CrossRef]
40. Tan, M.; Xu, X.; Zhang, W.; Wu, F.; Bo, X.; Qin, F.; Ju, S.; Song, Z.; Yang, T.; Li, J.; et al. Isolation and Insecticidal Activities of New Cyclic Peptides from Mangrove Endophytic Fungus *Aspergillus* Sp. GXNU-4QQY1a. *Fitoterapia* **2023**, *171*, 105693. [CrossRef]
41. Xu, J.; Kjer, J.; Sendker, J.; Wray, V.; Guan, H.; Edrada, R.; Lin, W.H.; Wu, J.; Proksch, P. Chromones from the Endophytic Fungus *Pestalotiopsis* sp. Isolated from the Chinese Mangrove Plant *Rhizophora mucronata*. *J. Nat. Prod.* **2009**, *72*, 662–665. [CrossRef]
42. Zhou, J.; Li, G.; Deng, Q.; Zheng, D.Y.; Yang, X.B.; Xu, J. Cytotoxic Constituents from the Mangrove Endophytic *Pestalotiopsis* sp. Induce G_0/G_1 Cell Cycle Arrest and Apoptosis in Human Cancer Cells. *Nat. Prod. Res.* **2018**, *32*, 2968–2972. [CrossRef] [PubMed]
43. Wei, C.W.; Sun, C.X.; Feng, Z.; Zhang, X.X.; Xu, J. Four New Chromones from the Endophytic Fungus *Phomopsis asparagi* DHS-48 Isolated from the Chinese Mangrove Plant *Rhizophora mangle*. *Mar. Drugs* **2021**, *19*, 348. [CrossRef] [PubMed]
44. Xu, Z.Y.; Wu, X.; Li, G.; Feng, Z.; Xu, J. Pestalotiopisorin B, a New Isocoumarin Derivative from the Mangrove Endophytic Fungus *Pestalotiopsis* sp. HHL101. *Nat. Prod. Res.* **2020**, *34*, 1002–1007. [CrossRef] [PubMed]
45. Zhang, X.X.; Li, G.; Deng, Q.; Xu, Z.Y.; Cen, J.R.; Xu, J. Vomifoliol Isolated from Mangrove Plant *Ceriops tagal* Inhibits the NFAT Signaling Pathway with CN as the Target Enzyme In Vitro. *Bioorganic Med. Chem. Lett.* **2021**, *48*, 128235. [CrossRef] [PubMed]
46. Feng, Z.; Zhang, X.; Wu, J.; Wei, C.; Feng, T.; Zhou, D.; Wen, Z.; Xu, J. Immunosuppressive cytochalasins from the mangrove endophytic fungus *Phomopsis asparagi* DHS-48. *Mar. Drugs* **2022**, *20*, 526. [CrossRef] [PubMed]
47. Feng, T.; Wei, C.; Deng, X.; Chen, D.; Wen, Z.; Xu, J. Epigenetic manipulation induced production of immunosuppressive chromones and cytochalasins from the mangrove endophytic fungus *Phomopsis asparagi* DHS-48. *Mar. Drugs* **2022**, *20*, 616. [CrossRef]
48. Stockert, J.C.; Blázquez-Castro, A.; Cañete, M.; Horobin, R.W.; Villanueva, Á. MTT Assay for Cell Viability: Intracellular Localization of the Formazan Product Is in Lipid Droplets. *Acta Histochem.* **2012**, *114*, 785–796. [CrossRef]
49. Guo, Z.; Chen, B.; Chen, D.; Deng, X.; Yuan, J.; Zhang, S.; Xiong, Z.; Xu, J. New Isocoumarin and Pyrone Derivatives from the Chinese Mangrove Plant *Rhizophora mangle*-Associated Fungus *Phomopsis* sp. DHS-11. *Molecules* **2023**, *28*, 3756. [CrossRef]
50. Bertrand, S.; Schumpp, O.; Bohni, N.; Bujard, A.; Azzollini, A.; Monod, M.; Gindro, K.; Wolfender, J.L. Detection of metabolite induction in fungal co-cultures on solid media by high-throughput differen tial ultra-high pressure liquid chromatography-time-of-flight mass spectrometry of ingerprinting. *J. Chromatogr. A* **2013**, *1292*, 219–228. [CrossRef]
51. Prabhu, G.; Bhat, D.; Bhat, R.M.; Selvaraj, S. A Critical Look at Bioproducts Co-Cultured Under Solid State Fermentation and Their Challenges and Industrial Applications. *Waste Biomass Valorization* **2022**, *13*, 3095–3111. [CrossRef]
52. Cartabia, M.; Girometta, C.E.; Milanese, C.; Baiguera, R.M.; Buratti, S.; Branciforti, D.S.; Vadivel, D.; Girella, A.; Babbini, S.; Savino, E.; et al. Collection and Characterization of Wood Decay Fungal Strains for Developing Pure Mycelium Mats. *J. Fungi* **2021**, *7*, 1008. [CrossRef] [PubMed]
53. Dashti, Y.; Grkovic, T.; Abdelmohsen, U.R.; Hentschel, U.; Quinn, R.J. Actinomycete Metabolome Induction/Suppression with N-Acetylglucosamine. *J. Nat. Prod.* **2017**, *80*, 828–836. [CrossRef] [PubMed]
54. Ding, B.; Yuan, J.; Huang, X.; Wen, W.; Zhu, X.; Liu, Y.; Li, H.; Lu, Y.; He, L.; Tan, H.; et al. New Dimeric Members of the Phomoxanthone Family: Phomolactonexanthones A, B and Deacetylphomoxanthone C Isolated from the Fungus *Phomopsis* sp. *Mar. Drugs* **2013**, *11*, 4961–4972. [CrossRef] [PubMed]
55. Stewart, S.; Ivy, M.A.; Anslyn, E.V. The Use of Principal Component Analysis and Discriminant Analysis in Differential Sensing Routines. *Chem. Soc. Rev.* **2014**, *43*, 70–84. [CrossRef]
56. Hamed, A.A.; Soldatou, S.; Qader, M.M.; Arjunan, S.; Miranda, K.J.; Casolari, F.; Pavesi, C.; Diyaolu, O.A.; Thissera, B.; Eshelli, M.; et al. Screening Fungal Endophytes Derived from Under-Explored Egyptian Marine Habitats for Antimicrobial and Antioxidant Properties in Factionalised Textiles. *Microorganisms* **2020**, *8*, 1617. [CrossRef]
57. Crüsemann, M.; O'Neill, E.C.; Larson, C.B.; Melnik, A.V.; Floros, D.J.; Da Silva, R.R.; Jensen, P.R.; Dorrestein, P.C.; Moore, B.S. Prioritizing Natural Product Diversity in a Collection of 146 Bacterial Strains Based on Growth and Extraction Protocols. *J. Nat. Prod.* **2017**, *80*, 588–597. [CrossRef]
58. Zhou, D.D.; Feng, T.; Xu, J. Mangrove endophytic fungi-derived dicerandrol A and its inhibitory effects and preliminary mechanism on HepG2 cells. *Chin. J. Antibiot.* **2022**, *47*, 481–487. [CrossRef]
59. Combès, A.; Ndoye, I.; Bance, C.; Bruzaud, J.; Djediat, C.; Dupont, J.; Nay, B.; Prado, S. Chemical Communication between the Endophytic Fungus Paraconiothyrium Variabile and the Phytopathogen Fusarium Oxysporum. *PLoS ONE* **2012**, *7*, e47313. [CrossRef]

60. Wang, Y.; Glukhov, E.; He, Y.; Liu, Y.; Zhou, L.; Ma, X.; Hu, X.; Hong, P.; Gerwick, W.H.; Zhang, Y. Secondary Metabolite Variation and Bioactivities of Two Marine Aspergillus Strains in Static Co-Culture Investigated by Molecular Network Analysis and Multiple Database Mining Based on LC-PDA-MS/MS. *Antibiotics* **2022**, *11*, 513. [CrossRef]
61. Oppong-Danquah, E.; Blümel, M.; Scarpato, S.; Mangoni, A.; Tasdemir, D. Induction of Isochromanones by Co-Cultivation of the Marine Fungus *Cosmospora* sp. and the Phytopathogen *Magnaporthe oryzae*. *Int. J. Mol. Sci.* **2022**, *23*, 782. [CrossRef]
62. Liu, S.; Wang, T.; Lu, Q.; Li, F.; Wu, G.; Jiang, Z.; Habden, X.; Liu, L.; Zhang, X.; Lukianov, D.A.; et al. Bioprospecting of Soil-Derived Actinobacteria Along the Alar-Hotan Desert Highway in the Taklamakan Desert. *Front. Microbiol.* **2021**, *12*, 604999. [CrossRef] [PubMed]
63. Van Santen, J.A.; Jacob, G.; Singh, A.L.; Aniebok, V.; Balunas, M.J.; Bunsko, D.; Neto, F.C.; Castaño-Espriu, L.; Chang, C.; Clark, T.N.; et al. The Natural Products Atlas: An Open Access Knowledge Base for Microbial Natural Products Discovery. *ACS Cent. Sci.* **2019**, *5*, 1824–1833. [CrossRef] [PubMed]

marine drugs

MDPI

Article

Indole Diketopiperazine Alkaloids from the Marine Sediment-Derived Fungus *Aspergillus chevalieri* against Pancreatic Ductal Adenocarcinoma

Dina H. El-Kashef [1,2], Deborah D. Obidake [1], Katja Schiedlauske [1], Alina Deipenbrock [1], Sebastian Scharf [3], Hao Wang [4], Daniela Naumann [5], Daniel Friedrich [5], Simone Miljanovic [1], Takin Haj Hassani Sohi [6], Christoph Janiak [6], Klaus Pfeffer [3] and Nicole Teusch [1,*]

1 Institute of Pharmaceutical Biology and Biotechnology, Heinrich Heine University Düsseldorf, 40225 Düsseldorf, Germany; dihas102@hhu.de (D.H.E.-K.)
2 Department of Pharmacognosy, Faculty of Pharmacy, Minia University, 61519 Minia, Egypt
3 Institute of Medical Microbiology and Hospital Hygiene, Medical Faculty, Heinrich Heine University, 40225 Düsseldorf, Germany
4 Hainan Key Laboratory for Research and Development of Natural Products from Li Folk Medicine, Institute of Tropical Bioscience and Biotechnology, Chinese Academy of Tropical Agricultural Sciences, Haikou 571101, China
5 Department of Chemistry and Biochemistry, University of Cologne, 50939 Cologne, Germany
6 Institute of Inorganic Chemistry and Structural Chemistry, Heinrich Heine University Düsseldorf, 40225 Düsseldorf, Germany
* Correspondence: nicole.teusch@hhu.de; Tel.: +49-(0)-211-81-14163

Abstract: A new prenylated indole diketopiperazine alkaloid, rubrumline P (**1**), was isolated along with six more analogues and characterized from the fermentation culture of a marine sediment-derived fungus, *Aspergillus chevalieri*, collected at a depth of 15 m near the lighthouse in Dahab, Red Sea, Egypt. In the current study, a bioassay-guided fractionation allowed for the identification of an active fraction displaying significant cytotoxic activity against the human pancreatic adenocarcinoma cell line PANC-1 from the EtOAc extract of the investigated fungus compared to the standard paclitaxel. The structures of the isolated compounds from the active fraction were established using 1D/2D NMR spectroscopy and mass spectrometry, together with comparisons with the literature. The absolute configuration of the obtained indole diketopiperazines was established based on single-crystal X-ray diffraction analyses of rubrumline I (**2**) and comparisons of optical rotations and NMR data, as well as on biogenetic considerations. Genome sequencing indicated the formation of prenyltransferases, which was subsequently confirmed by the isolation of mono-, di-, tri-, and tetraprenylated compounds. Compounds rubrumline P (**1**) and neoechinulin D (**4**) confirmed preferential cytotoxic activity against PANC-1 cancer cells with IC_{50} values of 25.8 and 23.4 μM, respectively. Although the underlying mechanism-of-action remains elusive in this study, cell cycle analysis indicated a slight increase in the sub-G1 peak after treatment with compounds **1** and **4**.

Keywords: *Aspergillus chevalieri*; indole diketopiperazine alkaloids; prenylation; genome sequencing; X-ray diffraction; pancreatic cancer; pancreatic ductal adenocarcinoma

Citation: El-Kashef, D.H.; Obidake, D.D.; Schiedlauske, K.; Deipenbrock, A.; Scharf, S.; Wang, H.; Naumann, D.; Friedrich, D.; Miljanovic, S.; Haj Hassani Sohi, T.; et al. Indole Diketopiperazine Alkaloids from the Marine Sediment-Derived Fungus *Aspergillus chevalieri* against Pancreatic Ductal Adenocarcinoma. *Mar. Drugs* **2024**, 22, 5. https://doi.org/10.3390/md22010005

Academic Editors: Bin-Gui Wang and Haofu Dai

Received: 30 November 2023
Revised: 14 December 2023
Accepted: 17 December 2023
Published: 20 December 2023

1. Introduction

Pancreatic ductal adenocarcinoma (PDAC), the most common form of pancreatic neoplasm, is considered the fourth leading cause of cancer-related mortality worldwide due to its high aggressiveness and malignancy with a poor survival rate [1,2]. Despite breakthroughs in diagnostic tools, surgical treatments, and chemotherapy, the overall survival rates of PDAC remain low, and the only available therapies are classical chemotherapeutics with an unselective cytotoxic profile. Consequently, there is an immense need to find new drugs or drug leads for curing this disease [2,3]. Many natural products have been reported to exert anticancer effects against pancreatic cancer [3–5]. As part of our ongoing search for discovering novel potential drug candidates for use against PDAC that originate from a natural source [6], a bioassay-guided isolation of secondary metabolites was conducted on a marine sediment-derived fungus, *Aspergillus chevalieri*, collected from the Red Sea at a depth of 15 m close to the lighthouse of Dahab, Egypt. Accordingly, seven prenylated indole diketopiperazine alkaloids, including a rare tetraprenylated derivative that has been reported only once before, were isolated from the bio-active fraction, identified, and further pharmacologically characterized [7,8]. Indole diketopiperazine alkaloids (DKPs) represent a group of secondary metabolites broadly distributed in filamentous fungi, especially in the genera *Aspergillus* and *Penicillium* [9]. Biosynthetically, in fungi, the core cyclic dipeptide nucleus of DKPs is mostly originated from the condensation of two tryptophan amino acids or one tryptophan molecule and another amino acid, assembled by non-ribosomal peptide synthetases (NRPSs) [8,10]. Different putative tailoring enzymes contribute to the structural diversity of DKP scaffolds, including oxidoreductases, hydrolases, methyl transferases, prenyltransferases, and ligases [11,12]. Prenylated indole diketopiperazines are constituted of an indole diketopiperazine backbone and isoprenoid moieties, in which prenylation is catalyzed by prenyltransferases acting as tailoring enzymes [7,8,13]. The subsequent prenylation cascade of echinulin and neoechinulin series of alkaloids in *Aspergillus ruber* has been demonstrated to be mediated via two prenyltransferases, EchPT1 and EchPT2. These two enzymes are involved in the prenylation and reverse prenylation cascades taking place at different positions of the indole DKP core structure of echinulin and neoechinulin congeners, resulting in the formation of mono-, di-, tri-, and tetraprenylated derivatives [7,8]. Indole DKPs are not only characterized by the diversities in their chemical structures but are also reported to exhibit significant biological activities, such as anti-inflammatory, anticancer, antiviral, anti-neurogenerative, and antioxidant activities [9,14]. Various indole DKPs have been evaluated for their anticancer activities against several cancer lines such as HL-60, P388, BEL-7402, A-549, PC12, and HeLa cells via induction of apoptosis and/inhibition of cell proliferation [14–18]. Owing to their improved lipophilicity, prenylated natural products are proven to possess better binding affinity to target proteins and better bioavailability compared to nonprenylated derivatives, creating a direct impact on the biological activity [19,20]. Therefore, prenylated DKPs have demonstrated to be possible candidates for drug development and discovery [21,22]. In the current study, we report the isolation of different prenylated indole diketopiperazine alkaloids, including a new triprenylated derivative rubrumline P (**1**) and a rare tetraprenylated known congener 1Q2 (**7**), from a marine sediment-derived fungus, *A. chevalieri*, through a bioassay-guided fractionation. In addition, the crystal structure of a known derivative, rubrumline I (**2**), is reported for the first time. Genome sequencing of the fungus revealed secondary metabolite regions presumably involved in the synthesis of the reported prenylated metabolites. Furthermore, an analysis of the isolated compounds' cytotoxic efficacy against pancreatic cancer cell line PANC-1 cells has been performed, followed by cell cycle analyses of the isolated compounds.

2. Results and Discussion

2.1. Bioactivity-Guided Fractionation of the Ethyl Acetate Extract of Aspergillus chevalieri

After cultivating *A. chevalieri* on a solid rice medium supplemented with 3.5% NaCl (mimicking the salinity of seawater), the crude ethyl acetate extract of the fungus was initially subjected to chromatographic separation via vacuum liquid chromatography. Subsequently, the produced VLC fractions were preliminary tested for their cytotoxic potential against the human pancreatic PANC-1 cancer cell line at a concentration of 100 μg/mL. Remarkable cytotoxic activity was observed for fractions 4 and 5 after 72 h compared to 1 μM of the microtubule inhibitor paclitaxel, part of the first-line treatment regimen for PDAC patients (Figure 1). Incubation of fraction 4 for 72 h reduced the viability of PANC-1 cells by about 99%. A comparable result was observed for fraction 5. Based on the bioassay result and HPLC profile of fraction 4 (100% EtOAc) compared to those of fraction 5 (90% DCM-MeOH), which was an oily fraction, fraction 4 was selected for further chromatographic separations to isolate and identify the active compounds.

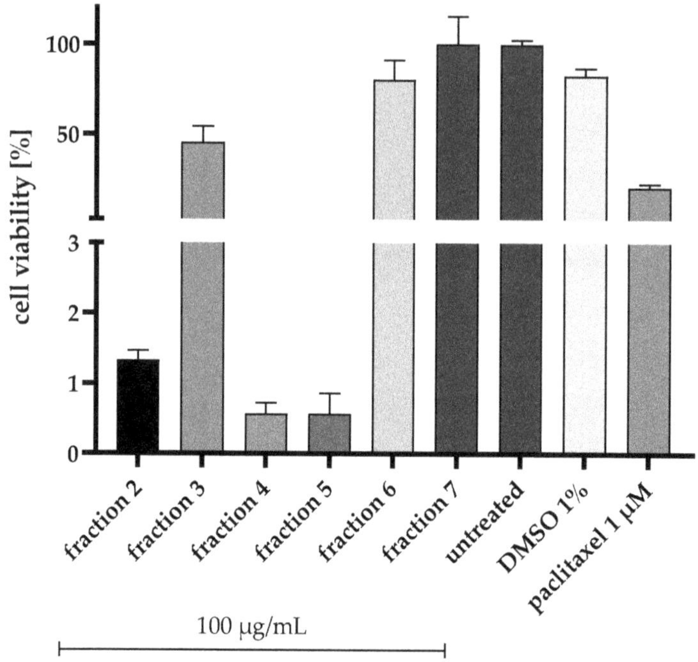

Figure 1. Results of the bioassay-guided fractionation: Cytotoxic effects of the vacuum liquid chromatography (VLC) fractions of *Aspergillus chevalieri* on PANC-1 cells. Cells were seeded into a 96-well plate at a concentration of 5×10^3 cells/well and treated for 72 h after seeding with the different VLC fractions at a concentration of 100 µg/mL. A solvent control (1% DMSO) and a positive control (1 µM paclitaxel) were incubated for 72 h. Cell viability is expressed as the percentage of untreated cells. Error bars indicate standard errors of the mean (SEM) (n = 1 with three technical replicates per subjected fraction).

2.2. Structure Elucidation

After our chromatographic fractionation of the crude extract of *A. chevalieri* on VLC, various separations were conducted on other stationary phases, followed by final purifications utilizing semi-preparative HPLC affording seven indole diketopiperazine alkaloids, compounds **1–7** (Figure 2), including one new alkaloid (**1**) in addition to six known mono-, di-, tri-, and tetraprenylated analogues identified as rubrumline I (**2**) [23], neoechinulin A (**3**) [24], neoechinulin D (**4**) [25], variecolorin G (**5**) [26], dehydroechinulin (**6**) [27], and 1Q2 (**7**) [8] through comparison of their 1D/2D NMR spectroscopic data with those in the literature.

Compound **1** was isolated as a beige-brown residue. Its molecular formula was determined to be $C_{30}H_{41}N_3O_4$ based on a positive HRESIMS analysis, indicating 12 degrees of unsaturation. The ^{1}H-NMR data (Table 1) and HSQC analysis exhibited signals for two meta-coupled aromatic protons (δ_H 6.93, d, J = 1.5 Hz, H-4; and δ_H 6.74, d, J = 1.5 Hz, H-6), six singlet methyls at δ_H 1.51, 1.50, 1.28, 1.23, 1.72, and 1.71, assigned for H_3-18, H_3-19, H_3-24, H_3-25, H_3-29, and H_3-30, respectively, a doublet methyl (δ_H 1.60, d, J = 7.0, H_3-20), one methoxy group (δ_H 3.28, 23-OCH_3), and three protons of a vinyl group (δ_H 6.05, dd, J = 17.4, 10.5 Hz, H-16; δ_H 5.14, d, J = 10.5 Hz, H-17a; and δ_H 5.12, d, J = 17.4 Hz, H-17b). Moreover, proton signals for two methylenes (H-21 and H-26), four methines—including one aliphatic proton (δ_H 4.27, q, J = 6.9 Hz; H-12), two olefinic protons (δ_H 7.23, s, H-8; and δ_H 5.35, m, H-27), and one oxygenated proton (δ_H 3.80, d, J = 8.7 Hz, H-22)—and three NH protons (NH-1, NH-11, and NH-14) were also detected. The APT and HMBC spectroscopic data (Table 1) assigned the presence of 30 carbons, allocated for: seven methyl groups and one methoxy group (CH_3-18, CH_3-19, CH_3-20, CH_3-24, CH_3-25, CH_3-29, CH_3-30, and 23-OCH_3); three methylenes, including one olefinic (CH_2-17) and two aliphatic (CH_2-21, and CH_2-26) groups; seven methines, with five olefinic/aromatic carbons (CH-4, CH-6, CH-8, CH-16, and CH-27), one oxygenated (CH-22), and one aliphatic (CH-12); and twelve quaternary carbons, accounting for two carbonyls (C-10, and C-13), eight olefinic/aromatic carbons (C-2, C-3, C-3a, C-5, C-7, C-7a, C-9, and C-28), one oxygenated (C-23), and one aliphatic carbon (C-15). Detailed inspection of the 1D and 2D NMR spectra of compound **1** revealed that compound **1** is an indole diketopiperazine derivative that is structurally similar to the known co-isolated rubrumline I (**2**), which was previously only characterized once from *Eurotium rubrum* F33, a marine sediment-derived fungus collected at a depth of 2067 m under the South Atlantic Ocean [23]. However, the signal for a methine group resonating at δ_H 7.06/δ_C 117.3 (CH-5) in rubrumline I was replaced by signals for an isoprenyl moiety in **1**. This assignment was suggested from the molecular composition of **1** ($C_{30}H_{41}N_3O_4$), which, compared to rubrumline I (**2**) ($C_{25}H_{32}N_3O_4$), bears an additional isoprenyl group (-C_5H_9) constituting the last remaining degree of unsaturation. Moreover, this assignment was further deduced based on COSY correlations between H_2-26 and H-27, together with the HMBC correlations from H_2-26 to C-4 (δ_C 116.6), C-5 (δ_C 134.5), C-6 (δ_C 123.8), and C-28 (δ_C 132.0) and from H_3-29/H_3-30 to C-28, and C-27 (δ_C 124.3) (Figure 3A), together with the NOE correlations between H_2-26 and H-4, and between H_2-26 and H-6. The geometry of the double bond at C-8 was assigned to be *Z* based on the NOESY correlations between H-4 and NH-14 and between H-8 and CH_3-18, together with the absence of cross peaks from H-8 to NH-14, which suggests the stable configuration of compound **1** (Figure 3B) that is also identical to the X-ray diffraction analysis structure of compound **2** (Figure 4). Moreover, the downfield shift of the olefinic proton H-8 (δ_H 7.23) implies a deshielding effect of the 10- carbonyl group, confirming the *Z* configuration of the double bond at C-8 [23,26,28]. Therefore, compound **1** was elucidated as a 5-prenyl analog of compound **2**, representing a new rubrumline derivative, for which the trivial name rubrumline P is proposed.

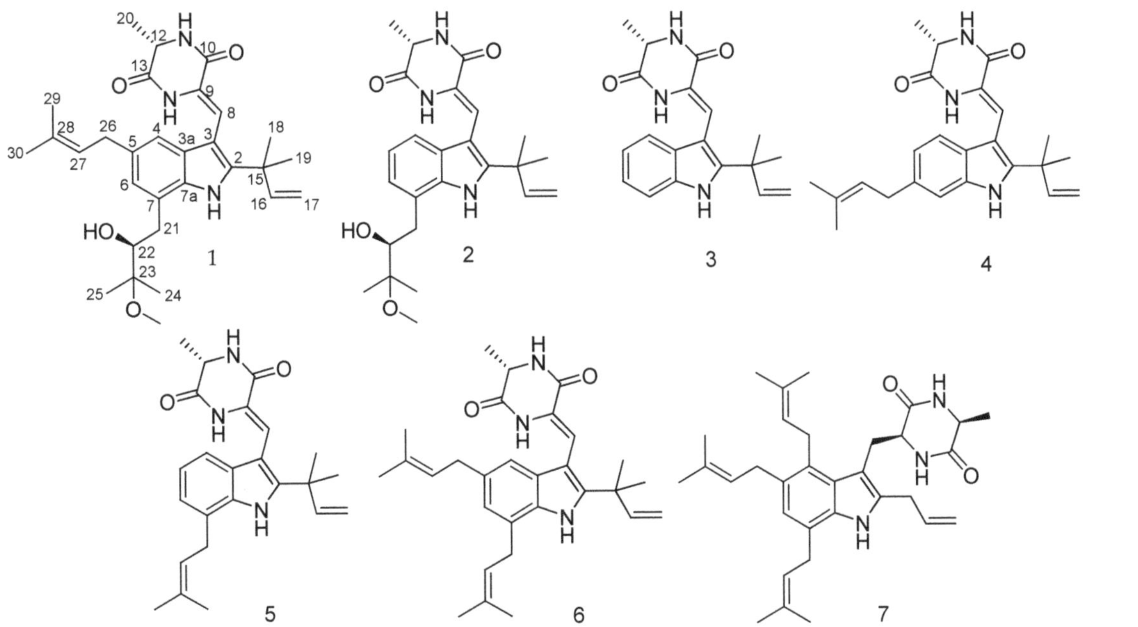

Figure 2. Structures of the isolated compounds from *Aspergillus chevalieri*.

Table 1. ^{1}H (600 MHz) and ^{13}C (150 MHz) NMR spectroscopic data recorded in $CDCl_3$ for compound **1** (δ in ppm).

No.	δ_C, Type	δ_H (Mult, *J* in Hz)	HMBC (from H to C)
2	144.7, C		
3	102.8, C		
3a	126.6, C		
4	116.6, CH	6.93, d (1.5)	3, 6, 7a, 21
5	134.5, C		
6	123.8, CH	6.74, d (1.5)	4, 7a, 26
7	123.5, C		
7a	132.9, C		
8	113.2, CH	7.23, s	2, 3a, 10
9	124.0, C		
10	160.4, C		
12	51.8, CH	4.27, q (6.9)	10, 13, 20
13	165.5, C		
15	39.4, C		
16	144.6, CH	6.05, dd (17.4, 10.5)	2, 15, 18, 19
17	112.7, CH_2	5.14, d (10.5) 5.12, d (17.4)	15, 16
18	27.6, CH_3	1.51, s	2, 15, 16, 19
19	27.5, CH_3	1.50, s	2, 15, 16, 18
20	21.1, CH_3	1.60, d (6.9)	12, 13
21	36.3, CH_2	2.85, d (15.0) 2.94, dd (15.0, 8.7)	7, 7a, 28
22	78.2, CH	3.80, d (8.7)	7, 26, 28, 29, 30
23	77.8, C		
24	20.5, CH_3	1.28, s	27, 28, 30
25	18.5, CH_3	1.23, s	27, 28, 29
26	34.6, CH_2	3.38, m	4, 5, 6, 22, 23
27	124.3, CH	5.35, m	24, 25
28	132.0, C		
29	25.9, CH_3	1.72, s	22, 23, 25
30	18.0, CH_3	1.71, s	22, 23, 24
1-NH		10.21, s	2, 3, 3a, 7a
11-NH		6.34, br s	10, 13
14-NH		7.47, br s	10, 12, 13
23-OCH_3	49.4, OCH_3	3.28, s	28

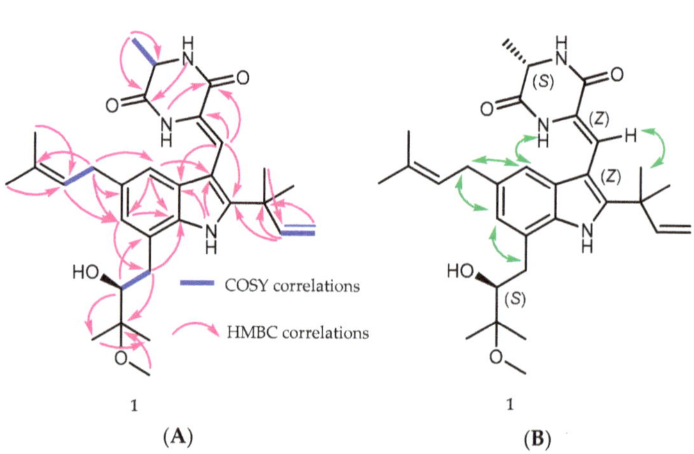

Figure 3. (**A**) Key ^{1}H-^{1}H COSY and HMBC correlations of compound **1**. (**B**) Key NOESY correlations of compound **1**.

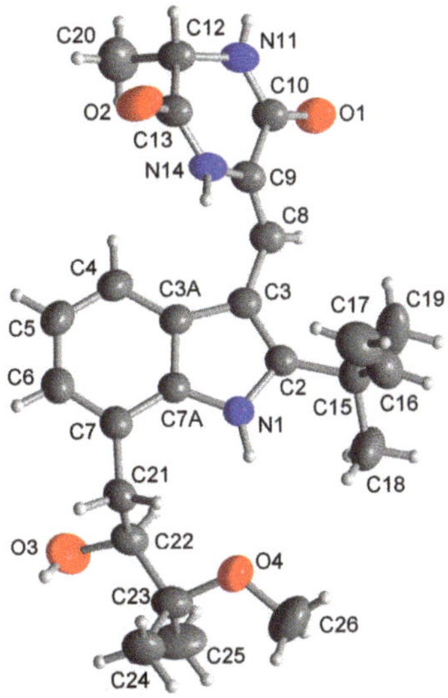

Figure 4. Molecular structure of **2** from a single-crystal X-ray structure determination analysis (50% thermal ellipsoids, H atoms with arbitrary radii). See Figure S9 in the Supplementary Materials for the packing diagram and a diagram of the hydrogen-bonding network.

The specific optical rotation of compound **1** is negative, which is consistent with the negative values reported for other known neoechinulin A series of indole diketopiperazine alkaloids with 12*S* and 22*S* configurations, proposing that the new compound **1** shares the same absolute configuration as the known derivatives [23,24,29]. Although the absolute configuration of the previously reported rubrumline derivatives was assigned using a modified Mosher's method and CD measurement, no crystal structure has been reported so far for these derivatives [23]. Herein, we present an independent assignment of the absolute configuration of the known rubrumline I (**2**) via single-crystal X-ray diffraction analysis through anomalous dispersion for the first time, allowing an unambiguous assignment of the absolute configuration of **2** as 12*S* and 22*S* (Figure 4). Based on NMR data, optical rotations and biogenetic considerations, the same absolute configurations, as elucidated for compound **2** via X-ray diffraction analysis, can be assumed for the other alkaloids as well.

2.3. Genome Sequencing

The genome assembly of *Aspergillus chevalieri* resulted in a total length of 29,419,542 bp, with an N50 fragment length of 1,488,407 bp and an average coverage of 3960×. The assembly consists of 192 contigs, of which 8 contigs contain more than 1,000,000 base pairs (Table 2). These eight contigs correspond to the eight chromosomes of the fungal genome as represented in the reference strain in the NCBI database (NCBI RefSeq assembly: GCF_016861735.1) [30].

Regions coding for potential secondary metabolites were found on six of these eight contigs (Figure 5). Most were gene clusters for non-ribosomal peptide synthetases (NRPSs), which were found on five contigs, followed by regions for type I polyketide synthases (T1PKSs), terpenes, and fungal ribosomally synthesized and post-translationally modified peptide (RiPP)-like secondary metabolites, on two contigs each. Regions for NI-siderophones were also found on one contig. In an additional analysis with InterProScan, a potential region for terpenoid cyclases and protein prenyltransferases was also found on a 21,917 bp-long contig, which does not belong to the core genome.

Table 2. Length and coverage of the *Aspergillus chevalieri* genome contigs and the identified secondary metabolite regions.

Contig No.	Length (bp)	Coverage (x-Fold)	Secondary Metabolite Regions
2	4,649,651	4036	NRPS-like; fungal-RiPP-like; terpene; NI-siderophore
187	3,079,001	4045	T1PKS
123	2,864,395	4042	NRPS; T1PKS
260	1,732,854	4039	NRPS
71	1,589,188	4047	NRPS; fungal-RiPP-like; terpene
284	1,488,407	4039	NRPS-like
328	1,357,148	4052	no secondary metabolite region found
315	1,177,976	4057	no secondary metabolite region found
149	21,917	3994	Terpenoid cyclases/Protein prenyltransferases

Analysis of the gene clusters in the genome of *A. chevalieri* revealed the presence of gene regions for the production of non-ribosomal peptide synthetases, polyketide synthases, and terpenes. There is also an evidence that this fungus is able to synthesize prenyltransferases. Both findings indicate that *A. chevalieri* has the ability to synthesize isoprenylated natural products. The synthesis of isoprenylated natural products through the formation of echinulin prenyltransferase 1 (EchPT1) has so far only been described for other members of the genus *Aspergillus* (*A. fumigatus*, *A. oryzae*, and *A. ruber*), but not yet for *Aspergillus chevalieri* [8,31,32].

2.4. Cytotoxicity and Cell Cycle Analyses in the Pancreatic Ductal Adenocarcinoma Cell Line PANC-1

Based on the remarkable cytotoxicity results from the bioactivity-guided screening of the starting fraction 4, the purified compounds **1** and **4** were characterized in full dose–response curves in the selected pancreatic cancer cell line. The respective cytotoxic efficacy levels are displayed in Table 3.

Table 3. Cytotoxic efficacy in PANC-1 cells represented by the respective IC_{50} values after 72 h of incubation. The arithmetic means of three independent biological repeats and their respective standard deviations are depicted.

Compound	IC_{50} (±SD) [µM]
	PANC-1
1	25.8 ± 1.3
4	23.4 ± 1.8

In contrast to the standard-of-care treatment paclitaxel, for which IC_{50} values in the low nanomolar range are reported [33], compounds **1** and **4** display only moderate cytotoxic efficacy levels, with IC_{50} values of 25.8 and 23.4 µM, respectively.

To determine whether compound **1** or compound **4** affects the cell cycle of PANC-1 cells, a cell cycle analysis was performed. The tumor cells were treated with 30 µM of compound **1** or compound **4** for 24 h, respectively, and stained according to the Nicoletti method. Treatment with the spindle toxin paclitaxel served as a positive control. Treatment with compound **1** or compound **4** did not result in any significant changes in the cell cycle compared to the solvent control DMSO (Figure 6). However, a slight increase in the sub-G1 peak was observed after treatment with compounds **1** and **4**, but this did not reach levels of statistical significance. In contrast, and as expected, treatment with 0.1 µM paclitaxel resulted in the arrest of the cell cycle in the G2 phase and a significant increase in the sub-G1 peak (Figure 6). Therefore, the cytotoxic effects of these compounds might be attributed to inducing apoptosis; however, the exact mechanism-of-action is still elusive.

Figure 5. Identified secondary metabolites regions for *Aspergillus chevalieri*. For each of the six genome assembly contigs on which regions for secondary metabolites were identified, the position of the region and the structure type of the metabolites encoded there are shown. In addition, the most similar metabolite cluster already known is indicated with the associated cluster type and the degree of similarity (%).

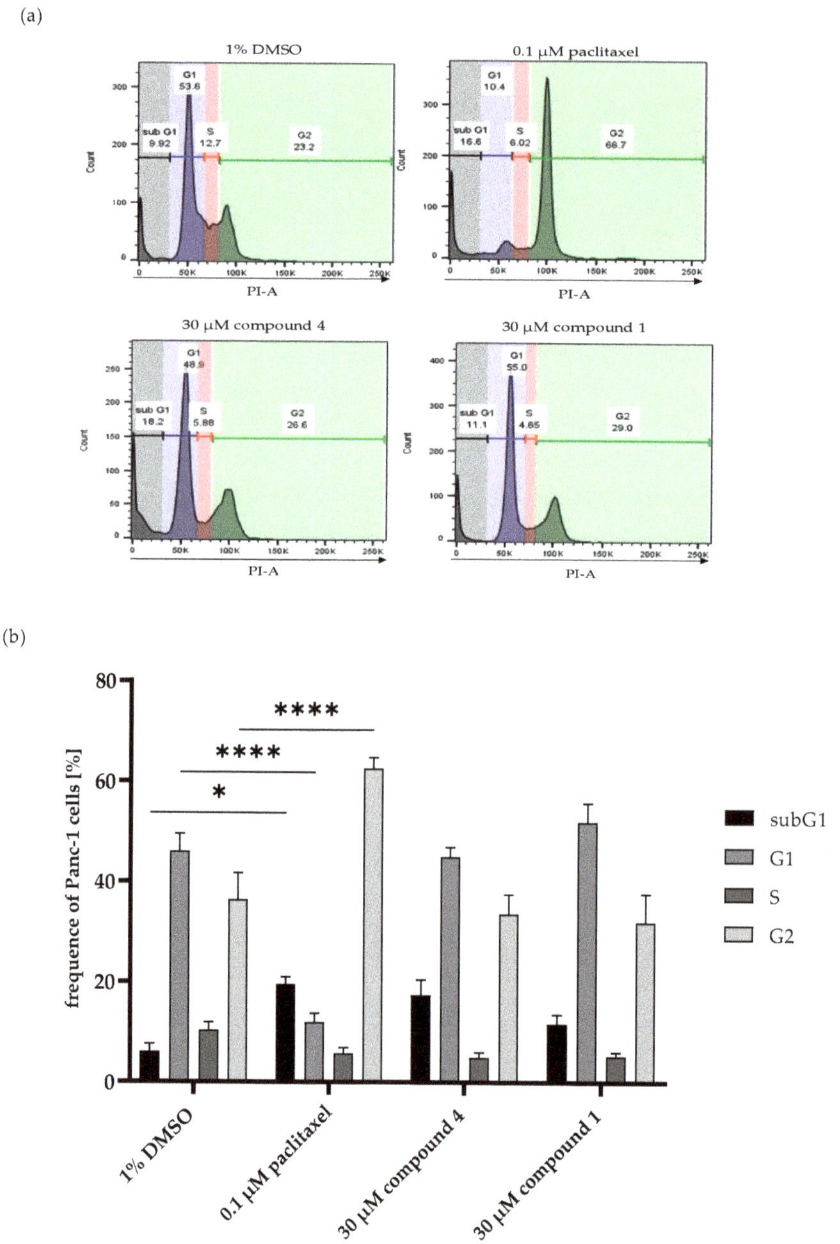

Figure 6. Detection of cell cycle phases in PANC-1 cells. The cells were treated with 30 μM of compound 1, 30 μM of compound 4, or 0.1 μM paclitaxel for 24 h. Cell cycle analysis was performed according to the Nicoletti method. (**a**) Percentage of PANC-1 cells in the different cell cycles after 24 h of treatment with compound 1, compound 4, or paclitaxel, shown in explementary histograms. (**b**) Quantification of cell cycle phases. The cell cycles of at least 10,000 cells were analyzed via flow cytometry. Error bars indicate the standard errors of the mean (n = three independent experiments, with * = $p \leq 0.05$, and **** = $p \leq 0.0001$).

3. Materials and Methods

3.1. General Compound Spectroscopic Analysis and Purification Procedures

Optical rotations were determined using a Jasco P-2000 polarimeter. The compounds were dissolved in optically pure solvents Uvasol® (spectroscopic-grade solvents, Merck, Darmstadt, Germany). The 1D and 2D NMR spectra were recorded in $CDCl_3$ and MeOD using Bruker Avance III 500 or 600 MHz NMR spectrometers (Bruker BioSpin GmbH, Rheinstetten, Germany). Low-resolution mass spectra (ESI) were measured with an Ion-Trap-API Finnigan LCQ Deca (Thermo Quest, Egelsbach, Germany) mass spectrometer, while high-resolution mass spectrometry (HRESIMS) data were recorded on an FTHRMS-Orbitrap (Thermo-Finnigan, Egelsbach, Germany) mass spectrometer. HPLC analysis was conducted using a Dionex UltiMate-3400 SD with an LPG-3400SD pump coupled to a photodiode array detector (DAD3000RS), with routine detections at 235, 254, 280, and 340 nm. The separation column was a Knauer Eurospher C18 analytical column (125 × 4 mm i.d., 5 m), using the solvent gradient MeOH-(0.1% HCOOH in H_2O), as follows: 0 min (10% MeOH); 5 min (10% MeOH); 35 min (100% MeOH); 45 min (100% MeOH). Purification of the compounds was performed using semipreparative HPLC on the VWR Hitachi Chromaster HPLC system (5160 pump; 5410 UV detector; Eurosphere-100 C18, 300 mm × 8 mm i.d., 10 m; Knauer, Germany) utilizing MeOH and H_2O as eluting solvents, employing a flow rate of 5 mL/min. Column chromatography included various stationary phases such as Sephadex LH-20 (0.25–0.1 mm mesh size, Merck) and Merck MN silica gel 60 M (0.04–0.063 mm mesh size). TLC plates precoated with silica gel F254 (Merck) were used for monitoring fractions. Detection of spots on TLC was performed through UV absorption at 254 and 365 nm, or by spraying the plates with an anisaldehyde spray reagent followed by heating.

3.2. Fungal Material

The fungus *Aspergillus chevalieri* LH15-100R2 was isolated from a marine sediment, which was collected from the Red Sea in November 2017 at a depth of 15 m near to the lighthouse of Dahab, Egypt. The fungus was identified as *A. chevalieri* (GenBank accession no. OR605556) through amplification and sequencing of the internal transcribed spacer region (ITS), including the 5.8S ribosomal DNA, followed by a BLAST search in NCBI, as described before [34]. This fungal strain is stored at −80 °C in the Institute of Pharmaceutical Biology and Biotechnology, Heinrich Heine University, Düsseldorf.

3.3. Fermentation, Extraction, and Isolation

Fermentation of the fungus was carried out in ten 1L Erlenmeyer flasks, each containing 100 g of rice (Oryza Milchreis), 3.5 g of NaCl, and 110 mL of demineralized water. After autoclaving at 121 °C for 20 min then cooling to room temperature, the fungus was transferred to the solid rice medium. The fungal culture was left to grow under static conditions for 23 days at room temperature until the fungus had totally overgrown the medium. Afterwards, the solid fermented fungus was extracted two times with EtOAc (each with 600 mL). After soaking overnight in EtOAc, the solid medium was cut into small pieces and then shaken for 8 h at 150 rpm. The combined extracts were concentrated via evaporation of EtOAc under reduced pressure, yielding around 15 g of EtOAc crude extract. The crude extract (15 g) was fractionated through VLC (vacuum liquid chromatography) on silica gel as a stationary-phase with a gradient elution of solvents consisting of mixtures of *n*-hexane/EtOAc and CH_2Cl_2/MeOH, to afford seven fractions (A. ch Fr. 1–7). Fraction A. ch. Fr. 4 (640 mg), eluted with 100% EtOAc, was further purified via Sephadex LH20 column chromatography using CH_2Cl_2-MeOH (1:1) as a mobile phase, yielding three subfractions (A. ch Fr. 4.1–4.3). Subfraction (A. ch Fr. 4.2) (205.6 mg) was purified using semi-preparative HPLC with a gradient of MeOH-H_2O containing 0.1% formic acid (65:35 to 95:05 in 21 min) to afford **1** (4.43 mg), **2** (1.04 mg), **3** (11.02 mg), **4** (40.48 mg), **5** (4.60 mg), **6** (3.01 mg), and **7** (1.47 mg) eluting at retention times; 33.5, 29.8, 25.7, 31.4, 32.0, 35.2, and 36.9 min, respectively

Rubrumline P (1): beige-brown residue; $[\alpha]^{20}{}_{D}$-14.216 (*c* 0.10, $CHCl_3$); UV (MeOH) λ_{max} 371.0, 341.0 and 210.8 nm; ^1H and ^{13}C NMR data, Table 1; HRESIMS *m/z* 508.3173 $[M + H]^+$ (calcd for $C_{30}H_{42}N_3O_4$, 508.3170).

3.4. Crystallographic Analysis of Compound 2

Crystals were obtained through solvent evaporation (MeOH). The data collection process was as follows: Single crystals were measured on a Rigaku XtaLAB Synergy-S, Dualflex, HyPix diffractometer with a micro-focus X-ray tube, with Cu-Kα radiation (λ = 1.54182 Å) at 301.6(1) K. Cell refinement, data collection, and data reduction were performed with CrysalisPro [35]. The Structure solution was conducted using SHELXT 2014/4 [36]. Structure refinement was performed with SHELXL 2017/1 [37], and implemented in the Olex2 software (version 1.5) package [38]. All non-hydrogen atoms were refined with anisotropic displacement parameters. All hydrogen atoms on C were positioned geometrically (with C–H = 0.93 Å for aromatic and vinyl CH, 0.98 Å for aliphatic CH, 0.97 Å for CH_2, and 0.96 Å for CH_3) and refined using riding models (AFIX 43, 93, 13, 23, and 137 with $U_{iso(H)} = 1.2\ U_{eq(C)}$ (CH, CH_2) and $1.5\ U_{eq(C)}$ (CH_3)). Protic hydrogen atoms on N and O were found and refined with $U_{iso(H)} = 1.5\ U_{eq(N/O)}$. The crystal structure was deposited into the Cambridge Crystallographic Data Center (CCDC no. 2308479).

3.5. DNA Isolation and Genome Sequencing

DNA isolation was performed according to the Quick DNA fungal/bacterial miniprep protocol provided by Zymo Research [39].

For long-read Nanopore sequencing, the NBE_9169_v114_revH_15Sep2022 protocol for ligation sequencing of genomic DNA was followed, using the SQK-NBD114 native barcoding kit and a PromethION Flow Cell R10 Version. Reads were assembled using Flye Version 2.9.2-b1794 [40] with asmCoverage-Mode. Antismash [41] and InterProScan Version 5.58-91.0 [42] were used to detect gene clusters and identify potential genes of interest.

3.6. Cell Culture and Cytotoxic Activity

The PDAC cell line PANC-1, purchased from the American Type Culture Collection (ATCC), was cultured in Dulbecco's modified Eagle medium (DMEM) (#41965039, Gibco, Grand Island, NY, USA) supplemented with 10% fetal bovine serum (FBS) (#10270-106, Gibco, Grand Island, NY, USA), as well as 1% penicillin-streptomycin (#15140122, Gibco, Grand Island, NY, USA). Cells were incubated at 37 °C in a humidified atmosphere supplemented with 5% CO_2. Cell viability was assessed using the PrestoBlue HS Cell Viability Assay (#P50201, Invitrogen^{TM}, Waltham, MA, USA) in 96-well plates (cat. no. 655090, Greiner Bio-One, Frickenhausen, Germany) by seeding 5000 cells per well. Compounds were incubated for 72 h. Paclitaxel (#sc-201439, Santa Cruz Biotechnology, Heidelberg, Germany) dissolved in DMSO (#A994.1, Carl Roth, Karlsruhe, Germany) was used as a positive control. Fluorescence values were recorded using the Tecan SPARK instrument (Tecan Group, Männedorf, Switzerland), according to the manufacturer's protocol.

3.7. Cell Cycle Analysis

For analysis of the cell cycles, 9×10^4 cells/mL of PANC-1 were cultured in 6-well plates (Greiner Bio-One). After treatment for 24 h and harvesting via trypsin, the cells were washed with D-PBS. For cell cycle analyses, the cells were suspended in 100 µL of hypotonic buffer (1% sodium citrate, 0.1% Triton X-100, and 50 µg/mL propidium iodide in double-distilled water). The suspension was incubated for 10 min at room temperature. Flow cytometry was performed with the BD FACSLyric flow cytometer (#87135, BD Biosciences, Heidelberg, Germany), whereby FlowJo software (version 10.8.1) was utilized for the analysis.

3.8. Statistical Analysis

GraphPad Prism version 8.4.3 (GraphPad Software, California, USA) was used to perform statistical analyses and graphical illustrations. Non-linear regression was performed to assess cell viability, obtaining IC_{50} values. To determine statistical significance, a one-way analysis of variance (ANOVA) was used, wherein a p-value below 0.05 was defined as statistically significant.

4. Conclusions

Following a bioassay-guided strategy for the isolation of presumable drugs or drug-led structures originating from a natural source for use against the human pancreatic adenocarcinoma cell line PANC-1, several indole diketopiperazine alkaloids, including a new triprenylated alkaloid, named rubrumline P (**1**), together with other known prenylated derivatives (**2–7**) were obtained from rice culture supplemented with 3.5% NaCl of the marine sediment-derived fungus, *Aspergillus chevalieri*. Two isolated indole DKPs, rubrumline P (**1**) and neoechinulin D (**4**), showed cytotoxic efficacy with IC_{50} values of 25.8 and 23.4 μM, respectively. In addition, a cell cycle analysis of the isolated secondary metabolites has been conducted. Most notably, our analysis of the fungal genome of *A. chevalieri*, which is described for the first time, revealed the ability of this fungus to synthesize prenyltransferases as tailoring enzymes involved in the production of prenylated compounds, as evidenced by the isolation of different prenylated indole DKPs from the fungus. This represents a starting point for the optimization of such prenylated alkaloids for future drug discoveries and development.

Supplementary Materials: The following supporting information can be downloaded at: https://www.mdpi.com/article/10.3390/md22010005/s1, Figures S1–S10: The UV spectrum, HRESIMS, 1D and 2D NMR spectra of compound **1**, and the packing diagram, determined via single-crystal x-ray diffraction, and hydrogen bond network of compound **2**; Tables S1–S5: crystal data and refinement, fractional atomic coordinates, atomic displacement parameters, geometric parameters, and hydrogen bond geometry for compound **2.**

Author Contributions: Investigation, D.H.E.-K., D.D.O., K.S., A.D., S.S., D.N., S.M., T.H.H.S. and H.W.; concept development, D.H.E.-K. and N.T.; resources, D.F., C.J, K.P. and N.T.; writing—original draft preparation, D.H.E.-K., K.S., A.D., S.S. and N.T.; writing—review and editing, C.J. and N.T. All authors have read and agreed to the published version of the manuscript.

Funding: This study received funding from Deutsche Forschungsgemeinschaft (DFG) under grant no. 440366605 (for the Rigaku diffractometer).

Institutional Review Board Statement: Not applicable.

Data Availability Statement: The CCDC number 2308479 for compound **2** contains the supplementary crystallographic data reported in this paper. These data can be obtained free of charge from the Cambridge Crystallographic Data Center via www.ccdc.cam.ac.uk/data_request/cif (accessed on 16 November 2023).

Acknowledgments: We would like to express our deep gratitude to Peter Proksch for his careful and critical revision of the manuscript and for his valuable and constant support as a mentor and friend. We wish to thank Dent. Abdel Rahman O. El Mekkawi, EFR, PADI IDC staff instructor, founder of I Dive Tribe, Dahab, South Sinai, Egypt, for collecting the sediment sample.

Conflicts of Interest: The authors declare no conflicts of interest.

References

1. Słodkowski, M.; Wroński, M.; Karkocha, D.; Kraj, L.; Śmigielska, K.; Jachnis, A. Current Approaches for the Curative-Intent Surgical Treatment of Pancreatic Ductal Adenocarcinoma. *Cancers* **2023**, *15*, 2584. [CrossRef] [PubMed]
2. Adamska, A.; Domenichini, A.; Falasca, M. Pancreatic Ductal Adenocarcinoma: Current and Evolving Therapies. *Int. J. Mol. Sci.* **2017**, *18*, 1338. [CrossRef] [PubMed]
3. He, X.; Wang, N.; Zhang, Y.; Huang, X.; Wang, Y. The therapeutic potential of natural products for treating pancreatic cancer. *Front. Pharmacol.* **2022**, *13*, 1051952. [CrossRef] [PubMed]

4. Zhang, Z.; Wang, J.; Liu, B.; Liu, Y.; Shi, X.; Li, W.; Xin, H.; Xin, J.; Hao, C. Anticancer effects of herbal medicines in pancreatic ductal adenocarcinoma through modulation of steroid hormone response proteins. *Sci. Rep.* **2022**, *12*, 9910. [CrossRef] [PubMed]
5. Kim, A.; Ha, J.; Kim, J.; Cho, Y.; Ahn, J.; Cheon, C.; Kim, S.-H.; Ko, S.-G.; Kim, B. Natural Products for Pancreatic Cancer Treatment: From Traditional Medicine to Modern Drug Discovery. *Nutrients* **2021**, *13*, 3801. [CrossRef]
6. Xie, B.; Hänsel, J.; Mundorf, V.; Betz, J.; Reimche, I.; Erkan, M.; Büdeyri, I.; Gesell, A.; Kerr, R.G.; Ariantari, N.P.; et al. Pseudopterosin and O-Methyltylophorinidine Suppress Cell Growth in a 3D Spheroid Co-Culture Model of Pancreatic Ductal Adenocarcinoma. *Bioengineering* **2020**, *7*, 57. [CrossRef] [PubMed]
7. Wohlgemuth, V.; Kindinger, F.; Li, S.M. Convenient synthetic approach for tri- and tetraprenylated cyclodipeptides by consecutive enzymatic prenylations. *Appl. Microbiol. Biotechnol.* **2018**, *102*, 2671–2681. [CrossRef] [PubMed]
8. Wohlgemuth, V.; Kindinger, F.; Xie, X.; Wang, B.G.; Li, S.M. Two Prenyltransferases Govern a Consecutive Prenylation Cascade in the Biosynthesis of Echinulin and Neoechinulin. *Org. Lett.* **2017**, *19*, 5928–5931. [CrossRef]
9. Ma, Y.-M.; Liang, X.-A.; Kong, Y.; Jia, B. Structural Diversity and Biological Activities of Indole Diketopiperazine Alkaloids from Fungi. *J. Agric. Food Chem.* **2016**, *64*, 6659–6671. [CrossRef]
10. Xu, W.; Gavia, D.J.; Tang, Y. Biosynthesis of fungal indole alkaloids. *Nat. Prod. Rep.* **2014**, *31*, 1474–1487. [CrossRef]
11. Giessen, T.W.; Marahiel, M.A. Rational and combinatorial tailoring of bioactive cyclic dipeptides. *Front. Microbiol.* **2015**, *6*, 785. [CrossRef] [PubMed]
12. Winkelblech, J.; Fan, A.; Li, S.-M. Prenyltransferases as key enzymes in primary and secondary metabolism. *Appl. Microbiol. Biotechnol.* **2015**, *99*, 7379–7397. [CrossRef] [PubMed]
13. Li, S.-M. Applications of dimethylallyltryptophan synthases and other indole prenyltransferases for structural modification of natural products. *Appl. Microbiol. Biotechnol.* **2009**, *84*, 631–639. [CrossRef] [PubMed]
14. Sharifi-Rad, J.; Bahukhandi, A.; Dhyani, P.; Sati, P.; Capanoglu, E.; Docea, A.O.; Al-Harrasi, A.; Dey, A.; Calina, D. Therapeutic Potential of Neoechinulins and Their Derivatives: An Overview of the Molecular Mechanisms Behind Pharmacological Activities. *Front. Nutr.* **2021**, *8*, 664197. [CrossRef] [PubMed]
15. Gao, H.; Zhu, T.; Li, D.; Gu, Q.; Liu, W. Prenylated indole diketopiperazine alkaloids from a mangrove rhizosphere soil derived fungus *Aspergillus effuses* H1-1. *Arch. Pharmacal Res.* **2013**, *36*, 952–956. [CrossRef] [PubMed]
16. Cai, S.; Sun, S.; Peng, J.; Kong, X.; Zhou, H.; Zhu, T.; Gu, Q.; Li, D. Okaramines S–U, three new indole diketopiperazine alkaloids from *Aspergillus taichungensis* ZHN-7-07. *Tetrahedron* **2015**, *71*, 3715–3719. [CrossRef]
17. Akashi, S.; Kimura, T.; Takeuchi, T.; Kuramochi, K.; Kobayashi, S.; Sugawara, F.; Watanabe, N.; Arai, T. Neoechinulin a impedes the progression of rotenone-induced cytotoxicity in PC12 cells. *Biol. Pharm. Bull.* **2011**, *34*, 243–248. [CrossRef]
18. Wijesekara, I.; Li, Y.-X.; Vo, T.-S.; Van Ta, Q.; Ngo, D.-H.; Kim, S.-K. Induction of apoptosis in human cervical carcinoma HeLa cells by neoechinulin A from marine-derived fungus *Microsporum* sp. *Process Biochem.* **2013**, *48*, 68–72. [CrossRef]
19. Terao, J.; Mukai, R. Prenylation modulates the bioavailability and bioaccumulation of dietary flavonoids. *Arch. Biochem. Biophys.* **2014**, *559*, 12–16. [CrossRef]
20. Botta, B.; Vitali, A.; Menendez, P.; Misiti, D.; Delle Monache, G. Prenylated flavonoids: Pharmacology and biotechnology. *Curr. Med. Chem.* **2005**, *12*, 717–739. [CrossRef]
21. Liu, R.; Zhang, H.; Wu, W.; Li, H.; An, Z.; Zhou, F. C7-Prenylation of Tryptophan-Containing Cyclic Dipeptides by 7-Dimethylallyl Tryptophan Synthase Significantly Increases the Anticancer and Antimicrobial Activities. *Molecules* **2020**, *25*, 3676. [CrossRef] [PubMed]
22. Li, S.-M. Prenylated indole derivatives from fungi: Structure diversity, biological activities, biosynthesis and chemoenzymatic synthesis. *Nat. Prod. Rep.* **2010**, *27*, 57–78. [CrossRef] [PubMed]
23. Chen, X.; Si, L.; Liu, D.; Proksch, P.; Zhang, L.; Zhou, D.; Lin, W. Neoechinulin B and its analogues as potential entry inhibitors of influenza viruses, targeting viral hemagglutinin. *Eur. J. Med. Chem.* **2015**, *93*, 182–195. [CrossRef] [PubMed]
24. Alhadrami, H.A.; Burgio, G.; Thissera, B.; Orfali, R.; Jiffri, S.E.; Yaseen, M.; Sayed, A.M.; Rateb, M.E. Neoechinulin A as a Promising SARS-CoV-2 Mpro Inhibitor: In Vitro and In Silico Study Showing the Ability of Simulations in Discerning Active from Inactive Enzyme Inhibitors. *Mar. Drugs* **2022**, *20*, 163. [CrossRef] [PubMed]
25. Dossena, A.; Marchelli, R.; Pochini, A. Neoechinulin D, a new isoprenylated dehydrotryptophyl metabolite from *Aspergillus amstelodami*. *Experientia* **1975**, *31*, 1249. [CrossRef]
26. Wang, W.-L.; Lu, Z.-Y.; Tao, H.-W.; Zhu, T.-J.; Fang, Y.-C.; Gu, Q.-Q.; Zhu, W.-M. Isoechinulin-type Alkaloids, Variecolorins A–L, from Halotolerant *Aspergillus variecolor*. *J. Nat. Prod.* **2007**, *70*, 1558–1564. [CrossRef] [PubMed]
27. Li, D.-L.; Li, X.-M.; Li, T.-G.; Dang, H.-Y.; Wang, B.-G. Dioxopiperazine Alkaloids Produced by the Marine Mangrove Derived Endophytic Fungus *Eurotium rubrum*. *Helv. Chim. Acta* **2008**, *91*, 1888–1893. [CrossRef]
28. Marchelli, R.; Dossena, A.; Pochini, A.; Dradi, E. The structures of five new didehydropeptides related to neoechinulin, isolated from *Aspergillus amstelodami*. *J. Chem. Soc. Perkin Trans.* **1977**, *1*, 713–717. [CrossRef]
29. Meng, L.-H.; Du, F.-Y.; Li, X.-M.; Pedpradab, P.; Xu, G.-M.; Wang, B.-G. Rubrumazines A–C, Indolediketopiperazines of the Isoechinulin Class from *Eurotium rubrum* MA-150, a Fungus Obtained from Marine Mangrove-Derived Rhizospheric Soil. *J. Nat. Prod.* **2015**, *78*, 909–913. [CrossRef]
30. Genome Assembly AchevalieriM1_assembly01. Available online: www.ncbi.nlm.nih.gov/datasets/genome/GCF_016861735.1 (accessed on 28 November 2023).

31. Li, W.; Xie, X.; Liu, J.; Yu, H.; Li, S.M. Prenylation of dimeric cyclo-L-Trp-L-Trp by the promiscuous cyclo-L-Trp-L-Ala prenyltransferase EchPT1. *Appl. Microbiol. Biotechnol.* **2023**, *107*, 6887–6895. [CrossRef]
32. Yu, X.; Li, S.-M. Chapter Thirteen—Prenyltransferases of the Dimethylallyltryptophan Synthase Superfamily. In *Methods in Enzymology*; Hopwood, D.A., Ed.; Academic Press: Cambridge, MA, USA, 2012; Volume 516, pp. 259–278.
33. Brumskill, S.; Barrera, L.N.; Calcraft, P.; Phillips, C.; Costello, E. Inclusion of cancer-associated fibroblasts in drug screening assays to evaluate pancreatic cancer resistance to therapeutic drugs. *J. Physiol. Biochem.* **2023**, *79*, 223–234. [CrossRef] [PubMed]
34. Kjer, J.; Debbab, A.; Aly, A.H.; Proksch, P. Methods for isolation of marine-derived endophytic fungi and their bioactive secondary products. *Nat. Protoc.* **2010**, *5*, 479–490. [CrossRef] [PubMed]
35. *CrysAlisPRO, v171.42*; Oxford Diffraction/Agilent Technologies UK Ltd.: Yarnton, UK, 2022.
36. Sheldrick, G.M. SHELXT–Integrated space-group and crystal-structure determination. *Acta Crystallogr. Sect. A Found. Adv.* **2015**, *71*, 3–8. [CrossRef]
37. Sheldrick, G.M. Crystal structure refinement with SHELXL. *Acta Crystallogr. Sect. C Struct. Chem.* **2015**, *71*, 3–8. [CrossRef] [PubMed]
38. Dolomanov, O.V.; Bourhis, L.J.; Gildea, R.J.; Howard, J.A.; Puschmann, H. OLEX2: A complete structure solution, refinement and analysis program. *J. Appl. Crystallogr.* **2009**, *42*, 339–341. [CrossRef]
39. Matsumoto, Y.; Nagamachi, T.; Yoshikawa, A.; Yamada, T.; Sugita, T. A joint PCR-based gene-targeting method using electroporation in the pathogenic fungus *Trichosporon asahii*. *AMB Express* **2022**, *12*, 91. [CrossRef] [PubMed]
40. Kolmogorov, M.; Bickhart, D.M.; Behsaz, B.; Gurevich, A.; Rayko, M.; Shin, S.B.; Kuhn, K.; Yuan, J.; Polevikov, E.; Smith, T.P.L.; et al. metaFlye: Scalable long-read metagenome assembly using repeat graphs. *Nat. Methods* **2020**, *17*, 1103–1110. [CrossRef]
41. Blin, K.; Shaw, S.; Kloosterman, A.M.; Charlop-Powers, Z.; van Wezel, G.P.; Medema, M.H.; Weber, T. antiSMASH 6.0: Improving cluster detection and comparison capabilities. *Nucleic Acids Res.* **2021**, *49*, W29–W35. [CrossRef]
42. Paysan-Lafosse, T.; Blum, M.; Chuguransky, S.; Grego, T.; Pinto, B.L.; Salazar, G.A.; Bileschi, M.L.; Bork, P.; Bridge, A.; Colwell, L.; et al. InterPro in 2022. *Nucleic Acids Res.* **2023**, *51*, D418–D427. [CrossRef]

marine drugs

Article

Cytosporones with Anti-Inflammatory Activities from the Mangrove Endophytic Fungus *Phomopsis* sp. QYM-13

Guisheng Wang [1,2,†], Yilin Yuan [3,†], Zhaokun Li [1], Junhao Zhu [1], Zhigang She [2,*] and Yan Chen [1,*]

1 School of Pharmacy, Anhui Medical University, Hefei 230032, China; wanggsh9@mail2.sysu.edu.cn (G.W.); lizhaokun0223@163.com (Z.L.); 19855683468@163.com (J.Z.)
2 School of Chemistry, Sun Yat-sen University, Guangzhou 510275, China
3 National R&D Center for Edible Fungus Processing Technology, Henan University, Kaifeng 475004, China; yilin@henu.edu.cn
* Correspondence: cesshzhg@mail.sysu.edu.cn (Z.S.) cychemistry@163.com (Y.C.)
† These authors contributed equally to this work.

Abstract: Six previously undescribed cytosporone derivatives (phomotones A-E (**1–5**) and phomotone F (**13**)), two new spiro-alkanol phombistenes A-B (**14–15**), and seven known analogs (**6–12**) were isolated from the mangrove endophytic fungus *Phomopsis* sp. QYM-13. The structures of these compounds were elucidated using spectroscopic data analysis, electronic circular dichroism (ECD), and ^{13}C NMR calculations. Compound **14** features an unprecedented 1,6-dioxaspiro[4.5]decane ring system. All isolates were evaluated for their inhibitory effect on nitric oxide (NO) in LPS-induced RAW264.7 cells. The results showed that compounds **1**, **6**, **8**, and **11** exhibited potent bioactivities by comparing with positive control. Then, compound **1** displayed the anti-inflammatory effect by inhibiting the MAPK/NF-κB signaling pathways. Molecular docking further revealed the possible mechanism of compound **1** interaction with ERK protein.

Keywords: mangrove endophytic fungus; *Phomopsis* sp.; cytosporone; anti-inflammatory

Citation: Wang, G.; Yuan, Y.; Li, Z.; Zhu, J.; She, Z.; Chen, Y. Cytosporones with Anti-Inflammatory Activities from the Mangrove Endophytic Fungus *Phomopsis* sp. QYM-13. *Mar. Drugs* **2023**, *21*, 631. https://doi.org/10.3390/md21120631

Academic Editors: Bin-Gui Wang and Haofu Dai

Received: 21 November 2023
Revised: 1 December 2023
Accepted: 5 December 2023
Published: 7 December 2023

1. Introduction

Cytosporones are a series of polyketide-derived octaketide phenolic lipids [1], the first metabolite of which is cytosporone A, isolated from an endophytic fungi *Cytospora* sp. in 2000 [2]. These compounds were likewise obtained from other fungi, like *Dothiorella*, *Pestalotiopsis*, *Diaporthe*, *Phomopsis*, *Aspergillus*, *Trichoderma*, and so on [3]. Furthermore, cytosporones were reported to have various kinds of biological activities, such as antimicrobial, antimalarial, cytotoxic, antivirus, anti-inflammatory, and allelopathic activity [4–6].

Mangrove endophytic fungi have attracted the attention of many researchers due to their ability to produce structurally novel and remarkably bioactive secondary metabolites [7]. To date, thousands of new metabolites have been isolated from mangrove endophytic fungi [8]. As part of our ongoing search for bioactive compounds from mangrove endophytic fungi, the strain *Phomopsis* sp. QYM-13, isolated from healthy leaves of *Kandelia candel*, was investigated. We recently reported twelve new cytochalasins, phomopchalasins D–O, including brominated and iodinated cytochalasins, with significant cytotoxicity from this fungus [9]. Subsequently, another eight new cytosporones phomotones A-E (**1–5**), phomotone F (**13**), and phombistenes A-B (**14–15**) (Figure 1) were obtained in our further research work. In bioassays, compounds **1**, **6**, **8**, and **11** exhibited significant anti-inflammatory activities. Herein, the isolation, structure elucidation, biological assays, molecular docking, and structure-activity relationship of these isolated compounds are described.

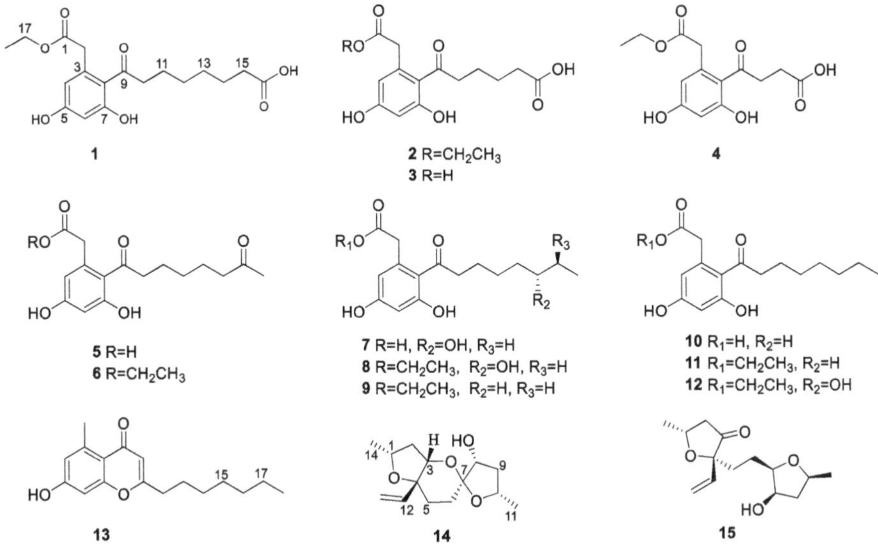

Figure 1. The structures of **1–15**.

2. Results

Compound **1** was isolated as a white solid with the molecular formula of $C_{18}H_{23}O_7$ based on the negative HRESIMS data *m/z* 351.14534 [M-H]$^-$. The ^{1}H NMR spectrum of **1** showed the presence of two aromatic protons (δ_H 6.20 (d, J = 2.2 Hz), 6.26 (d, J = 2.2 Hz)) and one methyl group (δ_H 1.24 (t, J = 7.1 Hz)). The ^{13}C NMR (Table 1) and HSQC spectra (Figure S3) of **1** revealed the coexistence of eighteen carbons attributable to one methyl, eight methylenes (one oxygenated), two methines, and seven unprotonated carbons (three carbonyl carbons and four olefinic carbons). These data suggested **1** to be a cytosporone class, and the ^{1}H and ^{13}C NMR data of **1** were similar to those of **12** **[10]**. The main difference involved the absence of the oxygenated methylene in **12** and the appearance of carboxyl carbon in **1**, which might mean the terminal hydroxyl group in **12** was oxidized to a carboxyl group in **1**. The deduction was further confirmed with the spin system of H_2-10/H_2-11/H_2-12/H_2-13/H_2-14/H_2-15 in the ^{1}H-^{1}H COSY spectrum (Figure 2), as well as the HMBC correlations (Figure 2) from H_2-10 to C-9 and from H_2-15 to C-16. Thus, the structure of **1** was assigned as shown in Figure 1.

Table 1. ^{1}H (500 MHz) and ^{13}C (125 MHz) NMR data for compounds **1–3** in MeOD-d_4.

	1		**2**		**3**	
No	δ_C	δ_H (J in Hz)	δ_C	δ_H (J in Hz)	δ_C	δ_H (J in Hz)
1	173.6		172.1		173.9	
2	40.5	3.58, s	39.0	3.60, s	38.9	3.57, s
3	137.0		135.7		135.8	
4	111.7	6.20, d (2.2)	110.4	6.21, d (2.2)	110.2	6.21, d (2.2)
5	161.3		159.9		160.2	
6	102.7	6.26, d (2.2)	101.3	6.28, d (2.2)	101.3	6.26, d (2.2)
7	159.8		158.4		157.0	
8	121.3		119.8		120.3	
9	208.9		206.9		207.1	

Table 1. *Cont.*

	1		2		3	
No	δ_C	δ_H (J in Hz)	δ_C	δ_H (J in Hz)	δ_C	δ_H (J in Hz)
10	45.1	2.91, t (7.4)	43.3	2.96, t (7.0)	43.2	2.95, m
11	25.4	1.62, m	23.6	1.68, m	23.6	1.65, m
12	30.1	1.35, m	24.5	1.65, m	24.4	1.62, m
13	30.1	1.33, m	33.6	2.32, t (7.0)	33.4	2.30, m
14	26.0	1.59, m	176.3		175.6	
15	35.1	2.28, d (7.4)	60.4	4.12, q (7.1)		
16	178.0		13.1	1.26, t (7.1)		
17	61.8	4.11, q (7.1)				
18	14.5	1.24, d (7.1)				

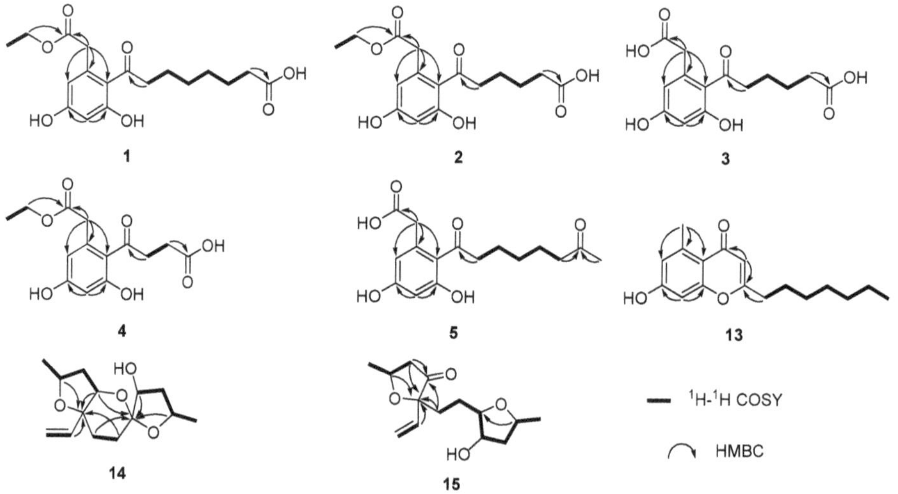

Figure 2. The key ^{1}H-^{1}H COSY and HMBC correlations of compounds **1–5** and **13–15**.

Compound **2** was isolated as a colorless oil with the molecular formula of $C_{16}H_{19}O_7$ based on the negative HRESIMS data m/z 323.11392 [M-H]$^-$. Detailed analysis of the NMR data (Table 1) indicated that **2** was structurally similar to **1**, with the difference being the absence of two methylenes. It was supported by its ^{1}H-^{1}H COSY cross-peaks of H_2-10/H_2-11/H_2-12/H_2-13 and HMBC correlations from H_2-10 to C-9 and from H_2-13 to C-14. Thus, the structure of **2** was established.

Compound **3** was isolated as a colorless oil with the molecular formula of $C_{14}H_{15}O_7$ based on the negative HRESIMS data m/z 295.08227 [M-H]$^-$. The ^{1}H and ^{13}C NMR data of **3** (Table 1) were similar to those of **2**, except for the absence of an ethyl group (δ_C 60.4, 13.1) in **3**. Then, the HMBC correlations (Figure 2) further confirmed the deduction above.

Compound **4** was isolated as a colorless oil with the molecular formula of $C_{14}H_{15}O_7$ based on the negative HRESIMS data m/z 295.08240 [M-H]$^-$. A comparison of the NMR data of **4** (Table 2) with those of **2** revealed that their structures were similar, with the only difference being the absence of two methylenes. The ^{1}H-^{1}H COSY cross-peaks of H_2-10/H_2-11 and HMBC correlations from H_2-10 to C-9 and from H_2-11 to C-12 further confirmed the structure.

Table 2. ^{1}H (500 MHz) and ^{13}C (125 MHz) NMR data for compounds **4–5** and **13**.

	4^{a}		**5**^{a}		**13**^{b}	
No	δ_C	δ_H (*J* in Hz)	δ_C	δ_H (*J* in Hz)	δ_C	δ_H (*J* in Hz)
1	172.3		176.2			
2	39.2	3.58, s	41.0	3.58, s	22.8	2.64, s
3	139.0		137.9		141.9	
4	110.5	6.19, d (2.2)	111.6	6.23, d (2.2)	116.9	6.60, d (2.2)
5	160.1		161.4		159.6	
6	101.5	6.27, d (2.2)	102.6	6.26, d (2.2)	101.0	6.62, d (2.2)
7	158.8		159.8		161.3	
8	119.3		121.3		114.8	
9	204.5		209.5		178.7	
10	38.7	3.20, t (6.8)	25.1	1.64, m	110.5	5.95, s
11	28.5	2.60, t (6.8)	35.3	1.30, m	167.3	
12	175.8		29.7	1.34, m	33.1	2.53, t (7.0)
13	60.4	4.11, q (7.1)	24.7	1.57, m	26.6	1.61, m
14	13.1	1.24, t (7.1)	44.2	2.49, t (7.4)	28.7	1.32, m
15			212.3	2.14, s	28.7	1.30, m
16			29.8		22.5	1.28, m
17					31.6	1.24, m
18					14.4	0.85, t (6.8)

^{a} Measured in MeOD-d_4, ^{b} measured in DMSO-d_6.

Compound **5** was isolated as a colorless oil with the molecular formula of $C_{16}H_{21}O_6$ based on the positive HRESIMS data m/z 309.1333 $[M + H]^+$. The ^{1}H and ^{13}C NMR data of **5** (Table 2) were similar to those of **6** **[10]**, except for the absence of an ethyl group (δ_C 61.5, 14.2) in **5**. The deduction was further confirmed by ^{1}H-^{1}H COSY and HMBC correlations (Figure 2).

For compound **13**, a yellow solid, the molecular formula $C_{17}H_{22}O_3$ with seven degrees of unsaturation was established using HRESIMS. The NMR (Table 2) information of **13** was similar to **10**, with the main difference being the presence of one methyl group (δ_H 2.64), one olefin proton (δ_H 5.95), and two olefin carbons (δ_C 167.3, 110.5) in **13**, and the disappearance of the carboxyl group at C-1 in **10**. The HMBC correlations from H_3-2 to C-3, C-4, and C-8, from H-10 to C-9, C-11, and C-12, together with the spin system of H_2-12/H_2-13/H_2-14/H_2-15/H_2-16/H_2-17/H_3-18 in the ^{1}H-^{1}H COSY spectrum (Figure 2) further confirmed the deduction. Thus, the structure of **13** was verified.

For compound **14**, a yellow solid, the molecular formula $C_{14}H_{22}O_4$ with four degrees of unsaturation was established using HRESIMS. The ^{1}H NMR spectrum (Table 3) showed two methyl groups at δ_H 1.34 (d, J = 6.2 Hz) and 1.30 (d, J = 6.4 Hz), and three olefin protons at δ_H 5.73 (ddd, J = 4.1, 10.7, 14.7 Hz), δ_H 5.07 (d, J = 10.7 Hz), and δ_H 5.30 (d, J = 17.1 Hz). The ^{13}C NMR and HSQC spectra of **14** revealed the existence of fourteen carbons attributable to two methyls, five methylenes (one olefinic carbon), five methines (one olefinic carbon), and two unprotonated carbons. From an analysis of the ^{1}H-^{1}H COSY spectrum, four independent coupling fragments could be inferred (Figure 2). Furthermore, the HMBC correlations from H-1, H-3, H-6, and H2-13 to C-4 and from H-3 and H-5 to C-7 indicated the furan ring and pyranoid ring were fused at positions C-3/C-4, and one group of terminal olefin was located at C-4. Then, the HMBC correlations from H-8 and H-10 to C-7 and the chemical shift at C-4 (δ_C 108.5) enabled the construction of a 1,6-dioxaspiro[4.5]decane ring system. Therefore, the planar structure of **14** was tentatively assigned.

Table 3. ^{1}H (500 MHz) and ^{13}C (125 MHz) NMR data for compounds **14–15** in $CDCl_3$.

	14		**15**	
No	δ_C	δ_H (*J* in Hz)	δ_C	δ_H (*J* in Hz)
1	73.0	4.33, dt (6.2,19.6)	70.2	4.43, ddd (3.3, 6.1,12.2)
2a	40.7	2.52, m	40.8	1.66, dd (2.9, 9.0)
2b		1.44, dd (4.8,13.6)		1.55, dd (2.5, 5.6)
3	72.9	4.24, d (5.8)	214.9	
4	81.8		86.4	
5	23.1	1.85, m	30.8	1.78, m
6	29.8	1.94, m	27.2	1.59, dd (5.0, 9.0)
7	108.5		74.0	3.71, ddd (3.1, 5.8, 9.0)
8	79.0		72.0	3.46, m
9a	39.4	2.21, ddd (5.9, 7.8, 13.5)	43.1	2.58, dd (6.1, 17.8)
9b		1.72, dd (3.8, 5.0)		2.27, dd (9.4, 17.8)
10	73.2	4.14, td (6.9, 12.6)	65.6	4.14, ddd (3.1, 6.3, 8.8)
11	22.4	1.34, d (6.2)	23.7	1.24, d (6.3)
12	140.7	5.73, ddd, (4.1, 10.7, 14.7)	137.1	5.78, dd (10.7,17.1)
13a	113.8	5.30, d (17.1)	116.2	5.45, dd (1.4,17.1)
13b		5.07, d (10.7)		5.23, dd (1.4,10.7)
14	22.6	1.30, d (6.4)	22.2	1.44, d (6.09)

Then, the relative configuration of **14** was determined through analysis of its ^{1}H NMR and NOESY correlations (Figure 3). The cross-peaks of H-3/H-8/H-10 and H-1/H_2-13/H-10 suggested that these protons were co-facial. However, the configuration at C-7 was difficult to determine due to the absence of a correlation between H_2-6 and H-8. Subsequently, the ^{13}C NMR calculations of (1*R**, 3*S**, 4*R**, 7*S**, 8*R**, 10*S**)-**14a** and (1*R**, 3*S**, 4*R**, 7*R**, 8*R**, 10*S**)-**14b** were carried out using the GIAO method at mPW1PW91-SCRF/6-311+G (d, p)/PCM (MeOH). The results showed that **14a** was the most likely candidate structure, with a better correlation coefficient (R^2 = 0.9988) and a high DP4+ probability of 100% (all data) probability (Figure 4). Thereafter, the ECD calculated was performed at the rB3LYP/6-311G level to determine the absolute configuration of **14**. The calculated curve matched well with its experimental ECD curve (Figure 5). Thus, the absolute configuration of **14** was assigned as 1*R*, 3*S*, 4*R*, 7*S*, 8*R*, 10*S*.

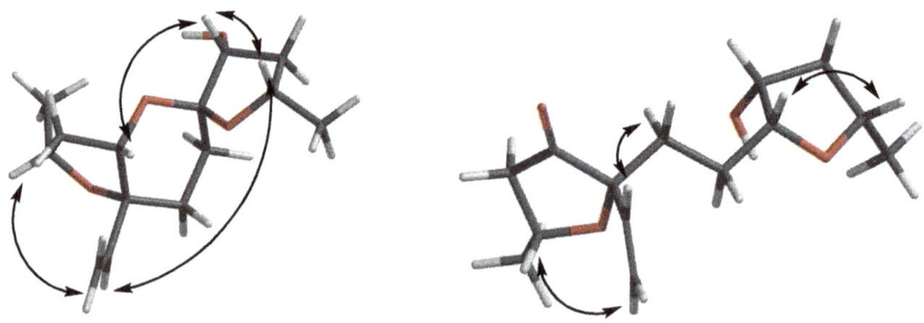

14 **15**

Figure 3. NOESY correlations of **14–15**.

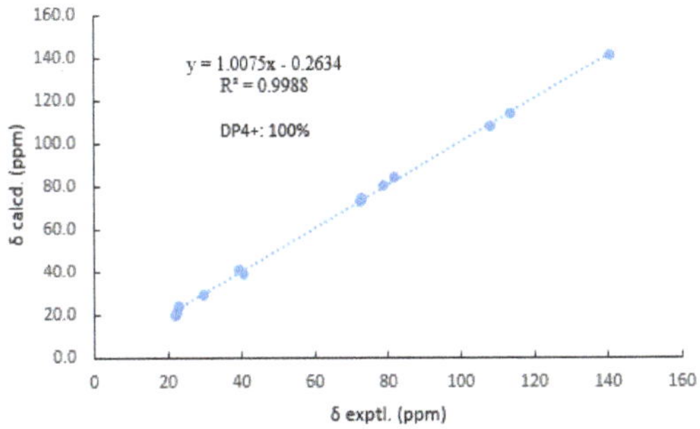

Figure 4. Comparisons of calculated and experimental ^{13}C NMR data of **14**.

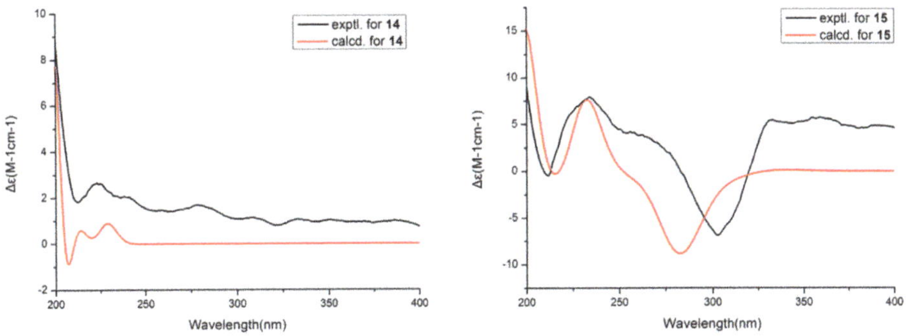

Figure 5. Experimental and calculated ECD spectra of **14** and **15**.

For compound **15**, a yellow solid, its molecular formula was determined as $C_{14}H_{22}O_4$ using HRESIMS data. The NMR (Table 3) data of **15** were similar to those of **14**, indicating that the structure of **15** closely resembled **14**. The main difference was that the oxygenated methine at C-3 in **14** was oxidated to a carbonyl group in **15**. Then, the spin systems of H_3-14/H-1/H_2-2 and H_2-5/H_2-6/H-7/H-8/H_2-9/H-10/H_3-11 from the ^1H-^1H COSY spectrum (Figure 2) and the HMBC correlations (Figure 2) from H-2 and H-5 to C-3 and C-4, and from H-10 to C-7, supported that the oxygen bridge bond between C-3 and C-7 was split. With the consideration of biogenetic origin, and the NOESY correlations (Figure 3) of H-1/H_2-13 and H-8/H-10, the relative configuration of **15** was established as 1*R**, 4*R**, 7*R**, 8*R**, and 10*S**. Then, the similarity of experimental and calculated ECD curves (Figure 5) allowed the assignment of the absolute configuration of **15** as 1*R*, 4*R*, 7*R*, 8*R*, and 10*S*.

The other compounds were identified as dothiorelone I (**6**) [10], (*S*)-dothiorelone Q (**7**) [11], dithiorelone B (**8**) [12], dithiorelone A (**9**) [13], cytoporone A (**10**) [1], cytoporone B (**11**) [1], a dothiorelone J (**12**) [10] by comparing the spectroscopic data to the literature.

Numerous inflammatory targets including iNOS, COX-1, COX-2, ICAM, IL-5, IL-17, JAK1, JAK2, SIRT2, and TNF-α were investigated through virtual screening for all compounds. The results indicated that iNOS showed stronger binding affinity with **1–13** than other targets (Table S2). Thereafter, all compounds were evaluated for their inhibitory activities against LPS-induced nitric oxide (NO) production in RAW 264.7 mouse macrophages. At non-cytotoxic concentrations, the results showed (Table 4) that compounds **1**, **6**, **8**, and **11** exhibited significant anti-inflammatory activities with IC_{50} values of 10.0, 12.0, 13.4, and 11.5 μM, respectively. Compounds **2**, **12**, and **13** showed potent anti-inflammatory activities compared with the positive control ($_L$-NMMA: 32.8 μM). Thereafter, the preliminary structure–activity relationship was discussed. The compounds **2**, **6**, **8**, and **11** displayed higher anti-inflammatory activity than **3**, **5**, **7**, and **10**, which indicated that the acetyl group at C-**1** is beneficial for activity. The carboxyl group in the side chain at C-16 may contribute to the activity by comparing **1** with **6**, **8**, **9**, **11**, and **12**. Moreover, the number of carbons in the side chain at C-8 may has not affected the anti-inflammatory activity by comparing the IC_{50} values of **1–12**.

Table 4. The anti-inflammatory activities of compounds **1–15**.

Comp.	**1**	**2**	**3**	**4**	**5**	**6**	**7**	**8**
IC_{50} (μM)	10.0 ± 0.3	17.2 ± 1.0	38.6 ± 0.3	42.1 ± 1.5	>50	12.0 ± 0.5	>50	13.4 ± 0.5
Comp.	**9**	**10**	**11**	**12**	**13**	**14**	**15**	L-NMMA [a]
IC_{50} (μM)	>50	47.0 ± 2.3	11.5 ± 0.3	28.2 ± 1.0	25.0 ± 1.2	25.6 ± 0.8	30.2 ± 0.3	32.8 ± 0.2

[a] positive control.

Cytosporones were reported to have anti-inflammatory activity. To explore the anti-inflammatory mechanism of these compounds, the inhibitory effects of inflammation-related iNOS and COX-2 for new compound **1** were measured using western blot. As a result, the protein expression of iNOS and COX-2 were apparently down-regulated after treatment of **1** with different concentrations (20.0, 10.0, and 5.0 μM) in a dose-dependent manner and dose-independent manner, respectively (Figure 6). These results were consistent with the project of target docking.

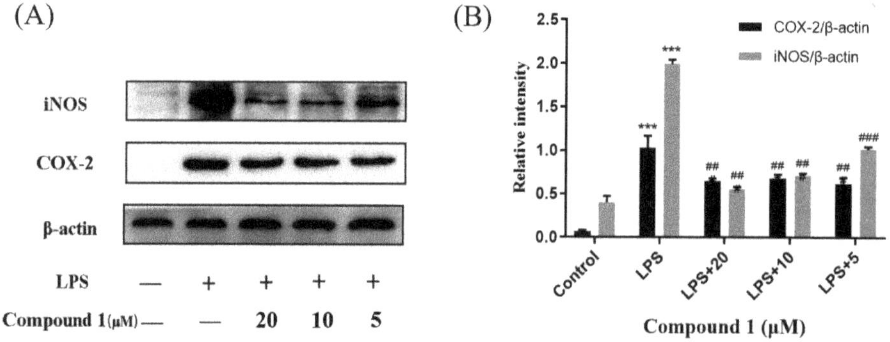

Figure 6. Influences of compound **1** on iNOS, COX-2, and β-actin protein expression were detected using western blotting (**A**); the ratio of the content of iNOS/β-actin and COX-2/β-actin (**B**). Data rendered are the mean ± SD, $n = 3$. In comparison to the control, *** $p < 0.001$. In comparison to the LPS group, ### $p < 0.001$, ## $p < 0.01$.

Mitogen-activated protein kinases (MAPK) and nuclear factor-κB (NF-κB) pathway were known to regulate the inflammatory response by modulating multiple pro-inflammatory cytokines in macrophages. So as to elucidate the mechanism of action by which compound **1** inhibited the level of NO, the MAPK and NF-κB signaling pathway of **1** was further investigated. As expected, the phosphorylation levels of p38, JNK, and ERK were significantly up-regulated after the treatment of RAW264.7 cells with LPS in the MAPK signaling pathway. Meanwhile, the phosphorylation levels of p38, JNK, and ERK were reduced after pretreatment of RAW264.7 cells with different concentrations of compound **1**. In addition, the western blotting methods have detected the expression of phosphorylation p65 in the NF-κB signaling pathway. The results implied that pretreatment with compound **1** obviously decreased the levels of phosphorylation p65 (Figure 7). Taken together, compound **1** displayed the anti-inflammatory effect by inhibiting the MAPK/NF-κB signaling pathways.

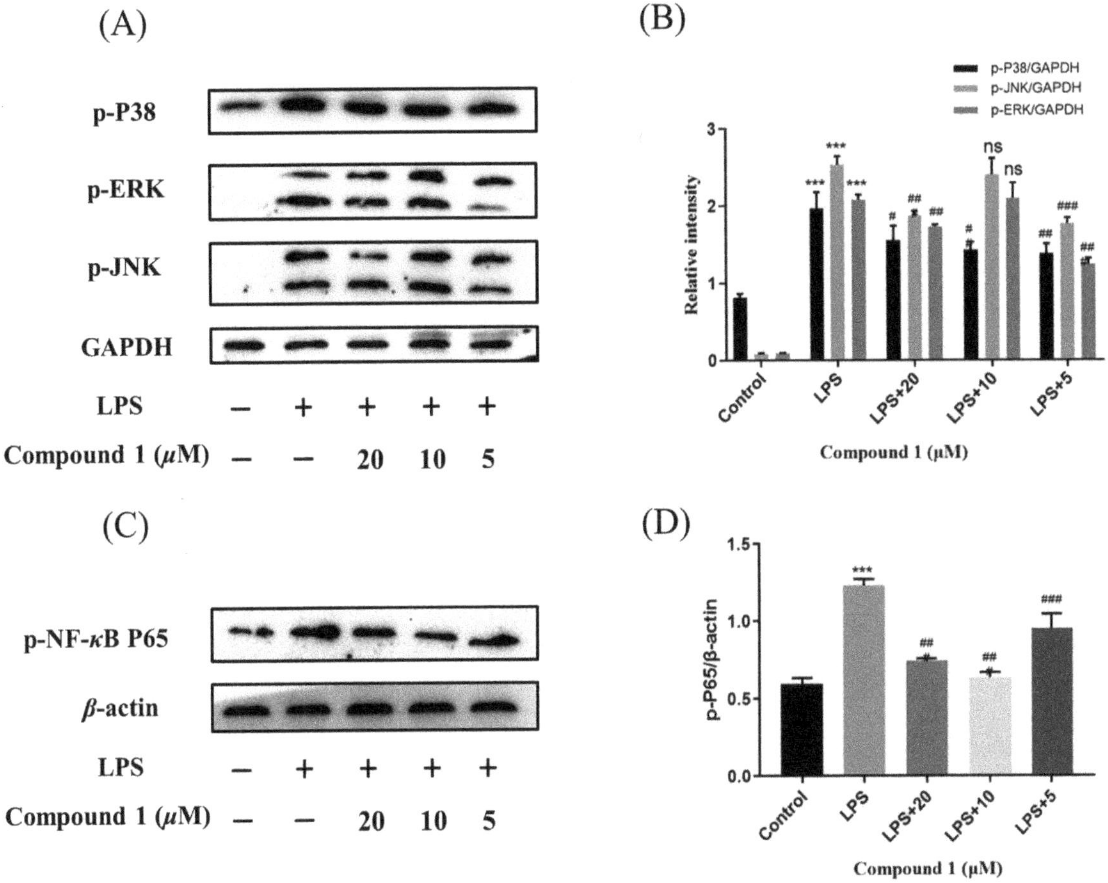

Figure 7. Influences of compound **1** on the MAPK and NF-κB pathway detected with western blotting. (**A**) The expression levels of p-JNK, p-ERK, p-P38, and GAPDH detected with Western blotting. (**B**) The proportion of p-JNK, p-ERK, and p-P38 to GAPDH content. (**C**) The expression levels of p-P65 and β-actin detected with western blotting. (**D**) The proportion of p-P65 to β-actin content. Data rendered are the mean ± SD, $n = 3$. In comparison to the control, *** $p < 0.001$. In comparison to the LPS, # $p < 0.05$, ## $p < 0.01$, ### $p < 0.001$.

In the MAPK signaling pathway, the levels of phosphorylation ERK were remarkably diminished compared with p38 and JNK. To further expound the role of phosphorylation ERK in compound **1** effects, molecular docking analysis was performed to investigate an inside interaction between compound **1** and ERK protein (PDB:5v60). As shown in Figure 8, the results indicated that **1** binds deeply in the active site pocket between Lys114, Tyr113, Asp111, Ser153, Glu33, Met38, and Ala35. Four hydrogen bonds are formed between the hydroxyl group at C-5 and C-7, and carbonyl at C-9 of **1** with Met108, Asp106, Gln105, and Cys166, respectively. In addition, multiple van der Waals-interacting residues such as Leu107, Ile31, Asn154, Lys54, Asp167, and Gly32 are generated between the ligand and the receptor protein, as well as π–π stacking interaction with Ala52, Val39, and Leu156. Thus, compound **1** could effectively activate the ERK signaling.

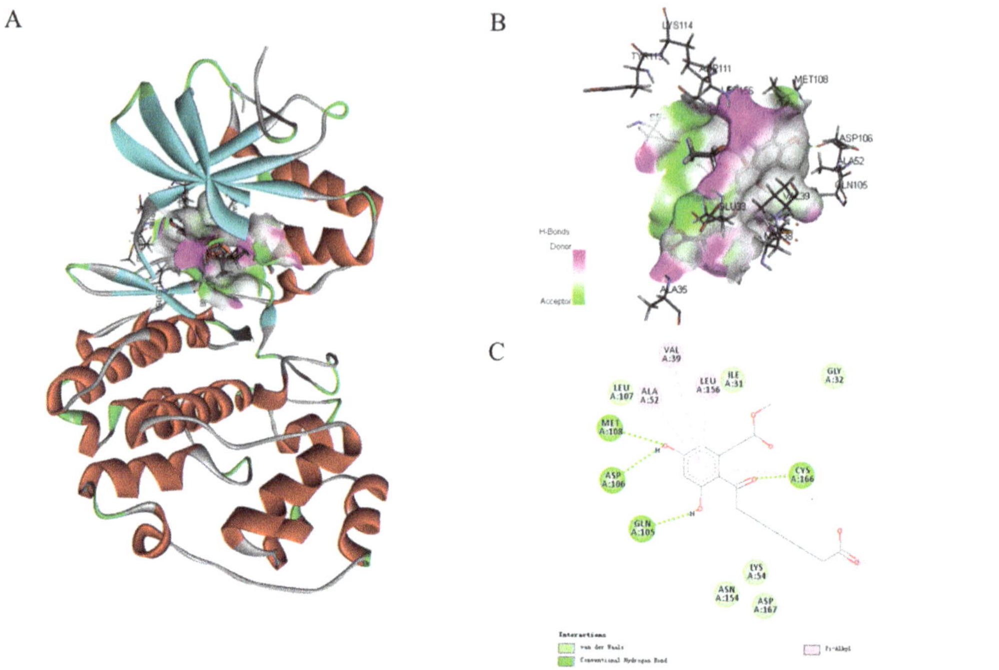

Figure 8. Molecular docking models for ERK (PDB:5v60) inhibition of compound **1** (**A**). Hydrogen bonding active pocket of compound **1** (**B**). 2D interaction diagrams of compound **1** (**C**).

3. Experimental Section

3.1. General Experimental Procedures

The HRESIMS data of compounds **1–4**, **5**, and **13–15** were obtained using LTQ Orbitrap LC-MS (Thermo Fisher Scientific, Bremen, Germany), Agilent G6230 Q-TOF LC/MS(Agilent, Santa Clara, CA, USA), and a Waters SYNAPT G2-Si (Waters Aisa Ltd., Singapore) mass spectrometer, respectively. All semi-preparative HPLC was performed on an Agilent 1260 liquid chromatography system(Agilent, Santa Clara, CA, USA) with an XSelect CSH C18 column (5 μm, 4.6 × 250 mm). The other procedures were the same as previously reported [9].

3.2. Fungal Material, Fermentation and Isolation

The strain *Phomopsis* sp. QYM-13 was described as previously reported [9]. Briefly, the mycelia of the fungus were inoculated into a 250 mL potato dextrose medium for 5 days to prepare the seed culture. Thereafter, the spore suspension was transferred into liquid mediums each containing 300 mL (3 g of potato extract, glucose 20 g/L, artificial sea salts 20 g/L, 100 × 1 L Erlenmeyer flasks) for 30 days at 25 °C. Then, the extracts (21.2 g) were obtained from the mycelia and broth as the methods previously described. It was subjected to silica gel column chromatography (CC, 200 mesh silica) eluting with petroleum ether/EtOAc (9:1 to 1:9) to yield 6 fractions (Fr. A-F). Fr. A was fractionated with Sephadex LH-20 CC (CH_2Cl_2/MeOH *v*/*v*, 1:1) to yield 2 fractions (Fr. A1-A2). Fr. A1 was further separated using silica gel CC (CH_2Cl_2/MeOH *v*/*v*, 97:3) and HPLC eluting with MeOH/H_2O (*v*/*v*, 86:14) to afford compounds **3** (3.4 mg) and **5** (4.6 mg). Fr. A2 was isolated using silica gel CC (CH_2Cl_2/MeOH *v*/*v*, 95:5) and further purified by semipreparative HPLC (MeOH-H_2O, 79:21) to give compounds **1** (2.2 mg) and **13** (6.7 mg). Fr. C was purified on Sephadex LH-20 CC (100% MeOH) to provide fractions C1-C3. Compound **2** (4.3 mg) and compound **14** (3.3 mg) were obtained from Fr. C3, which were purified using HPLC (MeOH-H_2O, 77:23). Fr. D was subjected to Sephadex LH-20 CC (CH_2Cl_2/

MeOH *v/v*, 1:1) to give fractions D1-D2. Fr. D1 was isolated using silica gel CC (CH_2Cl_2/MeOH *v/v*, 93:7) to yield compounds **6** (8.3 mg) and **4** (5.6 mg). Fr. D2 was purified using HPLC (MeOH-H_2O, 75:25) to give compounds **8** (3.2 mg) and **9** (2.8 mg). Fr. E was subjected to silica gel CC (CH_2Cl_2/MeOH *v/v*, 85:15) to give fractions E1-E3. Fr. E2 was purified using semipreparative HPLC (MeOH-H_2O, 73:27) to give compounds **7** (1.9 mg), **10** (3.5 mg), and **15** (2.8 mg). Fr. E3 was purified using HPLC eluting with MeOH-H_2O (*v/v*, 70:30) to give compounds **11** (2.6 mg) and **12** (3.3 mg).

Compound **1**: white solid, UV (MeOH) λ_{max} (logε): 220 (3.10), 232 (3.23), 271 (2.90), 288 (3.35) nm. IR (KBr): ν_{max} 3310, 1728, 1600, 1471, 1358, 1231 cm^{-1}. ^1H and ^{13}C NMR (500 MHz, MeOD-d_4) data, Table 1. HRESIMS *m*/*z* 351.14534 $[M-H]^-$ (calcd for $C_{18}H_{23}O_7$, 351.14531).

Compound **2**: colorless oil, UV (MeOH) λ_{max} (logε): 218 (3.15), 235 (3.20), 263 (2.89), 290 (3.41) nm. IR (KBr): ν_{max} 3286, 1726, 1610, 1465, 1235 cm^{-1}. ^1H and ^{13}C NMR (500 MHz, MeOD-d_4) data, see Table 1. HRESIMS *m*/*z* 323.11392 $[M-H]^-$ (calcd for $C_{16}H_{19}O_7$, 323.11390).

Compound **3**: colorless oil, UV (MeOH) λ_{max} (logε): 226 (3.08), 242 (2.80), 271 (3.0), 301 (3.38) nm. IR (KBr): ν_{max} 3310, 1727, 1600, 1463, 1362, 1230 cm^{-1}. ^1H and ^{13}C NMR (500 MHz, MeOD-d_4) data, see Table 1. HRESIMS *m*/*z* 295.08227 $[M-H]^-$ (calcd for $C_{14}H_{15}O_7$, 295.08229).

Compound **4**: colorless oil, UV (MeOH) λ_{max} (logε): 218 (3.0), 236 (3.06), 272 (2.85), 293 (3.23) nm. IR (KBr): ν_{max} 3286, 1731, 1605, 1469, 1370, 1231 cm^{-1}. ^1H and ^{13}C NMR (500 MHz, MeOD-d_4) data, see Table 2. HRESIMS *m*/*z* 295.08240 $[M-H]^-$ (calcd for $C_{14}H_{15}O_7$, 295.08242).

Compound **5**: colorless oil, UV (MeOH) λ_{max} (logε): 217 (3.10), 226 (2.90), 270 (2.89), 291 (3.32) nm. IR (KBr): ν_{max} 3288, 1731, 1600, 1463, 1365, 1230 cm^{-1}. ^1H and ^{13}C NMR (500 MHz, MeOD-d_4) data, see Table 2. HRESIMS *m*/*z* 309.1333 $[M + H]^+$ (calcd for $C_{16}H_{21}O_6$, 309.1328).

Compound **13**: yellow solid, UV (MeOH) λ_{max} (log ε) 228 (1.55) nm. IR (KBr) υ_{max} 3435, 2970, 1717 cm^{-1}. ^1H and ^{13}C NMR (500 MHz, DMSO-d_6) data, see Table 2. HRESIMS *m*/*z* 275.6262 $[M + H]^+$ (calcd for $C_{17}H_{23}O_3$, 275.6263).

Compound **14**: yellow solid, $[\alpha]_D^{25}$ = +8.5 (c 0.33, CH_3OH). UV (MeOH) λ_{max} (log ε) 228 (1.55) nm. IR (KBr) υ_{max} 3435, 2970, 1717 cm^{-1}. ^1H and ^{13}C NMR ($CDCl_3$) data, see Table 3. HRESIMS *m*/*z* 253.1436 $[M-H]^-$ (calcd for $C_{14}H_{21}O_4$, 253.1434).

Compound **15**: yellow solid, $[\alpha]_D^{25}$ = +7.5 (c 0.33, CH_3OH). UV (MeOH) λ_{max} (log ε) 224 (2.01) nm. IR (KBr) υ_{max} 3430, 2973, 1675 cm^{-1}. ^1H and ^{13}C NMR ($CDCl_3$) data, see Table 3. HRESIMS *m*/*z* 255.1587 $[M + H]^+$ (calcd for $C_{14}H_{23}O_4$, 255.1589).

3.3. ECD and NMR Calculation Methods

The ECD calculation was performed as described previously [7,14]. The conformers were subjected to geometric optimization at the level of B3LYP/6-31+G in the liquid phase. Thereafter, the optimized conformers were calculated using the TD-DFT method at the RB3LYP/6-311G (**14**) and PBEPBE/LAN12DZ (**15**) levels, respectively.

Typically, the Merck molecular force field in Spartan's 10 software was used for the conformational analysis of compound **14**. Conformers with populations exceeding 5% according to the Boltzmann distribution were optimized using the B3LYP/6-311+G (d, p) level in the polarizable continuum model (PCM) with methanol as the solvent. Subsequently, NMR calculations were performed using the gauge invariant atomic orbital (GIAO) method at the mPW1PW91-SCRF/6-311+G (d, p) level with PCM in methanol (Gaussian 09). Finally, the shielding constants were averaged using Boltzmann distribution theory for each stereoisomer, and their experimental and calculated data were analyzed using DP4+ probability.

3.4. Anti-inflammatory Assay

Cell culture: the RAW264.7 cells were cultured in Dulbecco's modified Eagle's medium (DMEM, Gibco, Grand Island, NY, USA) containing 10% fetal bovine serum (FBS, Grand Island, NY, USA) and 1% double antibodies (100 U/mL penicillin and 100 μg/mL streptomycin) at 37 °C with a 5% CO_2 humidified incubator.

Cell viability assay: the cell viability was measured using an MTT assay into 96-well plates. Approximately 3×10^6 cells/mL were inoculated overnight at 37 °C with 5% CO_2. Then, the cells were pretreated with different concentrations of L-NMMA or compound (5, 10, 20, 30, 40, and 50 μM) for 24 h. Thereafter, approximately 10 μL of MTT (0.5 mg/mL) was added to each well for 4 h. The absorbance was checked at 540 nm.

Bioassay of NO production: cells were inoculated overnight in 24-well plates with a density of 3×10^6 cells/mL (500 μL/well). Various concentrations of compounds were pretreated with LPS (1 μg/mL) for 24 h. The content of NO was measured according to the instructions of the Griess assay. The absorbance of the final product was measured at 540 nm.

Western blotting: cells (1×10^6 cells/well) were inoculated in 6-well plates with DMEM. Then, the cells were pretreated with compound **1** (20, 10, and 5 μM) and incubated for 24 h. The detailed operation process was according to the methods described previously [15].

3.5. Molecular Docking Studies

The molecular docking screening was performed using Sybyl-X 2.0 [16]. The three-dimensional structure of protein (iNOS: 2ORP; COX-1: 1HT8; COX-2: 5F19; ICAM: 6S8T; IL-17: 4HSA; IL-5: 3QT2; JAK1: 6N7A; JAK2: 6X8E; SIRT2: 4RMH; TNF-α: 2AZ5; p-ERK: 5V60) was obtained from the RCSB Protein Data Bank. The method used for the molecular docking screening was previously reported [17]. Briefly, the receptor compounds **1–13** were first optimized with the Gaussian View 5 program at DFT calculations. Then, the ligand substructures were extracted, and water molecules were removed. Subsequently, the target compounds were docked into the pocket of a receptor using Sybyl-X 2.0.

4. Conclusions

Six new cytosporone derivatives (phomotones A-D (**1–5**) and phomotone F (**13**)), two new spiro-alkanol phombistenes A-B (**14–15**), and seven known analogs (**6–12**) were isolated from the mangrove endophytic fungus *Phomopsis* sp. QYM-13. All compounds were evaluated for their inhibitory activities against LPS-induced nitric oxide (NO) production in RAW 264.7 macrophages after virtual screening using molecular docking with numerous inflammatory targets. The results showed that compounds **1**, **6**, **8**, and **11** exhibited promising anti-inflammatory activities. Thereafter, the mechanism of action suggested that compound **1** displayed the anti-inflammatory effect by inhibiting the MAPK/NF-κB signaling pathways. Furthermore, compound **1** could effectively activate the ERK signaling through hydrogen bonds, van der Waals, and π–π stacking. This research indicated that the cytosporone derivatives could be developed as anti-inflammatory therapeutic lead compounds.

Supplementary Materials: The following supporting information can be downloaded at https://www.mdpi.com/article/10.3390/md21120631/s1, HRESIMS, 1D and 2D NMR spectra of compounds **1–5** and **13–15**, the DP4+ evaluation of compound **14**, and the binding energy of compounds **1–13** with numerous inflammatory targets.

Author Contributions: G.W. and Y.Y. performed the separative experiments and structure identification; Z.L. and J.Z. carried out the anti-inflammatory activity; G.W. wrote the manuscript and Z.S. and Y.C. revised it. All authors have read and agreed to the published version of the manuscript.

Funding: We are grateful for the financial support from the National Natural Science Foundation of China (U20A2001, 42276114), the Key Scientific Research Project in Colleges and the Universities of Anhui Province (2022AH050706), and the Natural Science Foundation of Anhui Province (2308085QH302).

Informed Consent Statement: Not applicable.

Data Availability Statement: The data presented in this study are available in Supplementary Materials.

Conflicts of Interest: The authors declare no conflict of interest.

References

1. Brady, S.F.; Wagenaar, M.M.; Singh, M.P. The cytosporones, new octaketide antibiotics isolated from an endophytic fungus. *Org. Lett.* **2000**, *25*, 4043–4046. [CrossRef] [PubMed]
2. Beekman, A.M.; Barrow, R.A. Syntheses of Cytosporones A, C, J, K, and N, Metabolites from Medicinal Fungi. *Aust. J. Chem.* **2015**, *68*, 1583–1592. [CrossRef]
3. Zheng, C.J.; Huang, G.L.; Liao, H.X. Bioactive Cytosporone Derivatives Isolated from the Mangrove-derived Fungus *Dothiorella* sp. ML002. *Bioorg. Chem.* **2019**, *85*, 382–385. [CrossRef] [PubMed]
4. Yin, Y.; Yang, W.; Chen, T. Cytosporones W and X: Two Mutually Converting Epimers from a Mangrove Endophytic Fungus *Diaporthe* sp. ZJHJYZ-1. *ASC Omega* **2023**, *29*, 26628–26634. [CrossRef] [PubMed]
5. Jeremy, B.; Nida, M.; Whittney, N.B. Epigenetic Tailoring for the Production of Anti-Infective Cytosporones from the Marine Fungus *Leucostoma persoonii*. *Mar. Drugs* **2012**, *10*, 762–774.
6. Zamberlam, C.E.M.; Meza, A.; Leite, C.B. Total synthesis and allelopathic activity of cytosporones A–C. J. *Brazil. Chem. Soc.* **2012**, *23*, 124–131. [CrossRef]
7. Chen, Y.; Liu, Z.M.; She, Z.G. Ascomylactams A–C, cytotoxic 12-or 13-membered-ring macrocyclic alkaloids isolated from the mangrove endophytic fungus *Didymella* sp. CYSK-4, and structure revisions of phomapyrrolidones A and C. *J. Nat. Prod.* **2019**, *77*, 429–432. [CrossRef] [PubMed]
8. Chen, S.H.; Cai, R.L.; Liu, Z.M.; Cui, H.; She, Z.G. Secondary Metabolites from Mangrove-associated Fungi: Source, chemistry and bioactivities. *Nat. Prod. Rep.* **2022**, *39*, 560–595. [CrossRef] [PubMed]
9. Chen, Y.; Yang, W.; Zou, G.; Wang, G.; Kang, W.; Yuan, J.; She, Z.G. Cytotoxic bromine and iodine-containing cytochalasins produced by the mangrove endophytic fungus *Phomopsis* sp. QYM-13 using the OSMAC approach. *J. Nat. Prod.* **2022**, *85*, 1229–1238. [CrossRef] [PubMed]
10. Kornsakulkarn, J.; Somyong, W.; Supothina, S. Bioactive oxygen-bridged cyclooctadienes from endophytic fungus *Phomopsis* sp. BCC 45011. *Tetrahedron* **2016**, *47*, 9112–9116. [CrossRef]
11. Santos, A.L.; Ionta, M.; Horvath, R.; Soares, M.G.; Medeiros, L.S.D.; Uemi, M.; Kawafune, E.S.; Tangerina, M.M.P.; Ferreira, M.J.P.; Sartorelli, P. Molecular Network for Accessing Polyketide Derivatives from *Phomopsis* sp., an Endophytic Fungus of *Casearia Arborea* (*Salicaceae*). *Phytochem. Lett.* **2021**, *42*, 1–7. [CrossRef]
12. Takahashi, S.; Akita, Y.; Nakamura, T. Total synthesis of curvulone B and a proposed structure for dothiorelone B; determination of the configuration of curvulone B and structural revision of phomopsin A. *Tetrahedron* **2012**, *23*, 952–958. [CrossRef]
13. Izuchi, Y.; Koshino, H.; Hongo, Y. Synthesis and Structural Revision of Phomopsin B, a Novel Polyketide Carrying a 10-Membered Cyclic-Ether Ring. *Org. Lett.* **2011**, *13*, 3360–3363. [CrossRef] [PubMed]
14. Frisch, M.J.; Trucks, G.W.; Schlegel, H.B.; Scuseria, G.E.; Robb, M.A.; Cheeseman, J.R.; Scalmani, G.; Barone, V.; Mennucci, B.; Petersson, G.A. *Gaussian 16, Revision A. 03*; Gaussian. Inc.: Wallingford, CT, USA, 2016; Volume 3.
15. Chen, H.; Guo, S.; Song, W.; Hou, J.T. A stable NIR fluorescent probe for imaging lipid droplets in triple-negative breast cancer. *Sensor. Actuat. B-Chem.* **2024**, *398*, 134740. [CrossRef]
16. Joshi, S.D.; More, U.A.; Koli, D. Synthesis, evaluation and in silico molecular modeling of pyrroyl-1,3,4-thiadiazole inhibitors of InhA. *Bioorg. Chem.* **2015**, *59*, 151–167. [CrossRef] [PubMed]
17. Chen, Y.; Liu, H.J.; She, Z.G. Bioactive sesquiterpene derivatives from mangrove endophytic fungus *Phomopsis* sp. SYSU-QYP-23: Structures and nitric oxide inhibitory activities. *Bioorg. Chem.* **2020**, *107*, 104530. [CrossRef] [PubMed]

marine drugs

MDPI

Article

Rhizoaspergillin A and Rhizoaspergillinol A, including a Unique Orsellinic Acid–Ribose–Pyridazinone-*N*-Oxide Hybrid, from the Mangrove Endophytic Fungus *Aspergillus* sp. A1E3

Binbin Wu [1], Chenglong Xu [2], Jianjun Chen [2,*] and Guangying Chen [1,*]

[1] Key Laboratory of Tropical Medicinal Resource Chemistry of Ministry of Education, College of Chemistry and Chemical Engineering, Hainan Normal University, Haikou 571158, China; binbwu168@163.com

[2] Guangdong Provincial Key Laboratory of New Drug Screening, School of Pharmaceutical Sciences, Southern Medical University, 1838 Guangzhou Avenue North, Guangzhou 510515, China; xxuchenglong@163.com

* Correspondence: jchen21@smu.edu.cn (J.C.); chgying123@163.com (G.C.)

Abstract: Two new compounds, named rhizoaspergillin A (**1**) and rhizoaspergillinol A (**2**), were isolated from the mangrove endophytic fungus *Aspergillus* sp. A1E3, associated with the fruit of *Rhizophora mucronata*, together with averufanin (**3**). The planar structures and absolute configurations of rhizoaspergillinol A (**2**) and averufanin (**3**) were established by extensive NMR investigations and quantum-chemical electronic circular dichroism (ECD) calculations. Most notably, the constitution and absolute configuration of rhizoaspergillin A (**1**) were unambiguously determined by single-crystal X-ray diffraction analysis of its *tri*-pivaloyl derivative **4**, conducted with Cu Kα radiation, whereas those of averufanin (**3**) were first clarified by quantum-chemical ECD calculations. Rhizoaspergillin A is the first orsellinic acid–ribose–pyridazinone-*N*-oxide hybrid containing a unique β-oxo-2,3-dihydropyridazine 1-oxide moiety, whereas rhizoaspergillinol A (**2**) and averufanin (**3**) are sterigmatocystin and anthraquinone derivatives, respectively. From the perspective of biosynthesis, rhizoaspergillin A (**1**) could be originated from the combined assembly of three building blocks, viz., orsellinic acid, β-D-ribofuranose, and *L*-glutamine. It is an unprecedented alkaloid-*N*-oxide involving biosynthetic pathways of polyketides, pentose, and amino acids. In addition, rhizoaspergillinol A (**2**) exhibited potent antiproliferative activity against four cancer cell lines. It could dose-dependently induce G2/M phase arrest in HepG2 cells.

Keywords: *Aspergillus*; pyridazinone-*N*-oxide hybrid; sterigmatocystin; antitumor; cell cycle arrest

Citation: Wu, B.; Xu, C.; Chen, J.; Chen, G. Rhizoaspergillin A and Rhizoaspergillinol A, including a Unique Orsellinic Acid–Ribose–Pyridazinone-*N*-Oxide Hybrid, from the Mangrove Endophytic Fungus *Aspergillus* sp. A1E3. *Mar. Drugs* **2023**, *21*, 598. https://doi.org/10.3390/md21110598

Academic Editors: Bin-Gui Wang and Hao-Fu Dai

Received: 21 October 2023
Revised: 15 November 2023
Accepted: 16 November 2023
Published: 19 November 2023

1. Introduction

Pyridazines and pyridazinones are rare in nature but are common building blocks for heterocyclic organic synthesis [1–4]. Maleic hydrazide, i.e., 1,2-dihydro-3,6-pyradizinedione, is a synthesized selective herbicide and temporary plant growth regulator commonly used to prevent sprouting of potato tubers, onions, garlic, and radishes, etc., during storage. It can also inhibit crop growth and extend flowering periods [5]. Pyridaben, another example of a pyridazinone, is a broad-spectrum and contact killing acaricide. It is a mitochondrial electron transport inhibitor (METI) acaricide that promotes the formation of damaging oxygen and nitrogen radicals [6–8]. Pyridazine *N*-oxides are photoactivatable $O(^3P)$ precursors for applications in organic synthesis and chemical biology [9], whereas pyridazinone *N*-oxides are relatively stable. To the best of our knowledge, all the pyridazinone *N*-oxides are synthetic compounds. To date, no natural products of pyridazinone *N*-oxides have been reported.

Mangrove endophytic fungi of the genus *Aspergillus* can produce structurally unique metabolites with diverse bioactivities [10–19]. In order to search for bioactive natural compounds with new structures, two new compounds, named rhizoaspergillin A (**1**) and rhizoaspergillinol A (**2**), were isolated from the mangrove endophytic fungus *Aspergillus* sp.

A1E3, associated with the fruit of *Rhizophora mucronata*, together with averufanin (**3**) [20–25] (Figure 1). To our knowledge, rhizoaspergillin A (**1**) is the first alkaloid-*N*-oxide featuring the presence of an unprecedented orsellinic acid–ribose–pyridazinone-*N*-oxide hybrid scaffold containing a unique β-oxo-2,3-dihydropyridazine 1-oxide moiety, whereas rhizoaspergillinol A (**2**) and averufanin (**3**) are sterigmatocystin and anthraquinone derivatives, respectively. Herein, we report the isolation and structural identification of rhizoaspergillin A (**1**) and rhizoaspergillinol A (**2**), along with clarification of the absolute configuration of averufanin (**3**). The antiproliferative activities of compounds **1–3** were also evaluated against four cancer cell lines.

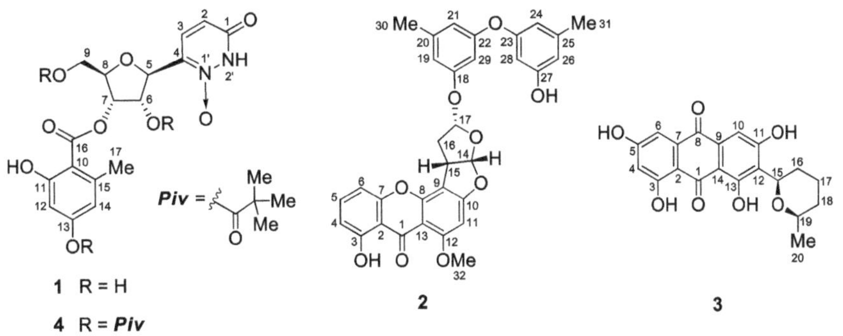

Figure 1. Structures of compounds **1–4**.

2. Results and Discussion

Rhizoaspergillin A (**1**) was obtained as an amorphous powder. The molecular formula $C_{17}H_{19}N_2O_9$ with ten degrees of unsaturation was determined by HR-ESIMS (*m*/*z*: calcd: 395.1085; found: 395.1084 $[M + H]^+$). The ^{13}C-NMR spectroscopic data and DEPT 135 experimental results for **1** revealed the presence of a methyl group, an oxygenated methylene group, eight methine groups (four oxygenated and four olefinic), and seven nonprotonated carbons (two carbonyl and five olefinic). According to the 1D and 2D NMR spectroscopic data for **1**, nine degrees of unsaturation are due to a carbon–carbon double bond, a carbon–nitrogen double bond, a tetrahydrofuran ring, an amide group, and a benzoate group. Thus, a pyridazinone ring should exist in the molecule.

The presence of a β-D-ribofuranose unit, being characterized by the corresponding NMR spectroscopic data [δ_H 5.96 (d, *J* = 6.4 Hz, H-5), 4.46 (dd, *J* = 11.0, 5.8 Hz, H-6), 6.00 (d, *J* = 5.6 Hz, 6-OH), 5.45 (dd, *J* = 5.2, 2.8 Hz, H-7), 4.26 (dd, *J* = 5.4, 2.8 Hz, H-8), 3.74 (br s, H_2-9), 5.38 (t, *J* = 4.8 Hz, 9-OH); δ_C 87.8 (CH, C-5), 72.0 (CH, C-6), 74.0 (CH, C-7), 83.0 (CH, C-8), 61.2 (CH_2, C-9)] (Table 1), was corroborated by ^{1}H–^{1}H COSY correlations between H-5/H-6, H-6/H-7, H-7/H-8, and H-8/H_2-9 (Figure 2a). The existence of an orsellinic acid moiety, i.e., a 2,4-dihydroxy-6-methylbenzoic acid unit, being evidenced by the corresponding NMR spectroscopic data [δ_H 6.24 (d, *J* = 2.0 Hz, H-12), 6.28 (d, *J* = 2.0 Hz, H-14), 2.47 (s, H_3-17), 10.85 (s, 11-OH), 10.17 (s, 13-OH); δ_C 106.3 (C, C-10), 162.6 (C, C-11), 100.9 (CH, C-12), 162.1 (C, C-13), 111.2 (CH, C-14), 142.4 (C, C-15), 169.1 (C, C-16), 23.1 (CH_3, C-17)] (Table 1), was confirmed by HMBC correlations between OH-11/C-10, OH-11/C-11, OH-11/C-12, OH-13/C-12, OH-13/C-13, OH-13/C-14, H_3-17/C-10, H_3-17/C-14, H_3-17/C-15, and H_3-17/C-16. Strong HMBC cross-peaks from protons of Me-17 (δ_H 2.47, s) to the nonprotonated C-15 (δ_C 142.4) placed it at C-15. Most notably, the key HMBC cross-peak from H-7 to the carbonyl C-16 connected the β-D-ribofuranose unit and the orsellinic acid moiety through the C-7–*O*–C-16 bond (Figure 2a).

Table 1. ^{1}H- and ^{13}C-NMR Data for **1** and its Derivative **4** in DMSO-d$_6$ and CDCl$_3$, respectively (δ in ppm and J in Hz).

Position	1 [a]		4 [b]	
	δ_H, Multi. (J)	δ_C, Type	δ_H, Multi. (J)	δ_C, Type
1		163.3, C		162.1, C
2	5.78 d (8.0)	102.6, CH	5.79 d (8.0)	103.4, CH
3	7.96 d (8.0)	140.7, CH	7.39 d (8.0)	139.4, CH
4		151.1, C		149.7, C
5	5.96 d (6.4)	87.8, CH	6.06 d (6.3)	87.8, CH
6	4.46 dd (11.0, 5.8)	72.0, CH	5.47 t (6.3)	72.6, CH
7	5.45 dd (5.2, 2.8)	74.0, CH	5.65 t (4.9)	72.0, CH
8	4.26 dd (5.4, 2.8)	83.0, CH	4.56 m	80.4, CH
9	3.74 2H, br s	61.2, CH$_2$	4.39 dd (12.6, 2.8) 4.47 dd (12.6, 3.5)	63.4, CH$_2$
10		106.3, C		108.6, C
11		162.6, C		164.9, C
12	6.24 d (2.0)	100.9, CH	6.53 br s	117.0, CH
13		162.1, C		156.2, C
14	6.28 d (2.0)	111.2, CH	6.62 br s	108.9, CH
15		142.4, C		142.9, C
16		169.1, C		170.1, C
17	2.47 s	23.1, CH$_3$	2.66 s	24.6, CH$_3$
2′	11.40 s		8.19 s	
18				176.2, C
19				38.9, C
20			1.35 s	27.0, CH$_3$
21			1.35 s	27.0, CH$_3$
22			1.35 s	27.0, CH$_3$
23				177.9, C
24				39.3, C
25			1.27 s	27.3, CH$_3$
26			1.27 s	27.3, CH$_3$
27			1.27 s	27.3, CH$_3$
28				177.3, C
29				38.8, C
30			1.09 s	26.8, CH$_3$
31			1.09 s	26.8, CH$_3$
32			1.09 s	26.8, CH$_3$
6-OH	6.00 d (5.6)			
9-OH	5.38 t (4.8)			
11-OH	10.85 s		11.08 s	
13-OH	10.17 s			

[a] ^{1}H- and ^{13}C-NMR data measured at 400 and 100 MHz, respectively; [b] ^{1}H- and ^{13}C-NMR data measured at 700 and 175 MHz, respectively.

The presence of a pyridazinone ring, being characterized by the corresponding NMR spectroscopic data [δ_H 11.40 (s, H-2′), 5.78 (d, J = 8.0 Hz, H-2), 7.96 (d, J = 8.0 Hz, H-3); δ_C 163.3 (C, C-1), 102.6 (CH, C-2), 140.7 (CH, C-3), 151.1 (C, C-4)] (Table 1), was corroborated by the ^{1}H–^{1}H COSY cross-peak between H-2/H-3 and HMBC correlations between H-2/C-1, H-3/C-1, H-3/C-2, and H-3/C-4 (Figure 2a). The Key HMBC correlation from H-5 to C-4 connected the β-D-ribofuranose unit and the above pyridazinone ring through the C-4–C-5 bond. In addition, the existence of a nitrogen–oxygen bond could be inferred by the molecular formula of **1** to be loaded on $N_{1'}$. Taken together, the planar structure of **1** was elucidated as shown (Figure 2a).

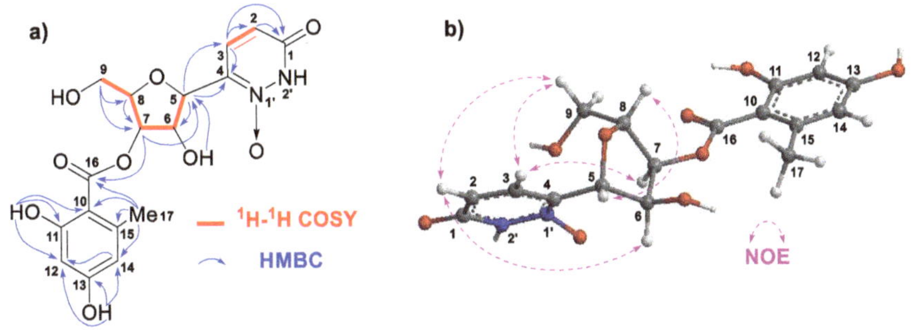

Figure 2. (**a**) Selected ^{1}H–^{1}H COSY and HMBC correlations for **1**. (**b**) Diagnostic NOE interactions for **1**.

The relative configuration of **1** was determined by NOE interactions (Figure 2b). Those between H-2/H-6, H-2/H$_2$-9, H$_2$-9/H-3, and H-3/H-7 revealed their cofacial relationships and were arbitrarily assigned as the α-oriented H-6 and H-7, whereas the diagnostic NOE interaction between H-5 and H-8 assigned their cofacial β-orientation. In order to reconfirm the constitution of **1** and establish its absolute configuration, single-crystal X-ray diffraction analysis was taken into account. However, it is impossible to obtain suitable crystals of **1** due to its poor solubility. Thus, derivatization reaction products were considered. Compound **1** was acylated by pivaloyl chloride to afford **4** (a *tri*-pivaloyl derivative of **1**, Figure 1). Suitable crystals of **4** were obtained in MeOH after considerable effort. Finally, the constitution and absolute configuration of **1** were established by single-crystal X-ray diffraction analysis of **4**, conducted with Cu Kα radiation [Flack parameter—0.12(9)] (Figure 3, CCDC 2291154). The absolute configuration of **1**, named rhizoaspergillin A, was unambiguously determined to be (5*S*,6*S*,7*S*,8*R*). To the best of our knowledge, rhizoaspergillin A is the first reported orsellinic acid–ribose–pyridazinone-*N*-oxide hybrid.

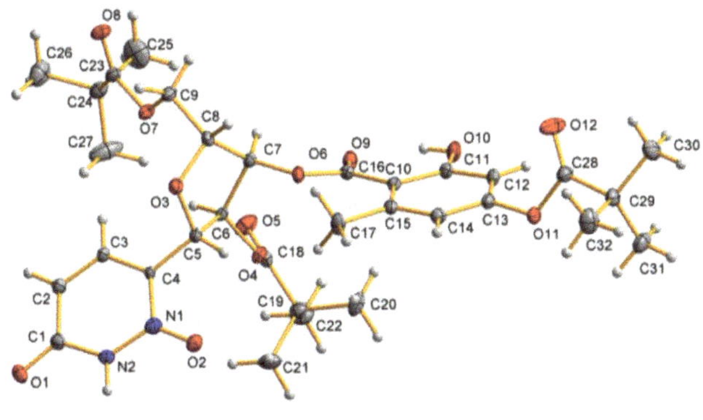

Figure 3. Oak Ridge Thermal-Ellipsoid Plot Program (ORTEP) illustration of the X-ray structure of **4** (Ellipsoids are given at the 30% probability level, Cu Kα).

Rhizoaspergillinol A (**2**) was isolated as a light white amorphous powder. The molecular formula $C_{32}H_{26}O_9$ was determined by the positive HR-ESIMS ion at m/z 555.1644 (calcd for $[M + H]^+$, 555.1650), indicating twenty degrees of unsaturation. According to ^{1}H- and ^{13}C-NMR spectroscopic data of **2** (Table 2), thirteen degrees of unsaturation are due to a keto-carbonyl function and twelve carbon–carbon double bonds. Therefore, the molecule has to be heptacyclic. The ^{13}C-NMR spectroscopic data and DEPT 135 experimental results for **2** revealed the presence of three methyl groups (a methoxy and two tertiary), a methylene group, thirteen methine groups (twelve olefinic and one oxygenated), and fifteen nonprotonated carbons (one carbonyl, five olefinic, and nine oxygenated).

Table 2. ^{1}H- and ^{13}C-NMR data for **2** and dihydrosterigmatocystin in $CDCl_3$ and DMSO-d_6, respectively (δ in ppm and J in Hz).

Position	2 ($CDCl_3$) [a]		Dihydrosterigmatocystin (DMSO-d_6) [26]	
	δ_H, Multi. (J)	δ_C, Type	δ_H, Multi. (J)	δ_C, Type
1		181.4, C		180.0, C
2		109.0, C		108.1, C
3		162.3, C		161.3, C
4	6.77 d (8.0)	111.4, CH	6.73 dd (8.5, 1.0)	110.5, CH
5	7.52 t (8.0)	135.8, CH	7.61 t (8.5)	136.0, CH
6	6.83 d (8.0)	105.9, CH	6.94 dd (8.5, 1.0)	106.1, CH
7		154.9, C		154.4, C
8		154.5, C		153.8, C
9		106.6, C		105.4, C
10		164.8, C		165.8, C
11	6.43 s	90.6, CH	6.60 s	90.2, CH
12		163.7, C		162.9, C
13		106.1, C		104.8, C
14	6.56 d (5.6)	112.3, CH	6.55 d (5.5)	113.4, CH
15	4.37 m	42.5, CH	4.25 m	43.3, CH
16	2.61 m, 2.73 m	37.4, CH_2	2.24 m, 2.45 m	30.7, CH_2
17	5.89 t (4.9)	103.3, CH	3.54 m, 4.10 m	67.2, CH_2
18		157.6, C		56.5, OCH_3
19	6.67 br s	112.0, CH		
20		140.9, C		
21	6.51 br s	114.0, CH		
22		157.8, C		
23		158.2, C		
24	6.400 d (2.0)	111.4, CH		
25		141.0, C		
26	6.399 d (2.0)	111.1, CH		
27		156.5, C		
28	6.29 t (2.0)	103.3, CH		
29	6.56 d (2.0)	105.1, CH		
30	2.30 s	21.5, CH_3		
31	2.27 s	21.7, CH_3		
32	4.01 s	56.9, CH_3		
3-OH	13.22 s		13.38 s	
27-OH	4.71 br s			

[a] ^{1}H- and ^{13}C-NMR data measured at 700 and 175 MHz, respectively.

The NMR spectroscopic data for **2** resembled those of dihydrosterigmatocystin [26,27], except for the presence of an additional 5,5′-oxybis(3-methylphenol) moiety, namely diorcinol [28–30], being characterized by the corresponding NMR spectroscopic data [δ_H 6.67 (br s, H-19), 6.51 (br s, H-21), 6.400 (d, J = 2.0 Hz, H-24), 6.399 (d, J = 2.0 Hz, H-26), 6.29 (t, J = 2.0 Hz, H-28), 6.56 (d, J = 2.0 Hz, H-29), 2.30 (s, H_3-30), 2.27 (s, H_3-31), 4.71 (br s, 27-OH); δ_C 157.6 (C, C-18), 112.0 (CH, C-19), 140.9 (C, C-20), 114.0 (CH, C-21), 157.8 (C, C-22), 158.2 (C, C-23), 111.4 (CH, C-24), 141.0 (C, C-25), 111.1 (CH, C-26), 156.5 (C, C-27), 103.3 (CH, C-28), 105.1 (CH, C-29), 21.5 (Me-30), 21.7 (Me-31)]. The existence of the diorcinol moiety was corroborated by HMBC correlations between H-19/C-18, H-19/C-21, H-19/C-29, H_3-30/C-19, H_3-30/C-20, H_3-30/C-21, H-21/C-22, H-24/C-23, H-24/C-28, H_3-31/C-24, H_3-31/C-25, H_3-31/C-26, 27-OH/C-26, 27-OH/C-27, and 27-OH/C-28. HMBC cross-peaks from protons of Me-30 (δ_H 2.30, s) to the nonprotonated C-20 (δ_C 140.9) and those from protons of Me-31 (δ_H 2.27, s) to the nonprotonated C-25 (δ_C 141.0) placed Me-30 at C-20 and Me-31 at C-25, respectively. In addition, the key HMBC correlation from H-17 (δ_H 5.89, t, J = 4.9 Hz) to the nonprotonated C-18 (δ_C 157.6, C) connected the dihydrosterigmatocystin unit and the diorcinol moiety through the C-17–*O*–C-18 bond. Therefore, the constitution of **2** was elucidated as shown (Figure 4).

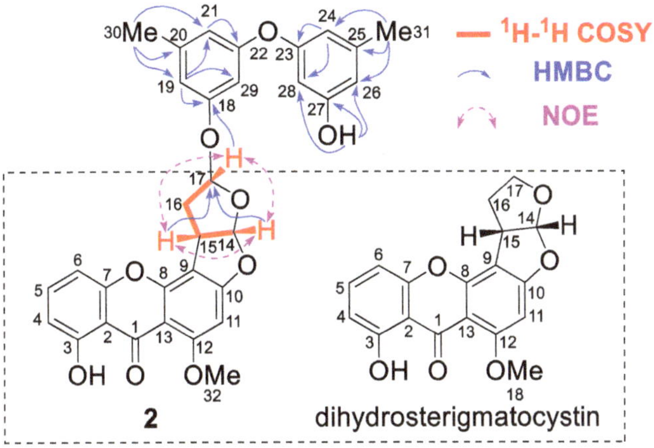

Figure 4. Selected ^{1}H–^{1}H COSY cross-peaks, HMBC correlations, and NOE interactions for **2**.

The relative configuration of **2** was determined by NOE interactions (Figure 4). Those between H-17/H-14 and H-17/H-15 revealed their cofacial relationship. Based on the previously reported absolute configuration of dihydrosterigmatocystin, the absolute configuration of C-17 in **2** was thereby assigned as *R*. In addition, the absolute configuration of **2** was reconfirmed by quantum-chemical electronic circular dichroism (ECD) calculations (Figure S1). The calculated ECD curve of (14*S*,15*S*,17*R*)-**2** showed good agreement with that of the experimental curve of **2** (Figure 5), whereas that of (14*R*,15*R*,17*S*)-**2** exhibited mirrored Cotton effects. Therefore, the absolute configuration of **2**, named rhizoaspergillinol A, was concluded to be (14*S*,15*S*,17*R*).

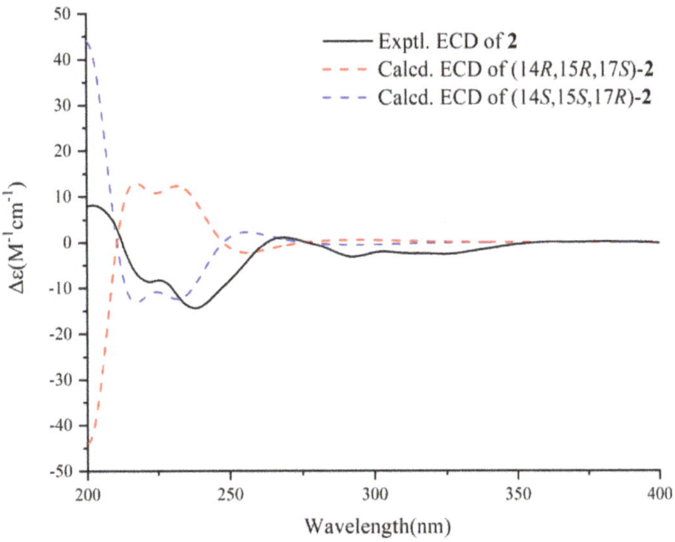

Figure 5. Experimental and calculated ECD curves for **2**.

Compound **3** was obtained as amorphous powder. The molecular formula $C_{20}H_{18}O_7$ was established by the positive HR-ESIMS ion at *m*/*z* 371.1136 (calcd for $[M + H]^+$, 371.1125), indicating twelve degrees of unsaturation. According to the NMR spectroscopic data for **3** (Table 3), eight degrees of unsaturation are due to two keto-carbonyl function, six carbon–carbon double bonds. Therefore, the molecule has to be tetracyclic. The presence of a

3,5,11,13-tetrahydroxyanthraquinone core, being characterized by the corresponding NMR spectroscopic data [δ_H 6.51 (d, *J* =1.2 Hz, H-4), 7.03 (br s, H-6), 7.00 (s, H-10); δ_C 188.5 (C, C-1), 108.4 (C, C-2), 165.5 (C, C-3), 108.0 (CH, C-4), 164.3 (C, C-5), 109.0 (CH, C-6), 134.7 (C, C-7), 181.0 (C, C-8), 133.1 (C, C-9), 109.0 (CH, C-10), 161.5 (C, C-11), 119.8 (C, C-12), 162.9 (C, C-13), 108.2 (C, C-14)], was confirmed by HMBC correlations between H-4/C-2, H-4/C-5, H-10/C-1, H-10/C-8, H-10/C-12, and H-10/C-14. The presence of a 20-methyltetrahydropyran moiety was confirmed by ^{1}H–^{1}H COSY cross-peaks between H-15/H_2-16, H_2-16/H_2-17, H_2-17/H_2-18, H_2-18/H-19, and H-19/H_3-20 and HMBC correlations between H-15/C-11, H-15/C-12, H-15/C-13, H-15/C-16, H-15/C-19, H_3-20/C-18, and H_3-20/C-19. HMBC correlations from H-15 to C-11, C-12, and C-13 connected the 3,5,11,13-tetrahydroxyanthraquinone core and the 20-methyltetrahydropyran moiety through the C-12–C-15 bond (Figure 6a).

Table 3. ^{1}H- and ^{13}C-NMR Data for **3** in DMSO-d_6 (δ in ppm and *J* in Hz).

Position	3 [a]	
	δ_H, multi. (*J*)	δ_C, type
1		188.5, C
2		108.4, C
3		165.5, C
4	6.51 d (1.2)	108.0, CH
5		164.3, C
6	7.03 br s	109.0, CH
7		134.7, C
8		181.0, C
9		133.1, C
10	7.00 s	109.0, CH
11		161.5, C
12		119.8, C
13		162.9, C
14		108.2, C
15	4.94 d (10.8)	73.3, CH
16	1.60 m, 1.93 m	28.3, CH_2
17	1.63 m, 1.87 m	23.4, CH_2
18	1.30 m, 1.65 m	32.5, CH_2
19	3.61 m	74.6, CH
20	1.18 d (6.0)	21.9, CH_3

[a] ^{1}H- and ^{13}C-NMR data measured at 400 and 100 MHz, respectively.

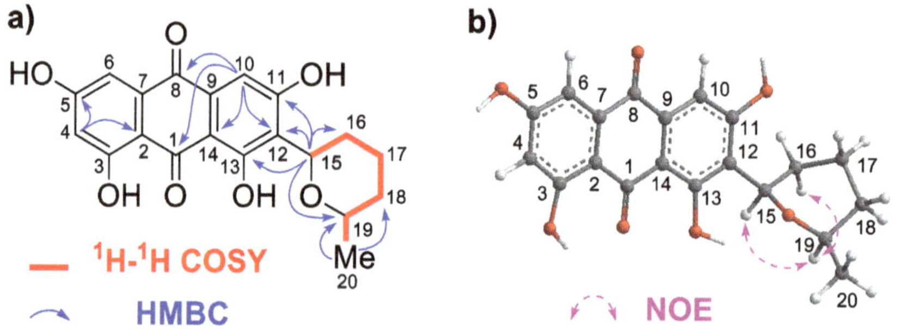

Figure 6. (**a**) Selected ^{1}H–^{1}H COSY and HMBC correlations for **3**. (**b**) Diagnostic NOE interactions for **3**.

The strong NOE interaction between H-15 and H-19 revealed their cofacial relationship (Figure 6b). The NMR spectroscopic data for **3** were the same as those of averufanin [21,22,25]. However, two absolute configurations, viz., (15*R*,19*R*) and (15*S*,19*S*), were confused in the literature [20–25]. In order to clarify the absolute configuration of **3**, quantum-chemical ECD calculations were employed (Figure S2). The calculated ECD curve of (15*R*,19*R*)-**3** showed good agreement with that of the experimental curve of **3**, whereas that of (15*S*,19*S*)-**3** exhibited mirrored Cotton effects (Figure 7). Therefore, the absolute configuration of **3**, i.e., averufanin, was concluded to be (15*R*,19*R*).

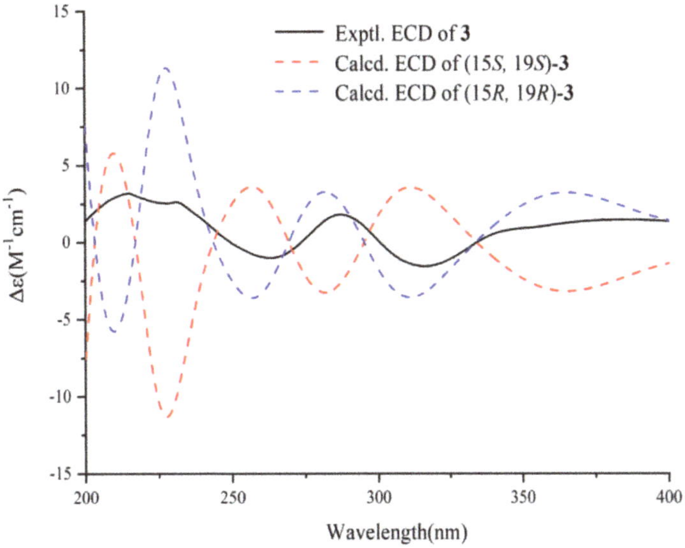

Figure 7. Experimental and calculated ECD curves for **3**.

The biosynthetic origin of **1** could be traced back to three building blocks, viz., 3-oxo-2,3-dihydropyridazine 1-oxide, β-D-ribofuranose, and orsellinic acid, among which the 3-oxo-2,3-dihydropyridazine 1-oxide moiety could be originated from one unit of *L*-glutamic acid, whereas the orsellinic acid unit could be biosynthesized from four units of acetyl coenzyme A (Scheme 1). The amidation of *L*-glutamic acid could generate *L*-glutamine, of which the decarboxylation and dehydrogenation would produce the intermediate (*E*)-4-aminobut-2-enamide. Subsequent cyclization and dehydrogenation could afford the heterocyclic intermediate pyridazin-3(2*H*)-one, of which the oxidation at the nitrogen atom would yield the crucial building block 3-oxo-2,3-dihydropyridazine 1-oxide. The dehydration of the anomeric center of the β-D-ribofuranose motif could produce a carbocation at C-1, of which nucleophilic attack at C-6 of the 3-oxo-2,3-dihydropyridazine 1-oxide moiety would generate a new carbon–carbon bond between C-1 of the β-D-ribofuranose motif and C-6 of the 3-oxo-2,3-dihydropyridazine 1-oxide moiety [31–35]. Finally, the esterification between 3-OH of the β-D-ribofuranose motif and the 7-carboxyl group of the orsellinic acid unit could produce rhizoaspergillin A (**1**). (Scheme 1)

The antiproliferative activity of **1**-**3** against four cancer cell lines were evaluated by using the standard MTT assay, with MS-275 (Entinostat, Figure 8) as the positive control [36]. MS-275 is a known histone deacetylase inhibitor as an anticancer agent. As summarized in Table 4 and Figure 9, **2** is the most potent one among the three compounds, with IC_{50} values of 8.83, 14.18, and 15.12 μM against HepG2, LLC, and B16-F10 cancer cell lines, respectively. Compounds **1** and **3** showed lower activities with IC_{50} values around or greater than 50.0 μM.

The most potent **2** was selected to evaluate its effects on the cell cycle of HepG2 cancer cells using flow cytometry. As shown in Figure 10A,B, after treatment with increasing concentrations of **2** (4.0, 8.0, and 16.0 μM) for 48h, the percentages of cells arrested at G2/M phase were increased from 14.83% to 25.91%, as compared to the control group (15.03%). Therefore, **2** could dose-dependently induce G2/M phase arrest in HepG2 cells.

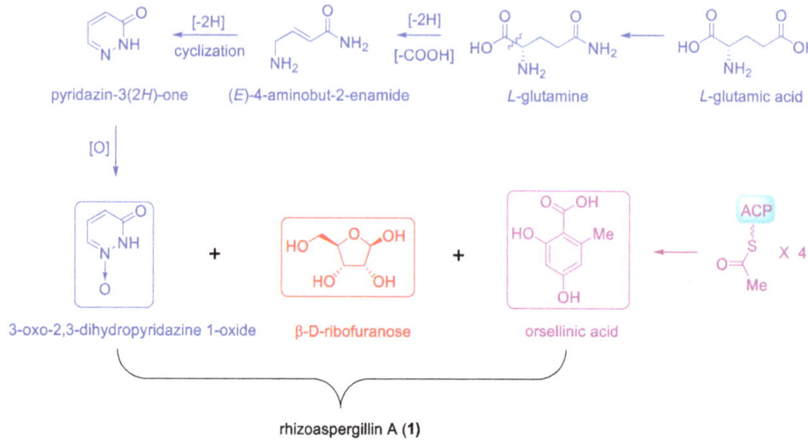

Scheme 1. Proposed biosynthetic origin of rhizoaspergillin A (**1**).

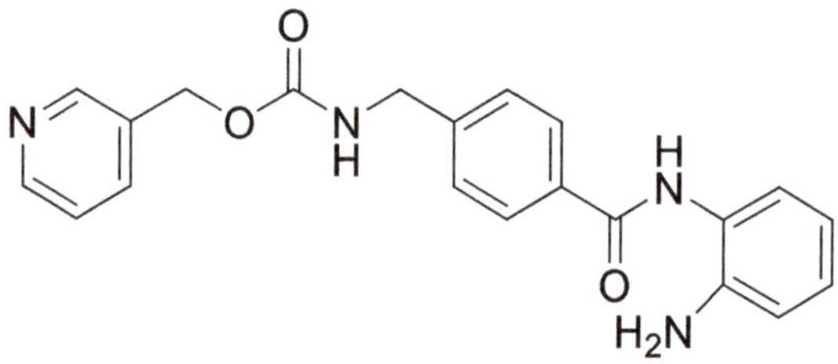

Figure 8. The structure of MS-275 (Entinostat).

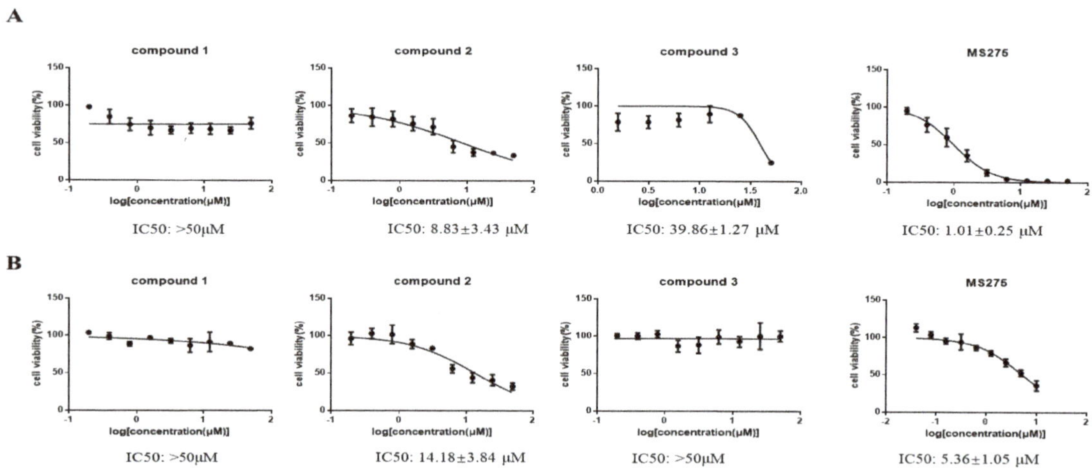

Figure 9. *Cont.*

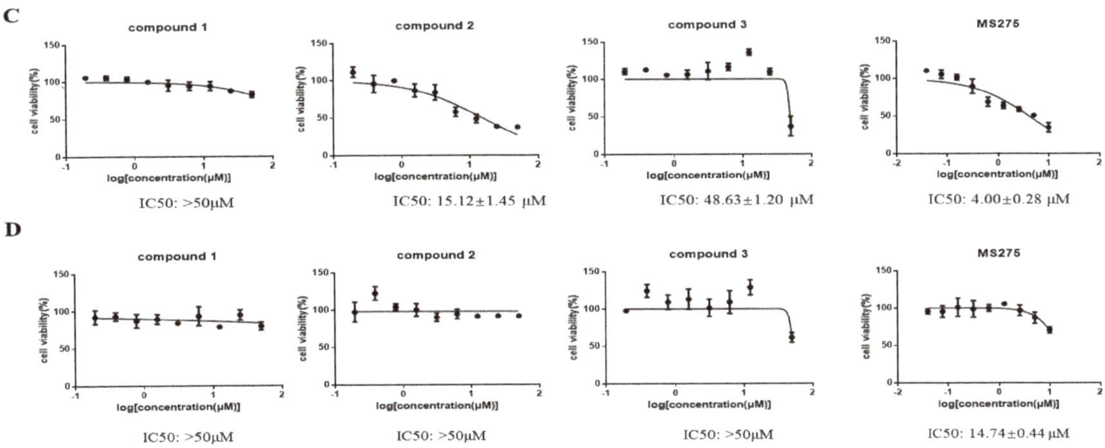

Figure 9. (**A**) MTT curve graph of **1–3** and MS-275 against HepG2 cells. (**B**) MTT curve graph of **1–3** and MS-275 against LLC cells. (**C**) MTT curve graph of **1–3** and MS-275 against B16-F10 cells. (**D**) MTT curve graph of **1–3** and MS-275 against MCF7 cells.

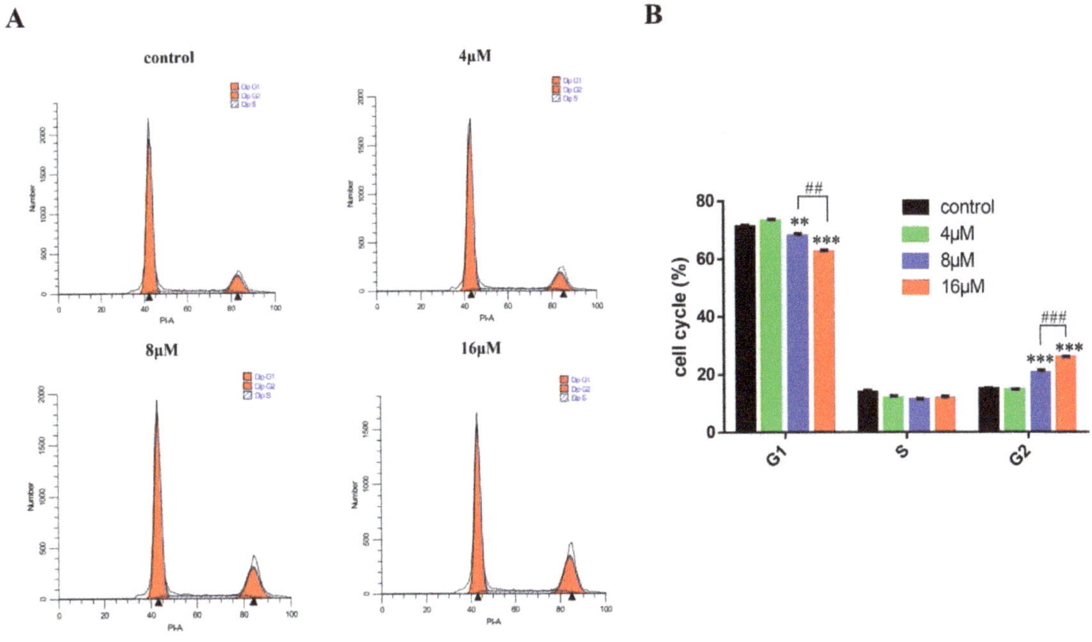

Figure 10. Cell cycle arrest induced by **2**. (**A**) HepG2 cells were incubated with varying concentrations of **2** (4.0, 8.0, and 16.0 μM) for 48 h. (**B**) Histograms showing the percentage of cell cycle distribution following **2** treatment (n = 3). The bar graphs are presented as mean ± SD. ***$p < 0.001$, ** $p < 0.01$ compared with the corresponding control group, ### $p < 0.001$, ## p <0.01 compared with the 16.0 μM-treated group, calculated by one-way ANOVA.

Table 4. In vitro antiproliferative activities of **1–3**.

Compounds	IC_{50} (μM) ± SD			
	HepG2	**LLC**	**B16-F10**	**MCF7**
1	>50.0	>50.0	>50.0	>50.0
2	8.83 ± 3.43	14.18 ± 3.84	15.12 ± 1.45	>50.0
3	39.86 ± 1.27	>50.0	48.63 ± 1.20	>50.0
MS-275 [a]	1.01 ± 0.25	5.36 ± 1.05	4.00 ± 0.28	14.74 ± 0.44

[a] MS-275 was used as a positive control.

3. Materials and Methods

3.1. General Experimental Procedures

HR-ESIMS spectra were obtained on a Bruker Daltonics Apex-Ultra 7.0 T (Bruker Corporation, Billerica, MA, USA). Optical rotations were recorded on an MCP500 modular circular polarimeter (Anton Paar GmbH, Graz, Austria) with a 0.5 cm cell at 25 °C and UV spectra were measured on a UV-2600 UV-Vis spectrophotometer (SHIMADZU) at room temperature. IR spectra were obtained on a SHIMADZU IRAffinity-1 Fourier transform infrared spectrometer and ECD spectra were recorded on a circular dichromatic spectrometer (Chirascan, Applied PhotoPhysics, Leatherhead, Surrey, UK). X-ray data were collected using an Agilent SuperNova with AtlasS2 X-ray single-crystal diffractometer with Cu Kα radiation. 1D and 2D NMR spectra were measured on a Bruker AV-400 or 700 MHz NMR spectrometer. High-performance liquid chromatography (HPLC) was performed on a Waters 2535 pump equipped with a 2998 photodiode array detector and C_{18} reversed-phase columns (YMC, Kyoto, Japan; 250 mm × 4.6 mm, length × i.d., 5 μm, for analysis; 250 mm × 10 mm, length × i.d., 5 μm, for preparation). Silica gel (Qingdao Haiyang Chemical Co., Ltd.; 100–200 and 200–300 mesh) and octadecylsilyl silica gel (YMC, Kyoto, Japan, ODS-A-HG, 12 nm, 50 μm) were used for column chromatography (CC).

3.2. Fungus Material

The fungal strain *Aspergillus* sp. A1E3 was isolated from the fruit of *Rhizophora mucronata*, collected from the Thai mangrove swamps of the Trang Province in February 2012. It was identified as *Aspergillus* sp. according to ITS rDNA sequence data. The strain was preserved in the School of Pharmaceutical Sciences, Southern Medical University.

3.3. Fermentation, Extraction and Isolation

The Fungal strain *Aspergillus* sp. A1E3 was inoculated into Erlenmeyer flasks (500 mL) containing 10‰ sea salt and potato glucose solution in a sterile environment at 25 °C for seven days to prepare the seed culture, which was then inoculated into 100 Erlenmeyer flasks (1000 mL) each containing rice solid medium (100 g rice and 150 mL water containing 10‰ sea salt) at room temperature in static conditions for 28 days. Then the fermentation material of *Aspergillus* sp. A1E3 was extracted three times with EtOAc, which was evaporated under vacuum. The resulting EtOAc extract was dissolved in EtOAc again and washed three times with water. The EtOAc solution was dried under reduced pressure to obtain the residue, which was then completely dissolved in $MeOH/H_2O$ (v/v, 1:1). After washing with double volume of *n*-hexane three times, the remaining aqueous methanol solution was dried to yield the resulting solid (70.11 g). The solid extract was then fractionated by silica gel column chromatography (200–300 mesh silica, 180 × 10 cm, i.d.) with a gradient mixture of $CHCl_3/MeOH$ (v/v, 100:0, 100:1, 50:1, 98:2, 30:1, 20:1, 10:1, 5:1, 3:1, 2:1, 1:1, 1:2) to afford 170 fractions. Fr. 42 and Fr. 43 were combined (3.89 g) and subjected to a C_{18} reversed-phase column (60 × 6 cm, i.d.), eluted with a gradient mixture of acetone/H_2O (v/v, 50:50 to 100:0) to afford 70 subfractions. Then subfractions 22 and 23 were combined and further purified by semi-preparative HPLC ($MeCN:H_2O$ 58:42) to yield **2** (1.0 mg, t_R = 90.0 min). Fr. 44 (1.01 g) was subjected to a C_{18} reversed-phase column (60 × 6 cm, i.d.), eluted with a gradient mixture of acetone/H_2O (v/v, 50:50 to

100:0) to give 50 subfractions, among which subfraction 46 (112.6 mg) was further purified by semi-preparative HPLC (CH_3CN:H_2O 65:35) to afford **3** (40.0 mg, t_R = 68.8 min). Fr. 145 (238.0 mg) was dissolved in chloroform and then filtered to give **1** (50.0 mg).

3.4. Spectroscopic Data of Compounds

Rhizoaspergillin A (**1**): white amorphous powder; $[\alpha]_D^{25}$—25.4 (*c* 0.10, MeOH); UV (MeOH) λ_{max} (logε) 213 (1.81) nm (Figure S3); IRv_{max} 3744, 3649, 3443, 3256, 2361, 2338, 1713, 1678, 1616, 1443, 1379, 1323, 1252, 1173, 1136, 1065 cm^{-1} (Figure S6); ^{1}H and ^{13}C NMR spectroscopic data (see Table 1); HR-ESIMS *m*/*z* 395.1084 $[M + H]^+$ (calcd for $C_{17}H_{20}N_2O_9$, 395.1085).

Rhizoaspergillinol A (**2**): light white amorphous powder; $[\alpha]_D^{25}$—57.6 (*c* 0.10, MeOH); UV (MeOH) λ_{max} (logε) 206 (1.52) nm (Figure S4); IRv_{max} 3439, 2918, 2851, 2359, 1647, 1620, 1582, 1460, 1233, 1155, 1040 cm^{-1} (Figure S7); ^{1}H and ^{13}C NMR spectroscopic data (see Table 2); HR-ESIMS *m*/*z* 555.1644 $[M + H]^+$ (calcd for $C_{32}H_{27}O_9$, 555.1650).

Averufanin (**3**): Deep red amorphous powder; $[\alpha]_D^{25}$ + 27.8 (*c* 0.05, MeOH); UV (MeOH) λ_{max} (logε) 208 (1.88) nm (Figure S5); IRv_{max} 3566, 3385, 3231, 2932, 2859, 2359, 2342, 1614, 1395, 1254, 1207, 1167, 1072, 1024, 995 cm^{-1} (Figure S8); ^{1}H and ^{13}C NMR spectroscopic data (see Table 3); HR-ESIMS *m*/*z* 371.1136 $[M + H]^+$ (calcd for $C_{20}H_{19}O_7$, 371.1125).

3.5. X-Ray Crystallographic Data of 4

$C_{32}H_{42}N_2O_{12}$, Mr 646.67. Colorless crystal from MeOH (mp 184.6–185.5). Temperature/K: 149.99(10). Crystal system: orthorhombic. Space group: P21212, a = 10.90168(15) Å, b = 11.02307(18) Å, c = 28.1523(4) Å, α: 90°, β: 90°, γ: 90°. Volume: 3383.06(9) Å3, Z: 4, ρcalcg: 1.270 cm^3, μ: 0.816 mm^{-1}, F(000): 1376.0. Crystal size: 0.26 × 0.05 × 0.03 mm^3. Radiation: CuKα (λ = 1.54184). 2^{Θ} range for data collection: 6.28 to 129.988°. Index ranges: $-12 \leq h \leq 12$, $-12 \leq k \leq 9$, $-33 \leq l \leq 33$. Reflections collected: 15534. Independent reflections: 5739 [R_{int} = 0.0313, R_{sigma} = 0.0343]. Data/restraints/parameters: 5739/35/426. Goodness-of-fit on F^2: 1.049. Final R indexes [I ≥ 2σ (I)]: R_1 = 0.0415, wR_2 = 0.1143. Final R indexes [all data]: R_1 = 0.0439, wR_2 = 0.1172. Largest diff. peak/hole: 0.73/−0.61 e $Å^{-3}$. Flack parameter: −0.12(9) (CCDC 2291154).

3.6. Synthesis of the Derivative 4

Compound **1** (20.0 mg) was dissolved in pyridine (0.5 mL). DMAP (3.0 mg) was then added, along with 50.0 μL of pivaloyl chloride under ice bath and stirred for 10 min. After the completion of the reaction (monitored by thin-layer chromatography, i.e., TLC), the reaction solution was evaporated under reduced pressure. The resulting residue was dissolved in methanol, purified by semi-preparative HPLC, and eluted with the mixture of MeCN/H_2O (*v*/*v*, 58:42, 3.0 mL/min) to afford **4** (5.0 mg). Suitable crystals of **4** were obtained in MeOH after considerable effort.

3.7. Cell Culture and Cytotoxicity (MTT) Assay

An MTT assay was used to determine the antiproliferative activities of compounds **1–3** and MS-275 (positive control) against four cancer cell lines, including a human lung cancer cell line (HepG2), mouse Lewis lung carcinoma (LLC) cells, a mouse skin melanoma cell line (B16-F10), and a human breast cancer cell line (MCF-7). Fetal bovine serum (FBS, 10%) in RPMI-1640 medium was used to culture MCF-7 cell lines and Fetal bovine serum (FBS, 10%) in DMEM medium was used to culture B16-F10, HepG2 and LLC cells. Cells were seeded into 96-well plates at a density of 5000 cells/well and incubated at 37 °C with 5% CO_2 overnight. The next day, the three compounds and MS-275 (InvivoChem) were dissolved in a complete medium to prepare different concentrations of solution, which were added to each well, followed by incubation at 37 °C for 48 h. Finally, 20 μL of MTT (5 mg/mL dissolved in PBS) was added to each well and incubated with cells at 37 °C for 4 h. Then, the complete medium was removed and the formazan crystals were dissolved in 100 μL of DMSO in each well. The absorbance was measured in a TECAN microplate reader at 490 nm. GraphPad Prism was utilized to calculate the IC_{50} values by a model of nonlinear regression to fit a normalized dose response. All of the experiments were performed independently three times.

3.8. Cell Cycle Analysis

HepG2 cells were seeded to a six-well plate at a density of 5×10^5 cells/well followed by incubation at 37 °C overnight. Then, different concentrations of **1–3** (4.0, 8.0, 16.0 μM) were added to the plate and incubated for 48 h. The collected cells were washed once with PBS. Then the cells were added 1.0 mL of DNA staining solution and 10.0 μL of permeabilization solution followed by incubation in the dark for 30.0 min. The DNA content of the samples was analyzed by flow cytometry.

4. Conclusions

In summary, two new compounds, named rhizoaspergillin A (**1**) and rhizoaspergillinol A (**2**), were obtained from the rice culture of the mangrove endophytic fungus *Aspergillus* sp. A1E3, together with the known compound averufanin (**3**). The planar structures and absolute configurations of rhizoaspergillinol A (**2**) and averufanin (**3**) were established by extensive NMR investigations and quantum-chemical electronic circular dichroism (ECD) calculations, whereas the constitution and absolute configuration of rhizoaspergillin A (**1**) were unambiguously determined by single-crystal X-ray diffraction analysis of its *tri*-pivaloyl derivative **4**, conducted with Cu Kα radiation. In addition, the absolute configuration of averufanin (**3**) was first clarified by ECD calculations. Most notably, rhizoaspergillin A (**1**) is the first reported orsellinic acid–ribose–pyridazinone-*N*-oxide hybrid containing a unique β-oxo-2,3-dihydropyridazine 1-oxide moiety, whereas rhizoaspergillinol A (**2**) is a sterigmatocystin derivative containing an additional diorcinol motif. From the perspective of biosynthesis, rhizoaspergillin A (**1**) could be originated from the combined assembly of three building blocks, viz., orsellinic acid, β-D-ribofuranose, and *L*-glutamine. It is an unprecedented alkaloid-*N*-oxide involving biosynthetic pathways of polyketides, pentose, and amino acids. Rhizoaspergillinol A (**2**) exhibited potent antiproliferative activity against a panel of cancer cell lines. Additionally, it could dose-dependently induce G2/M phase arrest in HepG2 cells. This work demonstrates that mangrove endophytic fungi of the genus *Aspergillus* harbor secondary metabolites with new structures.

Supplementary Materials: The following supporting information can be downloaded at: https://www.mdpi.com/article/10.3390/md21110598/s1, HR-ESIMS, 1D and 2D NMR spectra of compounds **1–4**, UV and IR spectra of compounds **1–3**, along with energy analyses of conformers for compounds **2** and **3**.

Author Contributions: B.W. carried out the chemistry work under the guidance of G.C., whereas C.X. performed the antitumor work under the guidance of J.C.; B.W. and C.X. prepared the first draft of the manuscript; G.C. and J.C. helped with all the methodological approaches and finalized the paper. All authors have read and agreed to the published version of the manuscript.

Funding: This research was funded by the National Natural Science Foundation of China (22177023), the Key Science and Technology Program of Hainan Province (ZDKJ202008), and Hainan Provincial Natural Science Foundation of China (221RC1054).

Institutional Review Board Statement: Not applicable.

Data Availability Statement: Raw data discussed in this study are available in the Supplementary Materials.

Acknowledgments: The authors gratefully thank Patchara Pedpradab (Rajamangala University of Technology Srivijaya, Trang Province, Thailand) for providing the mangrove fruits used in this work.

Conflicts of Interest: The authors declare no conflict of interest.

References

1. Chintakunta, V.K.; Akella, V.; Vedula, M.S.; Mamnoor, P.K.; Mishra, P.; Casturi, S.R.; Vangoori, A.; Rajagopalan, R. 3-*O*-Substituted benzyl pyridazinone derivatives as COX inhibitors. *Eur. J. Med. Chem.* **2002**, *37*, 339–347. [CrossRef] [PubMed]

2. Coelho, A.; Sotelo, E.; Novoa, H.; Peeters, O.M.; Blaton, N.; Raviña, E. Pyridazine derivatives. Part 38: Efficient Heck alkenylation at position 5 of the 6-phenyl-3(2*H*)-pyridazinone system. *Tetrahedron Lett.* **2004**, *45*, 3459–3463. [CrossRef]
3. Gong, Y.; Barbay, J.K.; Dyatkin, A.B.; Miskowski, T.A.; Kimball, E.S.; Prouty, S.M.; Fisher, M.C.; Santulli, R.J.; Schneider, C.R.; Wallace, N.H.; et al. Synthesis and biological evaluation of novel pyridazinone-based α_4 integrin receptor antagonists. *J. Med. Chem.* **2006**, *49*, 3402–3411. [CrossRef]
4. Rathish, I.G.; Javed, K.; Bano, S.; Ahmad, S.; Alam, M.S.; Pillai, K.K. Synthesis and blood glucose lowering effect of novel pyridazinone substituted benzenesulfonylurea derivatives. *Eur. J. Med. Chem.* **2009**, *44*, 2673–2678. [CrossRef]
5. Schoene, D.L.; Hoffmann, O.L. Maleic hydrazide, a unique growth regulant. *Science* **1949**, *109*, 588–589. [CrossRef]
6. Shipp, J.L.; Wang, K.; Ferguson, G. Residual toxicity of avermectin b1 and pyridaben to eight commercially produced beneficial arthropod species used for control of greenhouse pests. *Biol. Control* **2000**, *17*, 125–131. [CrossRef]
7. Navarro, A.; Bández, M.J.; Gómez, C.; Repetto, M.G.; Boveris, A. Effects of rotenone and pyridaben on complex I electron transfer and on mitochondrial nitric oxide synthase functional activity. *J. Bioenerg. Biomembr.* **2010**, *42*, 405–412. [CrossRef] [PubMed]
8. Sugimoto, N.; Osakabe, M. Cross-resistance between cyenopyrafen and pyridaben in the twospotted spidermite *Tetranychus urticae* (Acari: Tetranychidae). *Pest. Manag. Sci.* **2014**, *70*, 1090–1096. [CrossRef]
9. Basistyi, V.S.; Frederich, J.H. Pyridazine *N*-oxides as photoactivatable surrogates for reactive oxygen species. *Org. Lett.* **2022**, *24*, 1907–1912. [CrossRef]
10. Chen, S.H.; Cai, R.L.; Liu, Z.M.; Cui, H.; She, Z.G. Secondary metabolites from mangrove-associated fungi: Source, chemistry and bioactivities. *Nat. Prod. Rep.* **2021**, *39*, 560–595. [CrossRef]
11. Liu, X.; Fu, Y.; Zhou, Q.Q.; Wang, S.; Gao, L.; Lei, J.L.; Ke, A.B.; Li, Y.Y.; Zhang, X.X.; Huo, C.H.; et al. Aspergichromones A–E, five chromone derivatives with complicated polycyclic architecture from *Aspergillus deflectus*. *Org. Lett.* **2022**, *24*, 1610–1615. [CrossRef]
12. Liu, L.; Duan, F.F.; Gao, Y.; Peng, X.G.; Chang, J.L.; Chen, J.; Ruan, H.L. Aspersteroids A–C, Three Rearranged Ergostane-type Steroids from *Aspergillus ustus* NRRL 275. *Org. Lett.* **2021**, *23*, 9620–9624. [CrossRef]
13. Wang, L.; Yang, J.; Huang, J.P.; Li, J.; Luo, J.Y.; Yan, Y.J.; Huang, S.X. Bisaspochalasins A–C: Three cytochalasan homodimers with highly fused ring system from an endophytic *Aspergillus flavipes*. *Org. Lett.* **2020**, *22*, 7930–7935. [CrossRef]
14. Liu, L.; Wang, L.; Bao, L.; Ren, J.W.; Basnet, B.B.; Liu, R.X.; He, L.W.; Han, J.J.; Yin, W.B.; Liu, H.W. Versicoamides F–H, prenylated indole alkaloids from *Aspergillus tennesseensis*. *Org. Lett.* **2017**, *19*, 942–945. [CrossRef] [PubMed]
15. Kankanamge, S.; Khalil, Z.G.; Sritharan, T.; Capon, R.J. Noonindoles G–L: Indole diterpene glycosides from the Australian marine-derived fungus *Aspergillus noonimiae* CMB-M0339. *J. Nat. Prod.* **2023**, *86*, 508–516. [CrossRef]
16. Neuhaus, G.F.; Loesgen, S. Antibacterial drimane sesquiterpenes from *Aspergillus ustus*. *J. Nat. Prod.* **2021**, *84*, 37–45. [CrossRef] [PubMed]
17. Guo, Y.J.; Ding, L.; Ghidinelli, S.; Gotfredsen, C.H.; de la Cruz, M.; Mackenzie, T.A.; Ramos, M.C.; Sánchez, P.; Vicente, F.; Genilloud, O.; et al. Taxonomy driven discovery of polyketides from *Aspergillus californicus*. *J. Nat. Prod.* **2021**, *84*, 979–985. [CrossRef] [PubMed]
18. Wei, M.S.; Huang, L.P.; Li, Q.; Qiao, X.Y.; Zhao, Z.M.; Yin, J.; Fu, A.M.; Guo, J.R.; Hao, X.C.; Gu, L.H.; et al. Spectasterols, aromatic ergosterols with 6/6/6/5/5, 6/6/6/6, and 6/6/6/5 ring systems from *Aspergillus spectabilis*. *J. Nat. Prod.* **2023**, *86*, 1385–1391. [CrossRef] [PubMed]
19. Yang, W.C.; Chen, T.; Tan, Q.; Zang, Z.M.; Chen, Y.; Ou, Y.H.; Li, G.; Hu, D.; Wang, B.; Yao, H.L.; et al. Plasmodium-resistant indole diterpenoid biosynthesis gene cluster derived from *Aspergillus oryzae* was activated by exogenous P450 gene *Ast B*. *J. Nat. Prod.* **2023**, *86*, 1392–1401. [CrossRef] [PubMed]
20. Holkers, J.S.E.; Kagal, S.A.; Mulheirn, L.J.; White, P.M. Some new metabolites of *Aspergillus versicolor* and a revised structure for averufin. *Chem. Commun.* **1966**, *24*, 911–913. [CrossRef]
21. Castonguay, A. Synthesis of (±)-averufanin, noraverufanin and bis-deoxyaverufanin. *Tetrahedron* **1979**, *35*, 1557–1563. [CrossRef]
22. Sakai, K.; Ohte, S.; Ohshiro, T.; Matsuda, D.; Masuma, R.; Rudel, L.L.; Tomoda, H. Selective inhibition of Acyl-CoA: Cholesterol acyltransferase 2 isozyme by flavasperone and sterigmatocystin from *Aspergillus* Species. *J. Antibiot.* **2008**, *61*, 568–572. [CrossRef]
23. Shao, C.L.; Wang, C.Y.; Wei, M.Y.; Li, S.D.; She, Z.G.; Gu, Y.C.; Lin, Y.C. Structural and spectral assignments of six anthraquinone derivatives from the mangrove fungus (ZSUH-36). *Magn. Reson. Chem.* **2008**, *46*, 886–889. [CrossRef]
24. Chen, M.; Shao, C.L.; Kong, C.J.; She, Z.G.; Wang, C.Y. A new anthraquinone derivative from a gorgonian-derived fungus *Aspergillus* sp. *Chem. Nat. Compd.* **2014**, *50*, 617–620. [CrossRef]
25. Demirel1, D.; Ozkaya, F.C.; Ebrahim, W.; Sokullu, E.; Sahin, I.D. *Aspergillus Carneus* metabolite Averufanin induced cell cycle arrest and apoptotic cell death on cancer cell lines via inducing DNA damage. *Sci. Rep.* **2023**, *13*, 6460. [CrossRef]
26. Zhu, F.; Lin, Y.C. Three xanthones from a marine-derived mangrove endophytic fungus. *Chem. Nat. Compd.* **2007**, *43*, 132–135. [CrossRef]
27. Han, X.; Tang, X.L.; Luo, X.C.; Sun, C.X.; Liu, K.C.; Zhang, Y.; Li, P.L.; Li, G.Q. Isolation and identification of three new sterigmatocystin derivatives from the fungus *Aspergillus versicolor* guided by molecular networking approach. *Chem. Biodivers.* **2020**, *17*, e2000208. [CrossRef]
28. Ebrahim, W.; El-Neketi, M.; Lewald, L.I.; Orfali, R.S.; Lin, W.; Rehberg, N.; Kalscheuer, R.; Daletos, G.; Proksch, P. Metabolites from the fungal endophyte Aspergillus austroafricanus in axenic culture and in fungal–bacterial mixed cultures. *J. Nat. Prod.* **2016**, *79*, 914–922. [CrossRef] [PubMed]

29. Özkaya, F.C.; Ebrahim, W.; El-Neketi, M.; Tanrikul, T.T.; Kalscheuer, R.; Müller, W.E.G.; Guo, Z.Y.; Zou, K.; Liu, Z.; Proksch, P. Induction of new metabolites from sponge-associated fungus *Aspergillus carneus* by OSMAC approach. *Fitoterapia* **2018**, *131*, 9–14. [CrossRef] [PubMed]
30. Rodríguez-Urra, A.B.; Jiménez, C.; Nieto, M.I.; Rodríguez, J.; Hayashi, H.; Ugalde, U. Signaling the induction of sporulation involves the interaction of two secondary metabolites in *Aspergillus nidulans*. *ACS Chem. Biol.* **2012**, *7*, 599–606. [CrossRef] [PubMed]
31. Rasche, M.E.; White, R.H. Mechanism for the enzymatic formation of 4-(β-D-ribofuranosyl)aminobenzene 5′-phosphate during the biosynthesis of methanopterin. *Biochemistry* **1998**, *37*, 11343–11351. [CrossRef]
32. White, R.H. The conversion of a phenol to an aniline occurs in the biochemical formation of the 1-(4-aminophenyl)-1-deoxy-D-ribitol moiety in methanopterin. *Biochemistry* **2011**, *50*, 6041–6052. [CrossRef] [PubMed]
33. Xiang, Y.; Kotra, L.P.; Chu, C.K.; Schinazi, R.F. Synthesis and anti-HIV activities of 2′-deoxy-2′,2″-difluoro-β-*L*-ribofuranosyl-pyrimidine and -purine nucleosides. *Bioorg. Med. Chem. Lett.* **1995**, *5*, 743–748. [CrossRef]
34. Maeba, I.; Suzuki, M.; Hara, O.; Takeuchi, T.; Iijimar, T.; Furukawa, H. C-Nucleosides. 6. Synthesis of 5-methoxy-5-(2,3,5-*tri*-*O*-benzoyl-.beta.-D-ribofuranosyl)furan-2(5*H*)-one and its ring transformation. *J. Org. Chem.* **1987**, *52*, 4521–4526. [CrossRef]
35. Tomori, T.; Nagaoka, K.; Takeshita, L.; Shiozawa, T.; Miyatake, Y.; Masaki, Y.; Sekine, M.; Seio, K. Deoxynucleoside triphosphate containing pyridazin-3-one aglycon as a thymidine triphosphate substitute for primer extension and chain elongation by Klenow fragments. *J. Org. Chem.* **2018**, *83*, 8353–8363. [CrossRef] [PubMed]
36. Rosato, R.R.; Almenara, J.A.; Grant, S. The histone deacetylase inhibitor MS-275 promotes differentiation or apoptosis in human leukemia cells through a process regulated by generation of reactive oxygen species and induction of p21$^{CIP1/WAF11}$. *Cancer Res.* **2003**, *63*, 3637–3645.

marine drugs

MDPI

Article

Penidihydrocitrinins A–C: New Polyketides from the Deep-Sea-Derived *Penicillium citrinum* W17 and Their Anti-Inflammatory and Anti-Osteoporotic Bioactivities

Yong Zhang [1,†], Chun-Lan Xie [1,2,†], Yuan Wang [1], Xi-Wen He [1], Ming-Min Xie [1], You Li [1], Kai Zhang [1], Zheng-Biao Zou [1], Long-He Yang [1,*], Ren Xu [2,*] and Xian-Wen Yang [1,*]

1 Key Laboratory of Marine Genetic Resources, Technical Innovation Center for Utilization of Marine Biological Resources, Third Institute of Oceanography, Ministry of Natural Resources, 184 Daxue Road, Xiamen 361005, China; zhangyong@tio.org.cn (Y.Z.); xiechunlanxx@163.com (C.-L.X.); wy2012016284@163.com (Y.W.); hexiwen1224@163.com (X.-W.H.); xiemingmin@tio.org.cn (M.-M.X.); m220200718@st.shou.edu.cn (Y.L.); z18252730063@163.com (K.Z.); zhengbiaozou@njust.edu.cn (Z.-B.Z.)

2 School of Medicine, Xiamen University, South Xiangan Road, Xiamen 361005, China

* Correspondence: longheyang@tio.org.cn (L.-H.Y.); xuren526@xmu.edu.cn (R.X.); yangxianwen@tio.org.cn (X.-W.Y.); Tel.: +86-592-219-5319 (L.-H.Y. & X.-W.Y.); +86-592-288-0577 (R.X.)

† These authors contributed equally to this work.

Abstract: Three new polyketides (penidihydrocitrinins A–C, **1–3**) and fourteen known compounds (**4–17**) were isolated from the deep-sea-derived *Penicillium citrinum* W17. Their structures were elucidated by comprehensive analyses of 1D and 2D NMR, HRESIMS, and ECD calculations. Compounds **1–17** were evaluated for their anti-inflammatory and anti-osteoporotic bioactivities. All isolates exhibited significant inhibitory effects on LPS-stimulated nitric oxide production in murine brain microglial BV-2 cells in a dose-response manner. Notably, compound **14** displayed the strongest effect with the IC_{50} value of 4.7 μM. Additionally, compounds **6**, **7**, and **8** significantly enhanced osteoblast mineralization, which was comparable to that of the positive control, purmorphamine. Furthermore, these three compounds also suppressed osteoclastogenesis in a dose-dependent manner under the concentrations of 2.5 μM, 5.0 μM, and 10 μM.

Keywords: deep-sea; fungus; *Penicillium citrinum*; polyketides; anti-osteoporosis; anti-inflammation

Citation: Zhang, Y.; Xie, C.-L.; Wang, Y.; He, X.-W.; Xie, M.-M.; Li, Y.; Zhang, K.; Zou, Z.-B.; Yang, L.-H.; Xu, R.; et al. Penidihydrocitrinins A–C: New Polyketides from the Deep-Sea-Derived *Penicillium citrinum* W17 and Their Anti-Inflammatory and Anti-Osteoporotic Bioactivities. *Mar. Drugs* **2023**, *21*, 538. https://doi.org/10.3390/md21100538

Academic Editors: Bin-Gui Wang and Haofu Dai

Received: 23 September 2023
Revised: 10 October 2023
Accepted: 12 October 2023
Published: 14 October 2023

1. Introduction

Marine microbes are an ideal source to yield diverse secondary metabolites with unprecedented structures. As a matter of fact, about half of the new marine natural products are produced by marine microorganisms [1–3], especially those living in the deep sea under extremely tough environments such as low oxygen concentration, high salt, high hydrostatic pressure, and absence of light, which require various biochemical and physiological adaptations for survival [4,5]. These adaptions are accompanied by adjustments of gene regulation, resulting in the formation of different metabolic pathways to give birth to a large number of new secondary metabolites [4,5].

Polyketides are a class of structurally diverse natural products with carbon skeletons originating from the polymerization of short-chain carboxylic acids units, including acetate, propionate, butyrate, etc. [6], which are catalyzed by polyketide synthases (PKSs) [7]. Polyketides have attracted wide attention due to their promising bioactivities [8,9]. For example, salinosporamide A showed significant proteasomal chymotrypsin-like proteolytic inhibitory activity with an IC_{50} value of 1.3 nM [10]; microketide A exhibited remarkable antibacterial activities against *Pseudomonas aeruginosa*, *Nocardia brasiliensis*, *Kocuria rhizophila*, and *Bacillus anthraci* with an equal minimum inhibitory concentration (MIC) value (0.19 μg/mL) to that of ciprofloxacin [11]; and theissenone exhibited potent nitric oxide production inhibitory activity in murine brain microglial BV-2 cells with an IC_{50} value of 5.0 ± 1.0 μM [12]. As some of the most abundant fungi of the world, *Penicillium* species could generate a broad spectrum of unique polyketides. For instance, two new tricyclic polyketides, penijanthinones A and B, were isolated from *P. janthinellum* HK1-6 [13,14]; two new C-8 benzoyl-substituted azaphilones, pinazaphilones A and B, were obtained from *Penicillium* sp. HN29-3B1 [13,14]; chloctanspirones A and B, possessing an unprecedented bicyclo [2.2.2] octane-2-spiro cyclohexane skeleton, were discovered in *P. terrestre* [15].

As part of our continuing discovery of structurally novel and biologically interesting compounds from deep-sea-derived microorganisms [16–20], the crude extract of *Penicillium citrinum* W17 isolated from a deep-sea sediment sample (−5278 m) of the western Pacific Ocean revealed the rich chemical diversity of the secondary metabolites. Therefore, it was subjected to a systematic chemical investigation. As a result, three new polyketides (penidihydrocitrinins A–C, **1–3**) and fourteen known compounds (**4–17**) were obtained (Figure 1). Herein, we report the details of isolation, structure elucidation, and bioactivities of these isolates.

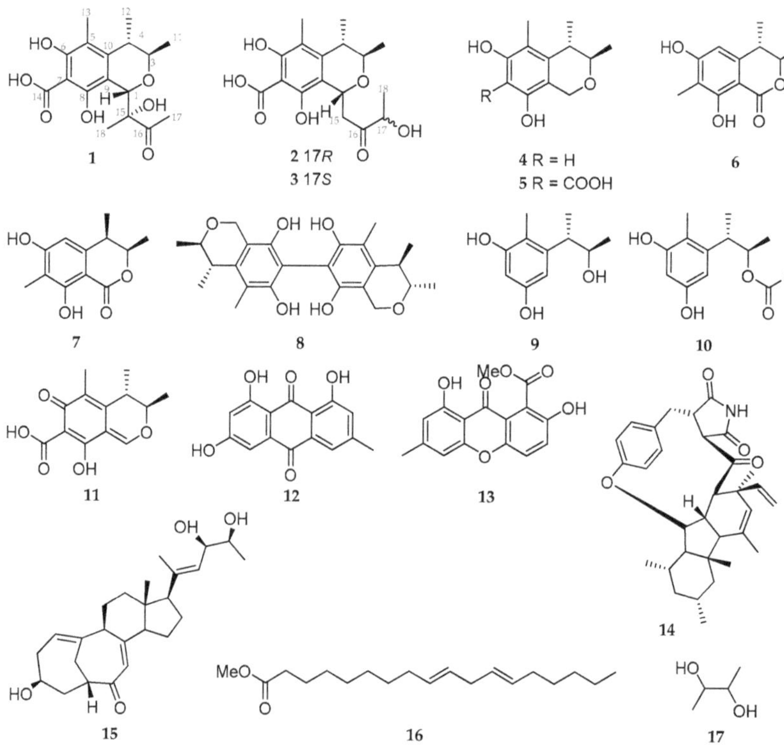

Figure 1. Compounds **1–17** isolated from the deep-sea-derived *Penicillium citrinum* W17.

2. Results and Discussion

The EtOAc-soluble extract of the deep-sea-derived *Penicillium citrinum* W17 was subjected to column chromatography (CC) on silica gel, ODS, and Sephadex LH-20, as well as preparative HPLC to afford 17 compounds (**1–17**).

Compound **1** was obtained as yellow amorphous powder. The molecular formula was determined as $C_{17}H_{22}O_7$ with seven degrees of unsaturation (DoU) on the basis of HRESIMS spectrum at m/z 337.1309 $[M - H]^-$ (Figure S1 of the Supplementary Materials). The ^1H NMR spectroscopic data (Table 1) showed two methyl doublets (δ_H 1.11 (d, J = 6.4 Hz, 11-Me) and 1.28 (d, J = 6.9 Hz, 12-Me)), three methyl singlets (δ_H 1.10 (s, 18-Me), 2.04 (s, 13-Me), and 2.30 (s, 17-Me)), and three methines (δ_H 2.62 (qd, J = 6.7 Hz, 1.8 Hz, H-4), 3.98 (qd, J = 6.9 Hz, 1.8 Hz, H-3), 5.17 (s, H-1)). The ^{13}C NMR spectroscopic data exhibited

17 carbon resonance signals, including five methyls (δ_C 10.2 (q, 13-Me), 18.3 (q, 11-Me), 20.1 (q, 12-Me), 20.7 (q, 18-Me), 25.0 (q, 17-Me)), three methines ((δ_C 36.9 (d, C-4), 74.0 (d, C-3), 75.4 (d, C-1)), and nine nonprotonated carbons (δ_C 83.2 (s, C-15), 102.2 (s, C-7), 110.3 (s, C-9), 114.0 (s, C-5), 143.8 (s, C-10), 156.7 (s, C-8), 160.0 (s, C-6), 178.2 (s, C-14), 211.3 (s, C-16)). The COSY correlations of 11-Me/H-3/H-4/12-Me, together with the HMBC correlations from H-4 to C-5 and C-10, from 13-Me to C-5, C-6, and C-10, from H-3 to C-10, and from H-1 to C-3, C-8, C-9, and C-10, constructed a dihydrocitrinin fragment. In combination with the HMBC cross peaks of 17-Me with C-16 and of 18-Me with C-1, C-15, and C-16, the planar structure of compound **1** was established as 2-hydroxy-3-butonyldihydrocitrinin (Figure 2).

Table 1. ^{1}H (400 Hz) and ^{13}C (100 Hz) NMR spectroscopic data of **1–3** (δ in ppm, J in Hz within parentheses).

No.	1 [a]		2 [b]		3 [b]	
	δ_C	δ_H	δ_C	δ_H	δ_C	δ_H
1	75.4 CH	5.17 s	64.8 CH	5.08 (d, 9.7)	64.8 CH	5.08 (d, 9.2)
3	74.0 CH	3.98 (qd, 6.9, 1.8)	71.7 CH	3.82 (qd, 6.6, 1.5)	71.7 CH	3.82 (qd, 6.6, 1.5)
4	36.9 CH	2.62 (qd, 6.7, 1.8)	35.1 CH	2.52 m	35.2 CH	2.51 m
5	114.0 C		109.5 C		109.6 C	
6	160.0 C		158.5 C		158.5 C	
7	102.2 C		101.8 C		101.8 C	
8	156.7 C		155.8 C		155.8 C	
9	110.3 C		110.7 C		110.9 C	
10	143.8 C		139.7 C		139.8 C	
11	18.3 CH_3	1.11 (d, 6.4)	18.2 CH_3	1.01 (d, 6.5)	18.3 CH_3	0.99 (d, 6.5)
12	20.1 CH_3	1.28 (d, 6.9)	20.1 CH_3	1.17 (d, 6.8)	20.2 CH_3	1.17 (d, 7.2)
13	10.2 CH_3	2.04 s	9.5 CH_3	1.93 s	9.6 CH_3	1.92 s
14	178.2 C		175.6 C		175.7 C	
15	83.2 C		43.1 CH_2	2.66 m 3.23 m	43.0 CH_2	2.72 m 3.16 m
16	211.3 C		211.9 C		212.1 C	
17	25.0 CH_3	2.30 s	72.2 CH	4.11 m	72.8 CH	4.02 m
18	20.7 CH_3	1.10 s	19.2 CH_3	1.21 (d, 7.0)	19.0 CH_3	1.15 (d, 7.2)
6-OH				14.64 s		14.62 s
8-OH				15.16 s		15.13 s

[a] Recorded in CD_3OD. [b] Recorded in DMSO-d_6.

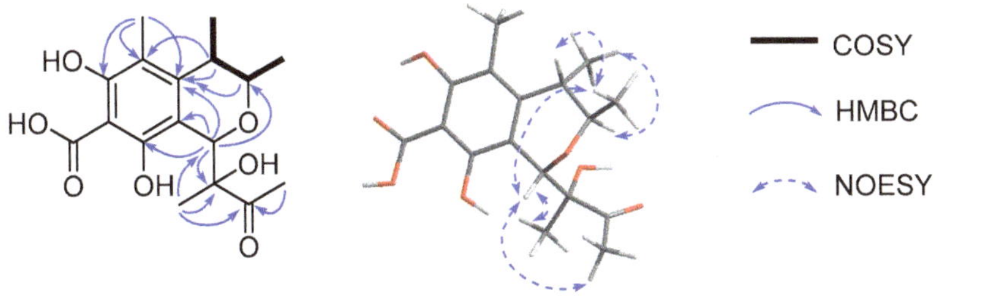

Figure 2. The key COSY, HMBC, and NOESY correlations of compound **1**.

The correlations of H-3/12-Me and 11-Me/H-1/H-4 were observed in the NOESY spectrum, indicating H-1, H-4, and 11-Me were on the same plane, opposite to H-3 and 12-Me (Figure 2). Although the NOESY correlations were found of H-1 to 17-Me and 18-Me, revealing the relative configuration of C-15, more solid evidence is needed to confirm the absolute configuration because of the flexible structure of the segment. Accordingly, the theoretical calculations of the electronic circular dichroism (ECD) spectrum of (1*S*,3*R*,4*S*,15*R*)-**1** and (1*S*,3*R*,4*S*,15*S*)-**1** were conducted along with their enantiomers of (1*R*,3*S*,4*R*,15*S*)-**1** and (1*R*,3*S*,4*R*,15*R*)-**1**. As shown in Figure 3, the calculated ECD spectrum of (1*S*,3*R*,4*S*,15*R*)-**1** matched well with the experimental one. On the basis of the above evidence, compound **1** was determined as (1*S*,3*R*,4*S*,15*R*)-2-hydroxy-3-butonyldihydrocitrinin and named penidihydrocitrinin A.

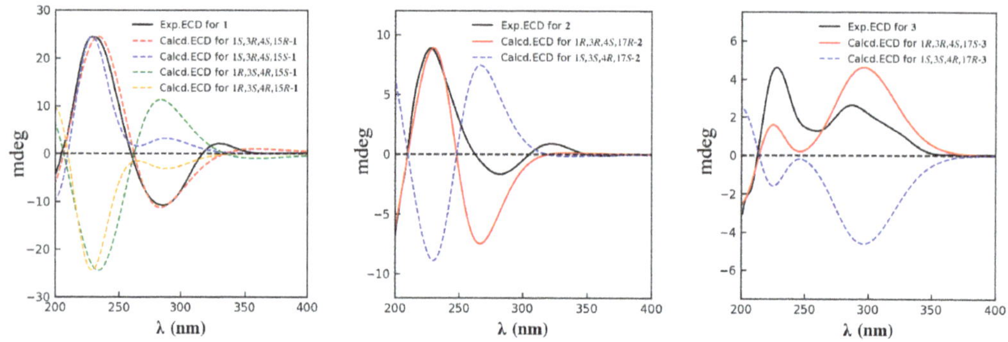

Figure 3. The calculated and experimental ECD spectra of compounds **1–3**.

Compound **2** was obtained as a colorless oil. Its molecular formula was assigned as $C_{17}H_{22}O_7$ on the basis of HRESIMS data at m/z 337.1348 $[M - H]^-$, suggesting seven degrees of unsaturation. Four methyls (δ_H 1.01 (d, J = 6.5 Hz, 11-Me), 1.17 (d, J = 6.8 Hz, 12-Me), 1.21 (d, J = 7.0 Hz, 18-Me), and 1.93 (s, 13-Me)) and four methines (δ_H 2.52 (m, H-4), 3.82 (qd, J = 6.6 Hz, 1.5 Hz, H-3), 4.11 (m, H-17), and 5.08 (d, J = 9.7 Hz, H-1)) were recognized in the ^1H NMR spectrum (Figure S9). The ^{13}C NMR spectrum in association with the HSQC spectrum indicated 17 carbon signals ascribed to four methyls at δ_C 9.5 (q, 13-Me), 18.2 (q, 11-Me), 19.2 (q, 18-Me), and 20.1 (q, 12-Me); one methylene at δ_C 43.1 (t, C-15); four methines at δ_C 35.1 (d, C-4), 64.8 (d, C-1), 71.7 (d, C-3), and 72.2 (d, C-17); and eight nonprotonated carbons at δ_C 101.8 (s, C-7), 109.5 (s, C-5), 110.7 (s, C-9), 139.7 (s, C-10), 155.8 (s, C-8), 158.5 (s, C-6), 175.6 (s, C-14), and 211.9 (s, C-16). The spin systems of 11-Me/H-3/H-4/12-Me, H-17/18-Me, and H-1/H-15 were observed in the COSY spectrum, constructing three segments (Figure 4). In the HMBC spectrum, correlations were found of 13-Me to C-5/C-6/C-10, 6-OH to C-5/C-6/C-7, 8-OH to C-7/C-8/C-9, H-4 to C-3/C-5/C-10/C-12, H-1 to C-9/C-10/C-15/C-16, H-3 to C-1, and 18-Me to C-16/C-17. Taking the COSY and HMBC correlations together, the planar structure of compound **2** was then constructed as 3-hydroxy-2-butonyldihydrocitrinin, an isomer of **1**. The relative configuration of **1** was regarded as the same as that of **2** on the basis of the key NOESY correlations of H-1/11-Me, H-3/12-Me, and H-4/11-Me (Figure 4). By the biosynthetic consideration, the absolute configuration of C-1, C-3, and C-4 in **2** and **1** should be the same. However, the stereochemistry of C-17 could not be determined. Therefore, the theoretical calculation of the ECD spectrum was performed. As a result, the experimental ECD spectrum matched well with that of (1*R*,3*R*,4*S*,17*R*)-**2** (Figure 4). Consequently, the structure of **2** was assigned as (1*R*,3*R*,4*S*,17*R*)-3-hydroxy-2-butonyldihydrocitrinin and named penidihydrocitrinin B.

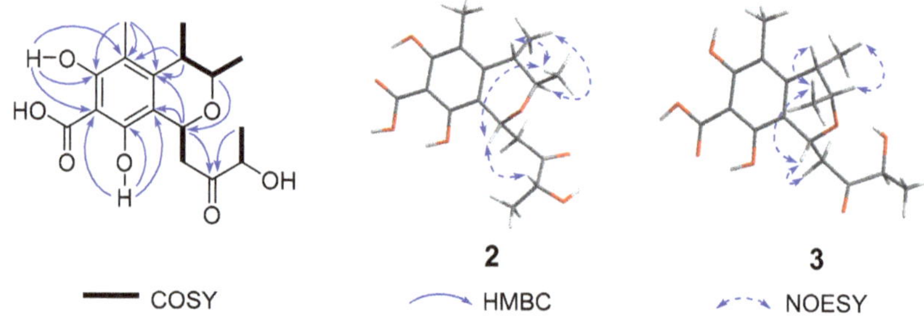

Figure 4. The key COSY, HMBC, and NOESY correlations of compounds **2** and **3**.

Compound **3** was isolated as a colorless oil. The molecular formula of $C_{17}H_{22}O_7$ was established based on its HRESIMS spectrum at *m/z* 337.1327 $[M - H]^-$ (calcd for $C_{17}H_{21}O_7$, 337.1287). Its ^1H and ^{13}C NMR spectroscopic data were very similar to those of **2**, except for the upshift of H-17 from δ_H 4.11 to δ_H 4.02 and H_3-18 from δ_H 1.21 to 1.15 and the downshift of C-17 from δ_C 72.2 to 72.8. This implied that compound **3** could be an epimer of **2** with *S*-configuration at the C-17 position. The assumption was evidenced by the positive Cotton effect (CE) at λmax 287 nm (Δε +0.54) in **3**, whereas a negative CE (Δε −0.34 at λmax 282 nm) in **2**. Final confirmation was obtained by comparison of the calculated and experimental ECD spectra of **3**, showing the calculated ECD spectrum of (1*R*,3*R*,4*S*,17*S*)-**3** was in good accordance with that of the experimental curve (Figure 4). Accordingly, compound **3** was defined as (1*R*,3*R*,4*S*,17*S*)-3-hydroxy-2-butonyldihydrocitrinin, and named penidihydrocitrinin C.

Compounds **1–3** were three novel polyketide adducts of citrinin and diacetyl. They might be biosynthesized by citrinin (**11**) and diacetyl, a widely found secondary metabolite in microorganisms [21] (Scheme 1). Noteworthily, the reduced derivative of diacetyl, 2,3-butanediol (**17**), was also co-isolated from the same extract of the strain.

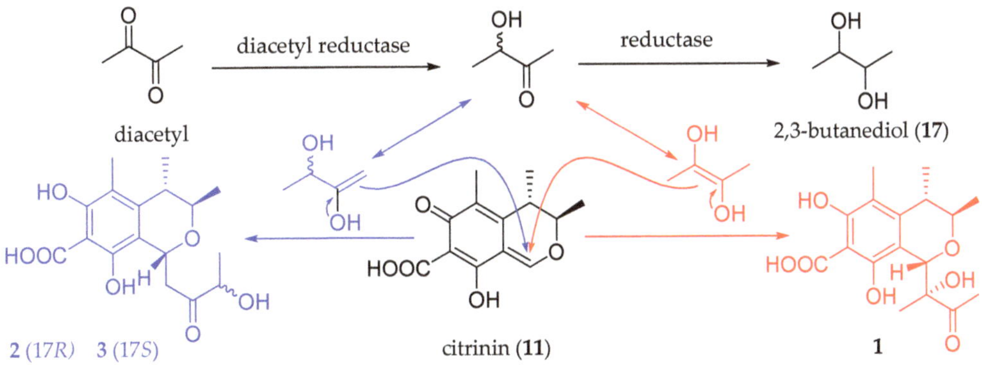

Scheme 1. The proposed biosynthetic pathway of penidihydrocitrinins A–C (**1–3**).

By comparison of the NMR and MS data with those published in the literature, 14 known compounds were identified as decarboxydihydrocitrinin (**4**) [22], dihydrocitrinin (**5**) [23], (3*R**,4*S**)-6,8-dihydroxy-3,4,7-trimethylisocoumarin (**6**) [24], sclerotinin C (**7**) [25], asperbiphenyl (**8**) [26], phenol A (**9**) [24], citrinin H2 (**10**) [27], citrinin (**11**) [28], emodin (**12**) [29], pinselin (**13**) [30], GKK1032 B (**14**) [31], neocyclocitrinol C (**15**) [32], (*Z,Z*)-9,12-ocadecadienoic acid methyl ester (**16**) [33], and 2,3-butanediol (**17**) [34].

Microglial activation plays a pivotal role in the pathogenesis of neurodegenerative diseases, orchestrating a complex interplay between inflammation and neuronal health [35]. While microglial cells function as the guardians of the central nervous system, their dysregulated activation can lead to chronic neuroinflammation and exacerbation of neuronal damage, contributing to the progression of disorders such as multiple sclerosis and Alzheimer's and Parkinson's diseases [36]. These diseases pose significant challenges to public health, which necessitate innovative therapeutic approaches. Thus, discovering new small molecules that can inhibit the dysregulated activation of microglial cells is essential for the targeted modulation of the immune response in the central nervous system, which can reduce inflammation and protect neurons from harm [37]. A growing amount of evidence demonstrates that secondary metabolites derived from marine resources are potential therapeutic strategies for microglial-mediated neuroinflammation.

Therefore, all isolates were tested for nitrite secretion in lipopolysaccharide (LPS)-induced BV-2 microglial cells. As a result, they all demonstrated a dose-dependent suppression of nitrite secretion induced by LPS, displaying inhibitory actions at concentrations of 3.0 μM, 10 μM, and 20 μM (Figure 5). Moreover, none exhibited cytotoxicity effects against BV-2 cells at 20 μM under the microscope. We further compared the inhibition of these compounds on nitrite production at a concentration of 10 μM (Figure 6), and the results showed that the inhibitory rates of compounds **1–7** on nitrite production were 31.1–53.5% which indicates that substitution of the C-1 position on isochroman weakened the anti-inflammatory activity of the compounds, suggesting that this substitution is related to anti-inflammatory activity. Compound **8** only inhibited nitrite production by 26.4% in LPS-stimulated BV-2 cells, demonstrating that the dimeric structure further weakened the anti-inflammatory effect. Meanwhile, the anti-inflammatory activity of compound **12**, emodin, was consistent with a previous report [38]. Notably, compound **14** (GKK1032 B) displayed the most potent nitrite inhibitory activity with an inhibitory ratio of 73.0 ± 1.6% at 10 μM (nitrite concentration: 12.2 ± 0.4 μM), compared to the LPS-treated group (nitrite concentration: 30.9 ± 0.4 μM) (Figures 5 and 6). Furthermore, this compound displayed an IC_{50} value of 4.7 μM. Although GKK1032 analogues were reported to exhibit antibacterial activities [39], it is the first time that they had anti-neuroinflammatory activity. These findings demonstrate the effects of marine-derived compounds in modulating microglial activation, suggesting their potential as therapeutic candidates for neuroinflammatory conditions and neurodegenerative diseases.

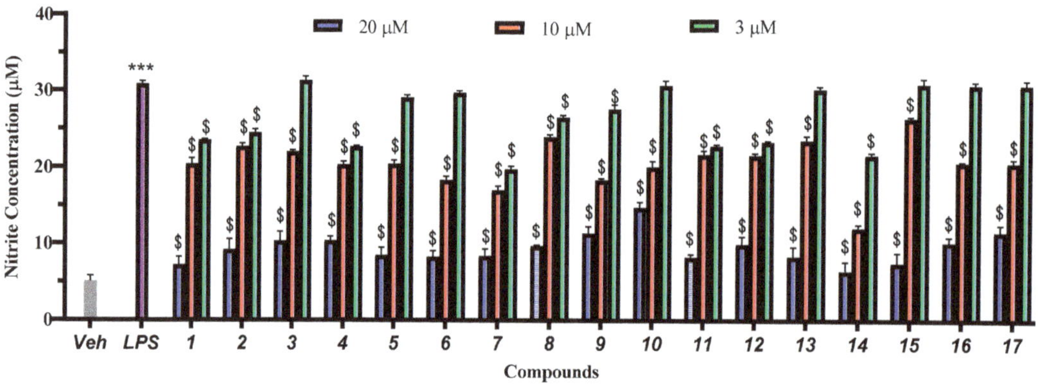

Figure 5. Effects of compounds **1–17** against nitrite production in LPS-induced BV-2 cells. *** $p < 0.0001$ vs. Veh; $ $p < 0.0001$ vs. LPS.

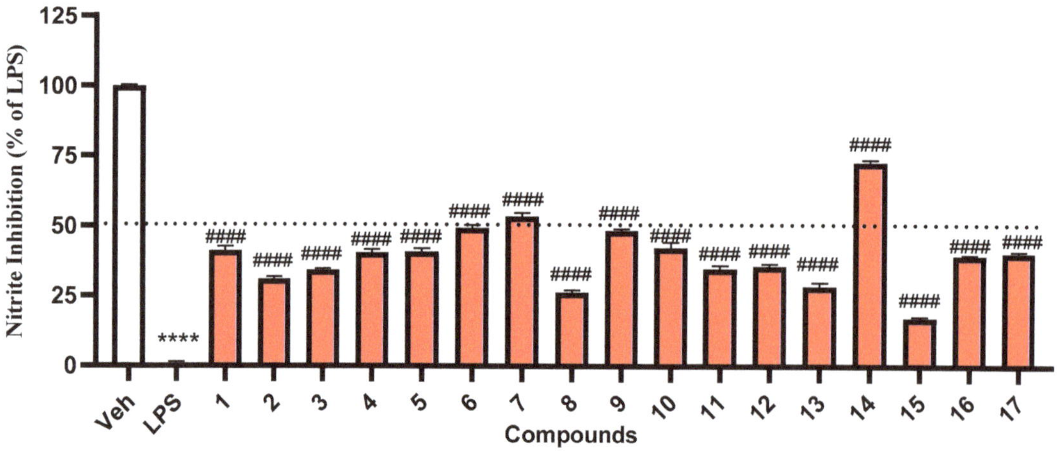

Figure 6. Inhibitory effects of compounds **1–17** (10 μM) on LPS-induced nitrite production in BV-2 cells. **** $p < 0.0001$ vs. Veh; #### $p < 0.0001$ vs. LPS.

Osteoporosis, a disease associated with aging, is characterized by excessive activation of osteoclasts or reduction of osteoblasts. Among women aged 65 or older, approximately 25% are affected by osteoporosis, with accelerated bone loss occurring after menopause. Therefore, promoting osteoblast differentiation and suppressing osteoclastogenesis are effective strategies for treating osteoporosis [40]. BMSCs are able to differentiate into osteoblasts, chondroblasts, and adipocytes [41], and bone regeneration achieved via osteogenic induction of MSCs could provide a rational therapeutic strategy for preventing age-related osteoporosis [42]. Accordingly, all 17 isolates were tested for both osteoblast and osteoclast activities. Firstly, compounds **1–17** were subjected to anti-proliferative tests on BMSCs. At the concentration of 10 μM, none showed cytotoxicity (Figure 7).

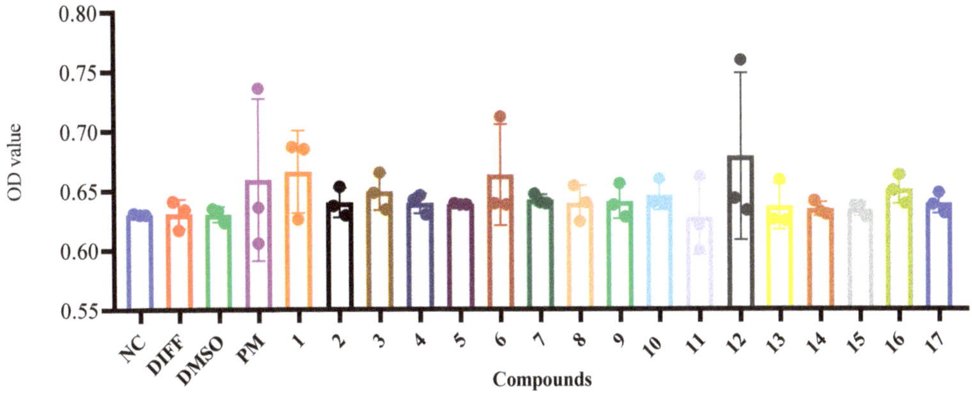

Figure 7. BMSC viability of compounds **1–17** was measured by the CCK-8 assay. The culture medium, osteogenesis differentiation medium (50 μg/mL ascorbic acid and 5 mM β-glycerophosphate), DMSO (0.1%), and purmorphamine (1 μM) were regarded as the negative control (NC), differentiation (DIFF), solvent (DMSO), and positive control (PM) groups, respectively.

Intracellular calcium deposition at a later stage served as a significant evaluation indicator for osteogenic activity. Interestingly, compounds **6**, **7**, and **8** exerted a noticeable enhancing effect on osteoblast mineralization within BMSCs (Figure 8).

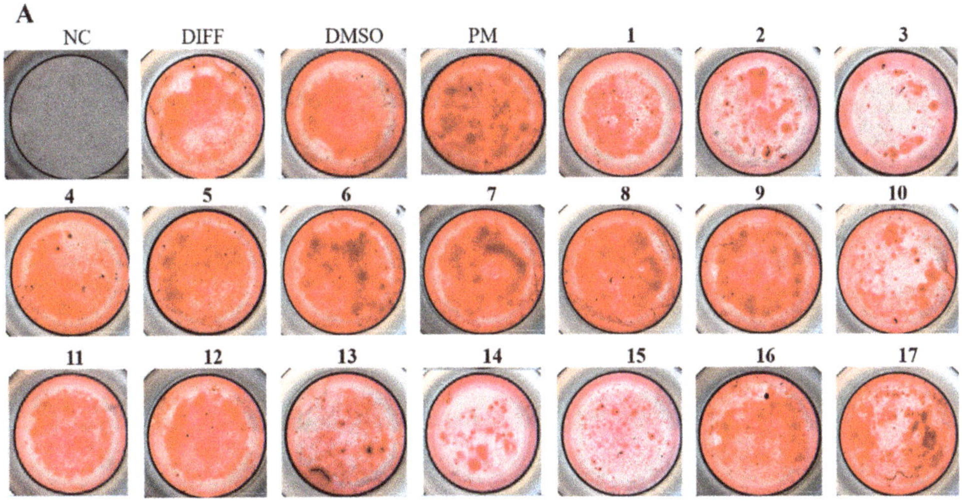

Figure 8. *Cont.*

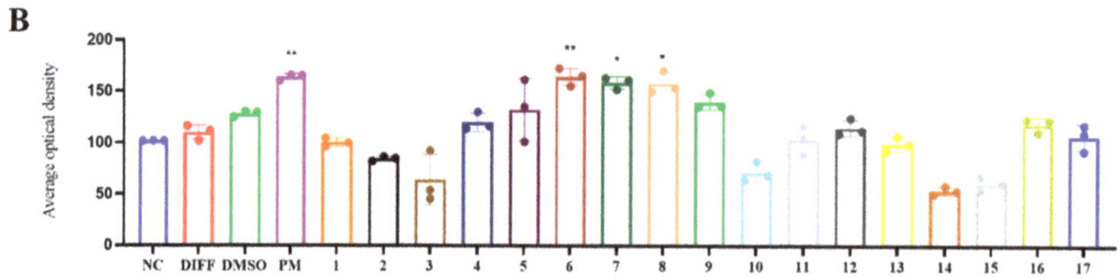

Figure 8. Preliminary screening results for the ability of compounds **1–17** to induce osteoblast activity among BMSCs. (**A**) Alizarin Red S staining. The BMSCs were cultures exposed to compounds **1–17** with osteogenic inducer. After 14 days, the cells were stained with Alizarin Red S and pictures were taken by Biotek cytation-5. The culture medium, osteogenesis differentiation medium (50 µg/mL ascorbic acid and 5 mM β-glycerophosphate), DMSO (0.1%), and purmorphamine (1 µM) were regarded as the negative control (NC), differentiation (DIFF), solvent (DMSO), and positive control (PM) groups, respectively. Scale bar = 1000 µm. (**B**) Quantification of Alizarin Red S staining based on average optical density. * $p < 0.05$, ** $p < 0.01$ vs. DMSO.

Furthermore, bioactive compounds **6**, **7**, and **8** also exhibited a distinct inhibitory effect on osteoclast activity, as evidenced by a significant reduction in tartrate-resistant acid phosphatase (TRAP)-positive cells (Figure 9). These findings indicate that compounds **6**, **7**, and **8** not only facilitated osteoblast mineralization but also exerted a substantial dose-dependent inhibitory effect on RANKL-induced osteoclasts.

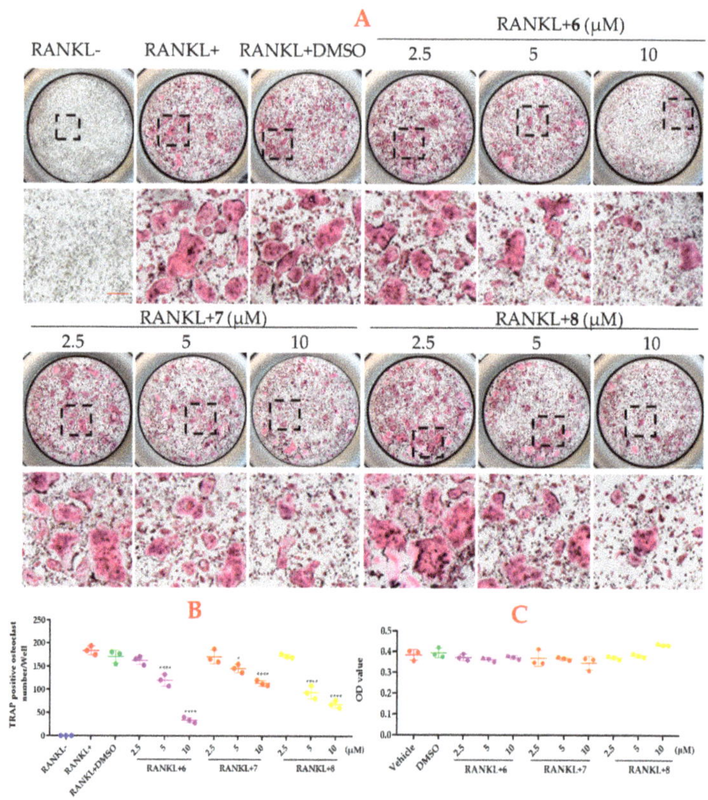

Figure 9. Compounds **6**, **7**, and **8** dose-dependently attenuated RANKL-activated osteoclastogenesis. (**A**) BMMs cultured with tested compounds (2.5 µM, 5.0 µM, and 10 µM) with the stimulation of

25 ng/mL RANKL (or not). After five days, these cells were fixed and stained with TRAP. (**B**) TRAP-positive multinucleated (there or more) cells were regarded as mature osteoclasts. Osteoclast numbers were quantified and analyzed (n = 3). * $p < 0.05$, **** $p < 0.0001$ vs. the RANKL+DMSO group. (**C**) BMM viability was measured by the CCK-8 assay. The culture medium and the solvent (0.1% DMSO) were regarded as Vehicle and DMSO groups, respectively.

3. Materials and Methods

3.1. General Experimental Procedures

NMR spectra of all compounds were recorded on a Bruker 400 MHz spectrometer (Bruker, Fällanden, Switzerland). Optical rotations were measured by an MCP 100 polarimeter (Anton Paar Trading Co., Ltd., Shanghai, China). The high-resolution electrospray ionization mass spectrometry (HRESIMS) results were acquired on a Xevo G2 Q-TOF mass spectrometer (Waters Corporation, Milford, MA, USA). ECD spectra were measured on a Chirascan spectrometer (Applied Photophysics, Surrey, UK). The semi-preparative high-performance liquid chromatography (semi-prep. HPLC) was performed on an Agilent instrument (1260) (Agilent Technologies, San Diego, CA, USA) with a semi-preparative chromatographic column (COSMOSIL 5 C18-MS-II, Nacalai Tesque, Japan). Column chromatography (CC) was performed on silica gel (Qingdao Marine Chemistry Co., Ltd., Qingdao, China), Sephadex LH-20 (Amersham Pharmacia Biotech AB, Uppsala, Sweden), and octadecyl silyl (ODS) (Daiso Chemical Co., Ltd., Osaka, Japan). Preparative thin-layer chromatography (Prep. TLC) was performed with silica gel precoated plates (Qingdao Marine Chemistry Co., Ltd., Qingdao, China).

3.2. Fungal Identification, Fermentation, and Extraction

The fungal strain W17 was isolated from a deep-sea sediment sample of the western Pacific Ocean at the depth of 5278 m. It was identified to be *Penicillium citrinum* as the 18S rRNA gene sequence alignment (OR398934) demonstrated that it showed great similarity (99.8%) to *Penicillium citrinum* NRRL 1841 (GenBank accession number NR_121224.1). The strain was preserved at the Key Laboratory of Marine Genetic Resources, Third Institute of Oceanography, Ministry of Natural Resources (Xiamen, China). The microbial strain was cultivated on a PDA plate medium at 25 °C for 3 days and the colony was inoculated into 250 mL Erlenmeyer flasks containing 50 mL PDB medium. Then, it was cultured in a rotary shaker (130 rpm) at 25 °C for 3 days as spore medium. After 3 days, the spore solution was inoculated in 120 Erlenmeyer flasks (1 L) with each containing 400 mL tap water, 80 g potato power, 8 g glucose, and 6 g marine salt. The fermentation was performed in a 130 rpm rotary shaker at 25 °C. After 7 days, the fermentation broth was extracted with EtOAc three times and concentrated under reduced pressure to give a crude extract (36.5 g).

3.3. Isolation and Purification

The crude extract (36.5 g) was separated into seven fractions (Fr.1−Fr.7) via CC over silica gel using a gradient of CH_2Cl_2−MeOH (100%→75%). Fraction Fr.4 (3.6 g) was subsequently separated by CC over ODS (MeOH-H_2O, 5%→100%), Sephadex LH-20 (MeOH), and semi-prep. HPLC with MeOH-H_2O (40%→100%) to yield **1** (5.7 mg) and **4** (15.4 mg). Fraction Fr.7 (1.5 g) was separated by CC over Sephadex LH-20 (MeOH), silica gel (CH_2Cl_2−MeOH, 10:1), and semi-prep. HPLC with MeOH−H_2O (30%→60%) to yield **2** (1.0 mg), **3** (1.0 mg), and **6** (45.9 mg). Compound **11** (43.0 mg) was purified from Fr.1 (5.0 g) by recrystallization in MeOH, while compounds **14** (14.0 mg) and **16** (49.0 mg) were isolated by CC over ODS (50%→100%), Sephadex LH-20 (MeOH), and prep. TLC (CH_2Cl_2-MeOH, 50:1). Fraction Fr.2 (2.0 g) was subjected to ODS (10%→100%) and Sephadex LH-20 (MeOH) to yield compound **12** (69.0 mg). Fraction Fr.3 was subjected to CC on ODS (10%→100%) and Sephadex LH-20 (MeOH). Final purification by semi-prep. HPLC (MeOH-H_2O, 60%→90%) afforded **9** (1.6 mg) and **10** (3.7 mg). Fraction Fr.5 (4.6 g) was separated by CC over ODS (10%→100%), Sephadex LH-20 (MeOH), and semi-prep. HPLC (MeOH-H_2O, 30%→60%) to give **5** (124.0 mg), **13** (7.0 mg), **15** (2.1 mg), and **17**

(190.0 mg). Compounds **7** (15.2 mg) and **8** (22.4 mg) were obtained by CC over ODS (10%→100%), Sephadex LH-20 (MeOH), and semi-prep. HPLC (MeOH−H_2O, 30%→60%) from Fr.6 (2.0 g).

Penidihydrocitrinin A (**1**): yellow amorphous powder; $[\alpha]_D^{25}$ +28.0 (*c* 0.10, MeOH); UV (MeOH) λmax (logε) 214 (3.16), 244 (2.64), 252 (2.65), 280 (1.85), 319 (2.16) nm; CD (MeOH) (Δε) 229 (+5.00), 233 (+0.53), 286 (−2.20), 330 (+0.42) nm; ^1H and ^{13}C NMR data, see Table 1; HRESIMS *m*/*z* 337.1309 $[M - H]^-$ (calcd for $C_{17}H_{21}O_7$ 337.1287).

Penidihydrocitrinin B (**2**): colorless oil; $[\alpha]_D^{25}$ +25.0 (*c* 0.10, MeOH); UV (MeOH) λmax (logε) 204 (2.78), 214 (2.84), 242 (2.28), 253 (2.32), 276 (1.40), 319 (1.90) nm; CD (MeOH) (Δε) 228 (+1.82), 282 (−0.34), 322 (+0.18) nm; ^1H and ^{13}C NMR data, see Table 1; HRESIMS *m*/*z* 337.1348 $[M - H]^-$(calcd for $C_{17}H_{21}O_7$ 337.1287).

Penidihydrocitrinin C (**3**): colorless oil; $[\alpha]_D^{25}$ +40.0 (*c* 0.10, MeOH); UV (MeOH) λmax (logε) 205 (2.71), 213 (2.72), 253 (2.24), 275 (1.39), 320 (1.83) nm; CD (MeOH) (Δε) 228 (+0.94), 260 (+0.26), 287 (+0.54) nm; ^1H and ^{13}C NMR data, see Table 1; HRESIMS *m*/*z* 337.1327 $[M - H]^-$ (calcd for $C_{17}H_{21}O_7$, 337.1287).

3.4. ECD Calculation

As reported previously [16], the conformational analysis was first conducted via random searching stochastically using the MMFF94 force field. All conformers were consecutively optimized at the PM6 and HF/6-31G(d) levels. Dominative conformers were further optimized at the B3LYP/6-31G(d) level in the gas phase. The optimized conformers possessed no imaginary frequencies and were true local minima. ECD calculations were conducted at the B3LYP/6-311G(d,p) level in MeOH with the IEFPCM model using time-dependent density functional theory (TD-DFT). The ECD spectrum was simulated by overlapping Gaussian functions for each transition.

3.5. BV-2 Cell Culture and Compound Treatment

BV-2 cell culture and compound treatment were carried out as previously reported [43,44]. Briefly, BV-2 microglial cells were cultured in DMEM supplemented with 10% fetal bovine serum (Thermofisher, Shanghai, China) and antibiotics (100 units/mL of penicillin and 100 μg/mL of streptomycin) in a humidified 5% CO_2 incubator at 37 °C. Cells were seeded into 24-well plates at a density of 2×10^4 cells per well and allowed to adhere overnight. On the subsequent day, the cells were treated with freshly prepared culture medium containing the specified concentrations of the investigated compounds for a duration of 30 min before exposure to LPS (1 μg/mL). A control group was treated with a vehicle solution (DMSO, 0.1%).

3.6. Quantification of Nitrite Levels

The concentration of nitrite present in the culture medium was assessed using the Griess Reagent Kit (Thermo Fisher, Shanghai, China). Briefly, 75 μL of cell culture supernatants was mixed with an equal volume of the Griess Reagent Kit and allowed to react for 30 min at room temperature. The absorbance of the resulting diazonium compound was measured at a wavelength of 560 nm. The concentration of nitrite production was calculated based on the nitrite standard solution.

3.7. Cell Extraction and Culture

The bone mesenchymal stem cells (BMSCs) and bone marrow monocytes (BMMs) were flushed from the femur of C57BL/6J mice aged 3 weeks and 6-week-old C57BL/6 mice, respectively, with the methods as previously described [16]. In brief, the BMSCs were carefully removed from the animals and cultured in α-MEM and induced with supplemented complete α-MEM (10% *v/v* FBS, 1% *v/v* penicillin/streptomycin (P/S)). The BMSCs were induced with 2 mM β-glycerophosphate and 50 μg/mL ascorbate, of which half were changed twice a week. The BMMs were cultured in complete α-MEM (10% *v/v* FBS, 1% *v/v* P/S, and 25 ng/mL M-CSF).

3.8. CCK-8 Assay

The in vitro cytotoxic bioassay was conducted using the CCK-8 method according to the previously reported protocols [16]. The BMSCs or BMMs were treated with or without compounds. After 48 h of culture, cells were treated with CCK-8 solution for 2 h before being scanned with a multimode scanner at 450 nm (Biotek, Winooski, VT, USA).

3.9. Osteoblast Differentiation and Mineralization

The BMSCs were digested and planted at a density of 2×10^4 cells/well into 96-well plates and cultivated overnight. Ascorbic acid (50 μg/mL) and β-glycerophosphate (5 mM) were added into the culture medium for osteogenic assay. The culture medium was changed every other day. AR S staining was performed after 14 days of differentiation. Alizarin Red staining was used to check the calcification conditions in cultures. Cells were fixed with 10% neutral-buffered formalin for 30 min and then 2% Alizarin Red S was used to incubate cells for 2 min at room temperature. Then, pictures were taken of the cells by a Biotek cytation-5.

3.10. Osteoclast Differentiation

BMMs were cultured for 7 days in the presence of M-CSF (25 ng/mL) and RANKL (25 ng/mL) for differentiation into mature osteoclasts. Media were refreshed every 2 days. For tartrate-resistant acid phosphatase (TRAP) staining, cells were fixed with 4% paraformaldehyde (PFA) and stained for TRAP activity. Under a microscope (Biotek cytation-5, USA), TRAP-positive multinucleated cells with three nuclei were counted as osteoclasts.

4. Conclusions

In summary, from the deep-sea-derived fungus *Penicillium citrinum* W17, 17 compounds were obtained. Penidihydrocitrinins A–C (**1**–**3**) are three novel citrinin and diacetyl adducts, which greatly enrich the chemical diversity of *Penicillium* species. Compound **14** displayed potent anti-inflammatory activity. Meanwhile, **6**, **7**, and **8** not only significantly promoted the osteoblast mineralization but also inhibited osteoclasts, providing potent drug leads for anti-osteoporosis.

Supplementary Materials: The following are available online at https://www.mdpi.com/article/10.3390/md21100538/s1, Figures S1–S21: One-dimensional and two-dimensional NMR spectra along with HRESIMS spectra of compounds **1**–**3**.

Author Contributions: X.-W.Y. designed the project; Y.Z. and Z.-B.Z. isolated and purified all compounds. C.-L.X. and R.X. conducted the anti-osteoporotic experiments; X.-W.H. and L.-H.Y. performed the anti-inflammatory bioassay. K.Z. identified the strain. Y.L. and M.-M.X. performed the fermentation. Y.Z., Y.W., and X.-W.Y. analyzed the data and wrote the paper, while critical revision of the publication was performed by all authors. All authors have read and agreed to the published version of the manuscript.

Funding: The work was supported by the National Key Research and Development Program of China (2022YFC2804800), the Xiamen Southern Oceanographic Center (22GYY007HJ07), and the National Natural Science Foundation of China (22177143).

Institutional Review Board Statement: Not applicable.

Informed Consent Statement: Not applicable.

Data Availability Statement: The data presented in this study are available in Supplementary Materials.

Acknowledgments: The authors wish to thank Xiao-Yong Zhang of South China Agricultural University for isolating the fungal strain and Ming-Ming Cao of Kunming Institute of Botany, Chinese Academy of Sciences for his constructive discussion on the biosynthetic pathway of the new compounds.

Conflicts of Interest: The authors declare no conflict of interest.

References

1. Carroll, A.R.; Copp, B.R.; Davis, R.A.; Keyzers, R.A.; Prinsep, M.R. Marine natural products. *Nat. Prod. Rep.* **2023**, *40*, 275–325. [CrossRef] [PubMed]
2. Voser, T.M.; Campbell, M.D.; Carroll, A.R. How different are marine microbial natural products compared to their terrestrial counterparts? *Nat. Prod. Rep.* **2022**, *39*, 7–19. [CrossRef] [PubMed]
3. Carroll, A.R.; Copp, B.R.; Davis, R.A.; Keyzers, R.A.; Prinsep, M.R. Marine natural products. *Nat. Prod. Rep.* **2022**, *39*, 1122–1171. [CrossRef] [PubMed]
4. Skropeta, D.; Wei, L. Recent advances in deep-sea natural products. *Nat. Prod. Rep.* **2014**, *31*, 999–1025. [CrossRef] [PubMed]
5. Sun, C.; Mudassir, S.; Zhang, Z.; Feng, Y.; Chang, Y.; Che, Q.; Gu, Q.; Zhu, T.; Zhang, G.; Li, D. Secondary metabolites from deep-sea derived microorganisms. *Curr. Med. Chem.* **2020**, *27*, 6244–6273. [CrossRef]
6. Chooi, Y.H.; Tang, Y. Navigating the fungal polyketide chemical space: From genes to molecules. *J. Org. Chem.* **2012**, *77*, 9933–9953. [CrossRef]
7. Niu, S.; Tang, X.X.; Fan, Z.; Xia, J.M.; Xie, C.L.; Yang, X.W. Fusarisolins A-E, polyketides from the marine-derived fungus *Fusarium solani* H918. *Mar. Drugs* **2019**, *17*, 125. [CrossRef]
8. Pojer, F.; Ferrer, J.L.; Richard, S.B.; Nagegowda, D.A.; Chye, M.L.; Bach, T.J.; Noel, J.P. Structural basis for the design of potent and species-specific inhibitors of 3-hydroxy-3-methylglutaryl CoA synthases. *Proc. Natl. Acad. Sci. USA* **2006**, *103*, 11491–11496. [CrossRef]
9. De Pascale, G.; Nazi, I.; Harrison, P.H.; Wright, G.D. *β*-Lactone natural products and derivatives inactivate homoserine transacetylase, a target for antimicrobial agents. *J. Antibiot.* **2011**, *64*, 483–487. [CrossRef]
10. Feling, R.H.; Buchanan, G.O.; Mincer, T.J.; Kauffman, C.A.; Jensen, P.R.; Fenical, W. Salinosporamide A: A highly cytotoxic proteasome inhibitor from a novel microbial source, a marine bacterium of the new genus *salinospora*. *Angew. Chem. Int. Ed.* **2003**, *42*, 355–357. [CrossRef]
11. Liu, Y.F.; Zhang, Y.H.; Shao, C.L.; Cao, F.; Wang, C.Y. Microketides A and B, polyketides from a gorgonian-derived *Microsphaeropsis* sp. fungus. *J. Nat. Prod.* **2020**, *83*, 1300–1304. [CrossRef] [PubMed]
12. Hsieh, M.H.; Hsiao, G.; Chang, C.H.; Yang, Y.L.; Ju, Y.M.; Kuo, Y.H.; Lee, T.H. Polyketides with anti-neuroinflammatory activity from *Theissenia cinerea*. *J. Nat. Prod.* **2021**, *84*, 1898–1903. [CrossRef] [PubMed]
13. Chen, M.; Zheng, Y.Y.; Chen, Z.Q.; Shen, N.X.; Shen, L.; Zhang, F.M.; Zhou, X.J.; Wang, C.Y. NaBr-induced production of brominated azaphilones and related tricyclic polyketides by the marine-derived fungus *Penicillium janthinellum* HK1-6. *J. Nat. Prod.* **2019**, *82*, 368–374. [CrossRef] [PubMed]
14. Liu, Y.; Yang, Q.; Xia, G.; Huang, H.; Li, H.; Ma, L.; Lu, Y.; He, L.; Xia, X.; She, Z. Polyketides with *α*-glucosidase inhibitory activity from a mangrove endophytic fungus, *Penicillium* sp. HN29-3B1. *J. Nat. Prod.* **2015**, *78*, 1816–1822. [CrossRef]
15. Li, D.; Chen, L.; Zhu, T.; Kurtán, T.; Mándi, A.; Zhao, Z.; Li, J.; Gu, Q. Chloctanspirones A and B, novel chlorinated polyketides with an unprecedented skeleton, from marine sediment derived fungus *Penicillium terrestre*. *Tetrahedron* **2011**, *67*, 7913–7918. [CrossRef]
16. He, Z.H.; Xie, C.L.; Wu, T.; Yue, Y.T.; Wang, C.F.; Xu, L.; Xie, M.M.; Zhang, Y.; Hao, Y.J.; Xu, R.; et al. Tetracyclic steroids bearing a bicyclo[4.4.1] ring system as potent antiosteoporosis agents from the deep-sea-derived fungus *Rhizopus* sp. W23. *J. Nat. Prod.* **2023**, *86*, 157–165. [CrossRef]
17. Hao, Y.J.; Zou, Z.B.; Xie, M.M.; Zhang, Y.; Xu, L.; Yu, H.Y.; Ma, H.B.; Yang, X.W. Ferroptosis inhibitory compounds from the deep-sea-derived fungus *Penicillium* sp. MCCC 3A00126. *Mar. Drugs* **2023**, *21*, 234. [CrossRef]
18. He, Z.H.; Xie, C.L.; Wu, T.; Zhang, Y.; Zou, Z.B.; Xie, M.M.; Xu, L.; Capon, R.J.; Xu, R.; Yang, X.W. Neotricitrinols A-C, unprecedented citrinin trimers with anti-osteoporosis activity from the deep-sea-derived *Penicillium citrinum* W23. *Bioorg. Chem.* **2023**, *139*, 106756. [CrossRef]
19. Xie, C.L.; Zhang, D.; Guo, K.Q.; Yan, Q.X.; Zou, Z.B.; He, Z.H.; Wu, Z.; Zhang, X.K.; Chen, H.F.; Yang, X.W. Meroterpenthiazole A, a unique meroterpenoid from the deep-sea-derived *Penicillium allii-sativi*, significantly inhibited retinoid X receptor (RXR)-*α* transcriptional effect. *Chin. Chem. Lett.* **2022**, *33*, 2057–2059. [CrossRef]
20. Niu, S.; Xie, C.L.; Xia, J.M.; Liu, Q.M.; Peng, G.; Liu, G.M.; Yang, X.W. Botryotins A-H, tetracyclic diterpenoids representing three carbon skeletons from a deep-sea-derived *Botryotinia fuckeliana*. *Org. Lett.* **2020**, *22*, 580–583. [CrossRef]
21. Wang, Y.; Sun, W.; Zheng, S.; Zhang, Y.; Bao, Y. Genetic engineering of *Bacillus* sp. and fermentation process optimizing for diacetyl production. *J. Biotechnol.* **2019**, *301*, 2–10. [CrossRef]
22. Wakana, D.; Hosoe, T.; Itabashi, T.; Okada, K.; de Campos Takaki, G.M.; Yaguchi, T.; Fukushima, K.; Kawai, K. New citrinin derivatives isolated from *Penicillium citrinum*. *J. Nat. Med.* **2006**, *60*, 279–284. [CrossRef]
23. Deruiter, J.; Jacyno, J.M.; Davis, R.A.; Cutler, H.G. Studies on aldose reductase inhibitors from fungi. I. Citrinin and related benzopyran derivatives. *J. Enzyme. Inhib.* **1992**, *6*, 201–210. [CrossRef]
24. Han, Z.; Mei, W.; Zhao, Y.; Deng, Y.; Dai, H. A new cytotoxic isocoumarin from endophytic fungus *Penicillium* SP. 091402 of the mangrove plant *Bruguiera sexangula*. *Chem. Nat. Compd.* **2010**, *45*, 805–807. [CrossRef]
25. Kuramata, M.; Fujioka, S.; Shimada, A.; Kawano, T.; Kimura, Y. Citrinolactones A, B and C, and sclerotinin C, plant growth regulators from *Penicillium citrinum*. *Biosci. Biotechnol. Biochem.* **2007**, *71*, 499–503. [CrossRef] [PubMed]
26. Wu, Z.J.; Ouyang, M.A.; Tan, Q.W. New asperxanthone and asperbiphenyl from the marine fungus *Aspergillus* sp. *Pest. Manag. Sci.* **2009**, *65*, 60–65. [CrossRef] [PubMed]

27. Hirota, M.; Menta, A.B.; Yoneyama, K.; and Kitabatake, N. A major decomposition product, citrinin H2, from citrinin on heating with moisture. *Biosci. Biotechnol. Biochem.* **2002**, *66*, 206–210. [CrossRef]
28. Chai, Y.J.; Cui, C.B.; Li, C.W.; Wu, C.J.; Tian, C.K.; Hua, W. Activation of the dormant secondary metabolite production by introducing gentamicin-resistance in a marine-derived *Penicillium purpurogenum* G59. *Mar. Drugs* **2012**, *10*, 559–582. [CrossRef]
29. Li, S.F.; Di, Y.T.; Wang, Y.H.; Tan, C.J.; Fang, X.; Zhang, Y.; Zheng, Y.T.; Li, L.; He, H.P.; Li, S.L.; et al. Anthraquinones and Lignans from *Cassia occidentalis*. *Helv. Chim. Acta* **2010**, *93*, 1795–1802. [CrossRef]
30. Kim, S.; Le, T.C.; Han, S.A.; Hillman, P.F.; Hong, A.; Hwang, S.; Du, Y.E.; Kim, H.; Oh, D.C.; Cha, S.S.; et al. Saccharobisindole, neoasterric methyl ester, and 7-chloro-4(1H)-quinolone: Three new compounds isolated from the marine bacterium *Saccharomonospora* sp. *Mar. Drugs* **2021**, *20*, 35. [CrossRef]
31. Oilawa, H. Biosynthesis of structurally unique fungal metabolite GKK1032A_2: Indication of novel carbocyclic formation mechanism in polyketide biosynthesis. *J. Org. Chem.* **2003**, *68*, 3552–3557.
32. Ren, J.M.; Yang, J.K.; Zhu, H.J.; Cao, F. Bioactive steroids from the marine-derived fungus *Aspergillus flavus* JK07-1. *Chem. Nat. Compd.* **2020**, *56*, 945–947. [CrossRef]
33. Biswas, A.; Sharma, B.K.; Willett, J.L.; Erhan, S.Z.; Cheng, H.N. Room-temperature self-curing ene reactions involving soybean oil. *Green Chem.* **2008**, *10*, 290–295. [CrossRef]
34. Uemura, Y.; Sugimoto, S.; Matsunami, K.; Otsuka, H.; Takeda, Y.; Kawahata, M.; Yamaguchi, K. Microtropins A–I: 6′-*O*-(2″*S*,3″*R*)-2″-ethyl-2″,3″-dihydroxybutyrates of aliphatic alcohol *β*-D-glucopyranosides from the branches of *Microtropis japonica*. *Phytochemistry* **2013**, *87*, 140–147. [CrossRef]
35. Dheen, S.T.; Kaur, C.; Ling, E.A. Microglial activation and its implications in the brain diseases. *Curr. Med. Chem.* **2007**, *14*, 1189–1197. [CrossRef]
36. Rawji, K.S.; Mishra, M.K.; Michaels, N.J.; Rivest, S.; Stys, P.K.; Yong, V.W. Immunosenescence of microglia and macrophages: Impact on the ageing central nervous system. *Brain* **2016**, *139*, 653–661. [CrossRef]
37. Costa, T.; Fernandez-Villalba, E.; Izura, V.; Lucas-Ochoa, A.M.; Menezes-Filho, N.J.; Santana, R.C.; de Oliveira, M.D.; Araujo, F.M.; Estrada, C.; Silva, V.; et al. Combined 1-deoxynojirimycin and ibuprofen treatment decreases microglial activation, phagocytosis and dopaminergic degeneration in MPTP-treated mice. *J. Neuroimmune Pharmacol.* **2021**, *16*, 390–402. [CrossRef]
38. Park, S.Y.; Jin, M.L.; Ko, M.J.; Park, G.; Choi, Y.W. Anti-neuroinflammatory effect of emodin in LPS-stimulated microglia: Involvement of AMPK/Nrf2 activation. *Neurochem. Res.* **2016**, *41*, 2981–2992. [CrossRef]
39. Song, T.; Chen, M.; Ge, Z.W.; Chai, W.; Li, X.C.; Zhang, Z.; Lian, X.Y. Bioactive penicipyrrodiether A, an adduct of GKK1032 analogue and phenol A derivative, from a marine-sourced fungus *Penicillium* sp. ZZ380. *J. Org. Chem.* **2018**, *83*, 13395–13401. [CrossRef]
40. El-Desoky, A.H.H.; Tsukamoto, S. Marine natural products that inhibit osteoclastogenesis and promote osteoblast differentiation. *J. Nat. Med.* **2022**, *76*, 575–583. [CrossRef]
41. Sun, Y.; Li, Q.F.; Yan, J.; Hu, R.; Jiang, H. Isoflurane preconditioning promotes the survival and migration of bone marrow stromal cells. *Cell Physiol. Biochem.* **2015**, *36*, 1331–1345. [CrossRef] [PubMed]
42. Guan, M.; Yao, W.; Liu, R.; Lam, K.S.; Nolta, J.; Jia, J.; Panganiban, B.; Meng, L.; Zhou, P.; Shahnazari, M.; et al. Directing mesenchymal stem cells to bone to augment bone formation and increase bone mass. *Nat. Med.* **2012**, *18*, 456–462. [CrossRef] [PubMed]
43. Yang, L.H.; Ou-Yang, H.; Yan, X.; Tang, B.W.; Fang, M.J.; Wu, Z.; Chen, J.W.; Qiu, Y.K. Open-ring butenolides from a marine-derived anti-neuroinflammatory fungus *Aspergillus terreus* Y10. *Mar. Drugs* **2018**, *16*, 428. [CrossRef]
44. Niu, S.; Yang, L.; Zhang, G.; Chen, T.; Hong, B.; Pei, S.; Shao, Z. Phenolic bisabolane and cuparene sesquiterpenoids with anti-inflammatory activities from the deep-sea-derived *Aspergillus sydowii* MCCC 3A00324 fungus. *Bioorg. Chem.* **2020**, *105*, 104420. [CrossRef] [PubMed]

marine drugs

MDPI

Article

New Fusarin Derivatives from the Marine Algicolous Fungus *Penicillium steckii* SCSIO41040

Yingying Song [1,2], Jianglian She [1,2], Weihao Chen [1], Jiamin Wang [1,2], Yanhui Tan [3], Xiaoyan Pang [1], Xuefeng Zhou [1,2], Junfeng Wang [1,2,*] and Yonghong Liu [1,2,*]

1 CAS Key Laboratory of Tropical Marine Bio-Resources and Ecology, Guangdong Key Laboratory of Marine Materia Medica, South China Sea Institute of Oceanology, Chinese Academy of Sciences, Guangzhou 510301, China; songyingying19@mails.ucas.ac.cn (Y.S.); shejianglian20@mails.ucas.ac.cn (J.S.); chenweihao17@mails.ucas.ac.cn (W.C.); wangjiamin20@mails.ucas.ac.cn (J.W.); xypang@scsio.ac.cn (X.P.); xfzhou@scsio.ac.cn (X.Z.)

2 University of Chinese Academy of Sciences, Beijing 100049, China

3 State Key Laboratory for Chemistry and Molecular Engineering of Medicinal Resources, School of Chemistry and Pharmaceutical Sciences, Guangxi Normal University, Guilin 541004, China; tyh533@126.com

* Correspondence: wangjunfeng@scsio.ac.cn (J.W.); yonghongliu@scsio.ac.cn (Y.L.)

Abstract: Five new fusarin derivatives, steckfusarins A–E (**1–5**), and two known natural products (**6**, **7**), were isolated and identified from the marine algicolous fungus *Penicillium steckii* SCSIO 41040. The new compounds, including absolute configurations, were determined by spectroscopic analyses and calculated electronic circular dichroism (ECD). All new compounds were evaluated for their antioxidant, antibacterial, antifungal, antiviral, cytotoxic, anti-inflammatory, antioxidant, cholesterol-lowering, acetyl cholinesterase (AChE) enzyme and 6-phosphofructo-2-kinase (PFKFB3) and phosphatidylinositol-3-kinase (PI3K) inhibitory activities. The biological evaluation results revealed that compound **1** exhibited radical scavenging activity against 2,2-diphenyl-1-picrylhydrazylhydrate (DPPH), with an IC_{50} value of 74.5 μg/mL. In addition, compound **1** also showed weak anti-inflammatory activity at a concentration of 20 μM.

Keywords: fusarin derivatives; *Penicillium steckii*; marine algicolous fungus

Citation: Song, Y.; She, J.; Chen, W.; Wang, J.; Tan, Y.; Pang, X.; Zhou, X.; Wang, J.; Liu, Y. New Fusarin Derivatives from the Marine Algicolous Fungus *Penicillium steckii* SCSIO41040. *Mar. Drugs* **2023**, *21*, 532. https://doi.org/10.3390/md21100532

Academic Editors: Bin-Gui Wang and Haofu Dai

Received: 28 September 2023
Revised: 9 October 2023
Accepted: 10 October 2023
Published: 12 October 2023

1. Introduction

Based on genome sequencing and mining and antimicrobial screening of crude extracts, *P. steckii* has shown strong activity against some pathogenic bacteria [1]. It has been recently reported that the secondary metabolites of *P. steckii* were mainly tanzawaic acids and alkaloids that demonstrated a broad range of significant biological activities, such as antimicrobial, anti-inflammatory and lipid-lowering activities [2–6].

Fusarins are a class of mycotoxins characterized by a 2-pyrrolidone moiety and a triene side chain of twelve carbon atoms [7] that are usually isolated from *Penicillium* sp. [8]. The compounds were biosynthesized by a hybrid polyketide synthase-nonribosomal peptide synthetase (PKS-NRPS) [9]. In the human neuroblastoma cell line SH-SY5Y, fusarin derivatives were first proven to effectively promote neurite growth and block the cell cycle in the G_0/G_1 phase [10]. Fusarin derivatives inhibit mammalian DNA polymerases and human DNA topoisomerase II in vitro [11]. Additionally, they induce the formation of 6-thioguanine-resistant mutants in V79 cells, as well as the asynchronous replication of polyoma DNA sequences subsequent to the presence of a microsomal activation system [12,13]. Fusarins have been reported to be unstable after prolonged exposure to ultraviolet light and elevated temperatures [14]; additionally, light can induce the rearrangement of fusarins, leading to the simultaneous appearance of fusarin analogues [15].

During our continuous study exploring new bioactive natural products from marine microorganisms [16,17], five new fusarin derivatives (**1–5**) (Figures S1–S4) and two known natural products (**6**, **7**) (Figure 1) were obtained from the fungus *Penicillium steckii* SCSIO 41040, which was isolated from a green algae *Botryocladia* sp. of the South China Sea. Herein, we describe the fermentation, isolation, structural determination and biological activities of these compounds.

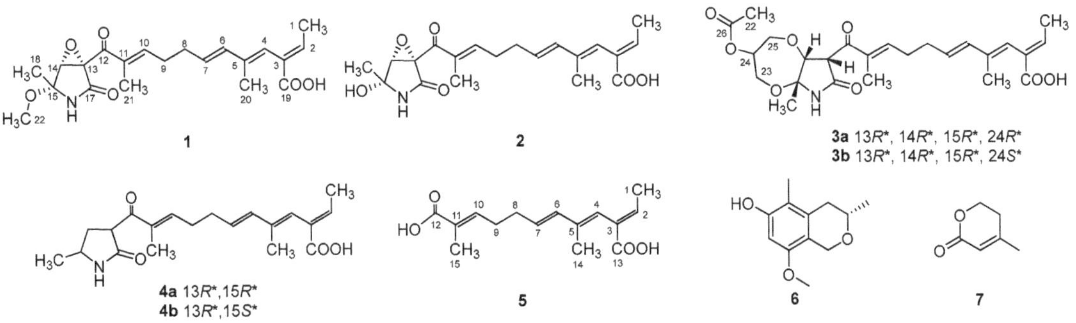

Figure 1. Chemical structures of compounds **1**–**7**.

2. Results and Discussion

Compound **1** was obtained as a yellow oil, and the molecular formula $C_{21}H_{27}NO_6$ was assigned to it based on its HRESIMS (m/z 390.1912 $[M + H]^+$, calcd for $C_{21}H_{28}NO_6^+$, 389.1911), which required nine degrees of unsaturation. The ^1H NMR and ^{13}C NMR data (Table 1) indicated five olefinic methines ($\delta_{C/H}$ 138.5/6.80, CH-2; 123.5/5.94, CH-4; 134.9/6.25, CH-6; 128.4/5.73, CH-7; 148.0/6.86, CH-10), five methyls ($\delta_{C/H}$ 15.6/1.66, CH_3-1; 18.7/1.39, CH_3-18; 14.2/1.56, CH_3-20; 10.8/1.75, CH_3-21; 49.5/3.26, CH_3-22), two methylenes ($\delta_{C/H}$ 31.0/2.26, CH_2-8; 28.8/2.42, CH_2-9), one oxygenated methine ($\delta_{C/H}$ 62.4/4.28, CH-14) and eight quaternary carbonyls (δ_C 131.0, C-3; 136.7, C-5; 137.5, C-11; 190.0, C-12; 62.2, C-13; 87.4, C-15; 169.9, C-17; 167.9, C-19). Seven degrees of unsaturation could be accounted for by the presence of three carbonyl signals (a ketone resonance (δ_C 190.0), an amide (δ_C 169.9) and a carboxyl (δ_C 167.9)) and the eight sp^2 carbon signals of four olefinic bonds. Therefore, the remaining unsaturation indicated the presence of two rings. HMBC correlations from H-14 to C-15 and NH-16 to C-13/C-14/C-15/C-17 established the presence of the 2-pyrrolidone ring structure. The highly deshielded chemical shift of C-13 and CH-14 inferred that both were oxygenated, which prompted us to speculate on the existence of an epoxide ring in the 2-pyrrolidone ring. Additionally, the location of the methyl and methoxyl groups were determined via HMBC correlations from H_3-18 to C-14/C-15 and H_3-22 to C-15. According to the 2D NMR spectra, a 2-ethylidene-4,10-dimethyl-11-oxoundeca-3,5,9-trienoic acid side chain was located at C-13 of the 2-pyrrolidone ring, supported by COSY correlations of H_3-1/H-2 and H-6/H-7/H_2-8/H_2-9/H-10 and HMBC correlations from H_3-1 to C-3, H-2 to C-19, H-4 to C-2/C-6, H_3-20 to C-4/C-5 and H_3-21 to C-10/C-11/C-12. The *E*-configuration of the $\Delta^{6(7)}$ double bond was determined by the coupling constant (J = 15.6 Hz). NOESY correlations indicated an *E*-configuration for the other double bonds in the side chains of H_3-1/H-4, H-4/H-6, H_2-9/H_3-21 and the cofacial orientation of H_3-18/H-14. The presence of an epoxide ring indicated the homoconfiguration of H-13/H-14. The absolute configurations of compound **1** were determined by ECD calculations. The experimental ECD spectra were compared with two calculated spectra, (13*R*, 14*R*, 15*R*)-**1** and (13*S*, 14*R*, 15*R*)-**1**. The results showed that the experimental ECD spectra of **1** matched well with the calculated spectrum for (13*R*, 14*R*, 15*R*)-**1**, confirming the absolute configurations of **1** as depicted (Figure 2).

Table 1. ^{1}H and ^{13}C NMR spectroscopic data for compounds **1–3** and **5** (700, 175 MHz, DMSO-d_6).

No.	1		2		3		5	
	δ_C, Type	δ_H (J in Hz)	δ_C, Type	δ_H (J in Hz)	δ_C, Type	δ_H (J in Hz)	δ_C, Type	δ_H (J in Hz)
1	15.6 CH_3	1.66 (d, 7.3)	15.6 CH_3	1.65 (d, 7.1)	14.5 CH_3	1.64 (dd, 7.2, 0.9)	15.6 CH_3	1.65 (d, 7.1)
2	138.5 CH	6.80 (q, 7.3)	138.4 CH	6.80 (q, 7.1)	138.8 CH	6.78 (q, 7.2)	138.1 CH	6.78 (q, 7.1)
3	131.0 C		131.0 C		131.1 C		131.3 C	
4	123.5 CH	5.94 s	123.5 CH	5.93 s	123.5 CH	5.92 s	123.6 CH	5.94 s
5	136.7 C		136.7 C		137.0 C		136.5 C	
6	134.9 CH	6.25 (d, 15.6)	134.9 CH	6.25 (d, 15.6)	134.7 CH	6.26 (dd, 15.6, 7.2)	134.5 CH	6.23 (d, 15.6)
7	128.4 CH	5.73 (dt, 15.6, 7.0)	128.5 CH	5.74 (dt, 15.6, 7.0)	129.2 CH	5.79 (dt, 15.6, 7.0)	128.3 CH	5.72 (dt, 15.6, 7.0)
8	40.0 CH_2	2.26 (q, 7.2)	31.0 CH_2	2.25 (q, 7.3)	31.3 CH_2	2.32 m	28.1 CH_2	2.26 m
9	28.7 CH_2	2.42 (q, 7.2)	28.9 CH_2	2.40 (q, 7.3)	29.4 CH_2	2.40 m	31.3 CH_2	2.45 m
10	148.0 CH	6.86 (td, 6.8, 0.7)	147.5 CH	6.92 (dt, 7.2, 1.6)	147.4 CH	6.93 (td, 6.8, 0.9)	140.6 CH	6.66 (td, 6.9, 0.7)
11	135.5 C		135.4 C		137.5 C		128.9 C	
12	190.0 C		190.2 C		197.0 C		169.0 C	
13	62.2 C		62.6 C		48.8 CH	4.93 (d, 8.4)	168.1 C	
14	64.4 CH	4.28 (d, 2.3)	64.1 CH	4.1 (d, 2.3)	77.2 CH	4.33 (d, 8.4)	14.3 CH_3	1.55 s
15	87.4 C		83.0 C		82.9 C		12.4 CH_3	1.74 s
16-NH		8.82 s		8.63 s		8.89 s		
17	169.9 C		169.4 C		170.5 C			
18	18.7 CH_3	1.39 s	22.3 CH_3	1.39 s	20.0 CH_3	1.48 s		
19	167.9 C		168.0 C		168.2 C			
20	14.2 CH_3	1.56 s	14.3 CH_3	1.55 s	15.8 CH_3	1.54 s		
21	10.8 CH_3	1.75 s	10.9 CH_3	1.75 s	11.5 CH	1.72 s		
22	49.5 CH_3	3.26 s						

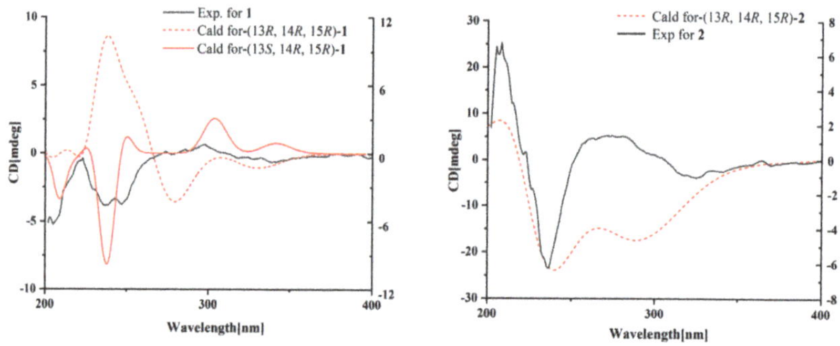

Figure 2. Experimental and calculated ECD spectra for compounds **1** and **2**.

Compound **2** was isolated as a yellow oil. The HRESIMS (m/z 347.1609 [M − H]$^-$, calcd for $C_{20}H_{24}NO_6^-$, 374.1609) data suggested a molecular formula of $C_{20}H_{25}NO_6$, revealing nine degrees of unsaturation. The NMR signals (Table 1) of **2** resembled those of **1**, except for the absence of a methoxyl group at $\delta_{C/H}$ 49.5/3.26. Large coupling (J = 15.6 Hz) between H-6 (δ_H 6.25) and H-7 (δ_H 5.74) showed that the $\Delta^{6(7)}$ olefinic bond was in *E*-configuration. NOESY data of H_3-1 (δ_H 1.65)/H-4 (δ_H 5.93), H-4 /H-6 and H_2-9 (δ_H 2.40)/H_3-21 (δ_H 1.75), indicated that the $\Delta^{2(3)}$, $\Delta^{4(5)}$ and $\Delta^{10(11)}$ olefinic bonds were all in *E*-configuration. The NOESY spectrum of **2** showed key correlations of H_3-18/H-14, indicating that they were co-facial. The high similarity between the calculated ECD curve of (13*R*, 14*R*, 15*R*)-**2** and its experimental curve unambiguously confirmed the absolute configuration of **2** (Figure 2) [18–20].

Compound **3** was obtained as a pink oil. Its molecular formula of $C_{25}H_{33}NO_8$ was determined via HRESIMS (m/z 476.2290 [M + H]$^+$, calcd for $C_{25}H_{34}NO_8^+$, 476.2279), which required ten degrees of unsaturation. Proton network systems of H-13 (δ_H 4.93)/H-14 (δ_H 4.33) and H_2-23 (δ_H 3.51, 3.67)/H-24 (δ_H 4.16)/H_2-25 (δ_H 3.34, 3.99) were detected in the COSY spectrum. Moreover, HMBC correlations from H-13 to C-17 (δ_C 170.5), NH-16 (δ_H 8.89) to C-13 (δ_C 48.8)/C-14 (δ_C 77.2)/C-15 (δ_C 82.9)/C-17, H_3-18 (δ_H 1.48) to C-14/C-15 and H_2-23 to C-15 established the presence of a glycerol moiety and the 2-pyrrolidone ring structure, which were connected via ether linkages to form a 8a-methyl-2-oxohexahydro-

1*H*,5*H*-[1,4]dioxepino [2,3-*b*]pyrrol-6-yl acetate ring system (Figure 3). A detailed analysis of its NMR spectroscopic features implied that it was closely structurally related to dothilactaene B [15]. A set of signals at CH_3-22 and C-26 (Tables 1 and 2), which were not present in the ^1H and ^{13}C NMR spectra of dothilactaene B, were observed in the spectrum of **3**. Meanwhile, compound **3** lacked a methoxy group at $\delta_{C/H}$ 51.9/3.55. Through the analyses of the HMBC spectrum, the acetic acid group was confirmed by the HMBC correlations from H_3-22 (δ_H 1.98) to C-26 (δ_C 170.8) and H_2-25 to C-26. The *E*-configuration of the $\Delta^{6(7)}$ double bond was determined by the coupling constant (J = 15.6 Hz). NMR data analyses of **3** showed a similar multiplicity pattern in the glycerol moiety, indicating that they are two components and have the same plane structure. The resonances of the glycerol moiety suggested them to be diastereoisomers presenting in a ratio of 3:1 (**3a**:**3b**), which may be formed by Michael addition [21]. NOESY correlations (Figure 4) of **3a** indicated the *E*-configuration for the other double bonds in the side chain and the cofacial orientation of H-13, H-14, CH_3-18 and H-24. The NOESY spectrum of **3b** was the same as **3a**, except the signal of H-24 was the opposite. These data permitted assignment of the (13*R**, 14*R**, 15*R**, 24*R**)-**3a** and (13*R**, 14*R**, 15*R**, 24*S**)-**3b** relative configurations [15,19].

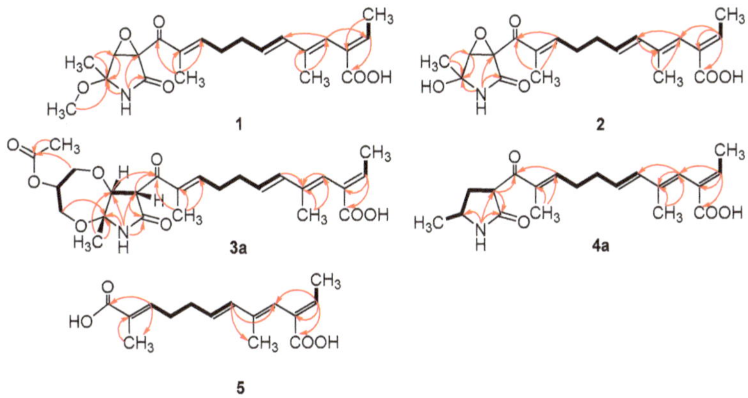

Figure 3. Key COSY (bold lines) and HMBC (arrows) correlations of compounds **1–5**.

Table 2. Partial ^1H and ^{13}C NMR spectroscopic data for **3** (700, 175 MHz, DMSO-d_6).

No.	3a		3b	
	δ_C, Type	δ_H (J in Hz)	δ_C, Type	δ_H (J in Hz)
22	20.7 CH_3	1.98 s	21.0 CH_3	2.00 s
23	61.2 CH_2	3.51 m 3.67 m	62.8 CH_2	3.34 m 3.99 m
24	65.6 CH	4.16 m	69.4 CH	3.63 m
25	63.1 CH_2	3.34 m 3.99 m	65.9 CH_2	3.87 m 4.02 m
26	170.8 C		170.9 C	

Compound **4** was obtained as a yellow oil. Its HRESIMS ($[M + H]^+$, 346.2016; calcd for $C_{20}H_{28}NO_4^+$, 346.2013) data were in agreement with the molecular formula $C_{20}H_{27}NO_4$. Its ^1H NMR and ^{13}C NMR data (Table 3) closely resembled the data for **2**, except for the absence of the hydroxyl group and an epoxy system at the 2-pyrrolidone ring structure. To data, all fusarin derivatives reported from natural sources have had a free hydroxyl group at C-15; however, compound **4** has no such substituent [22–24]. The relative configuration of **4** was confirmed by its NOESY correlations and coupling constants (Figure 4). The NOESY correlations of H_3-1 (δ_H 1.64)/H-4 (δ_H 5.95), H-4/H-6 (δ_H 6.26), H_2-9 (δ_H 2.38)/H_3-21 (δ_H 1.72) and the large coupling constant of H-6/H-7 (J = 15.6 Hz) suggested the *E* geometry of all double bonds. Each pair of ^1H and ^{13}C NMR signals showed a similar multiplicity pattern, suggesting that compound **4** was an inseparable mixture of two geometric isomers presenting in a ratio of 1:1 (**4a**:**4b**). Even under mild conditions, the stereo center of C-15 was easy to be epimerized through ring opening [19,20]. Comparing **4a** and **4b** (Table 2), the main difference occurred at the configuration at C-13, C-15 and C-18, which was proven by the carbon chemical shifts of C-13 ($\Delta\delta_C$ 1.1 ppm), C-15 ($\Delta\delta_C$ 1.1 ppm) and C-18 ($\Delta\delta_C$ 0.2 ppm). The NOESY spectrum of **4a** showed cross-peaks between H-13 (δ_H 4.33) and H_3-18 (δ_H 1.11) that indicated H-13 and H-15 (δ_H 1.48) were on the opposite side, whereas no relevant signal between H-13 and H_3-18 was found in **4b**. Accordingly, the relative configurations of **4a** and **4b** could be assigned as (13*R**, 15*R**)-**4a** and (13*R**, 15*S**)-**4b**.

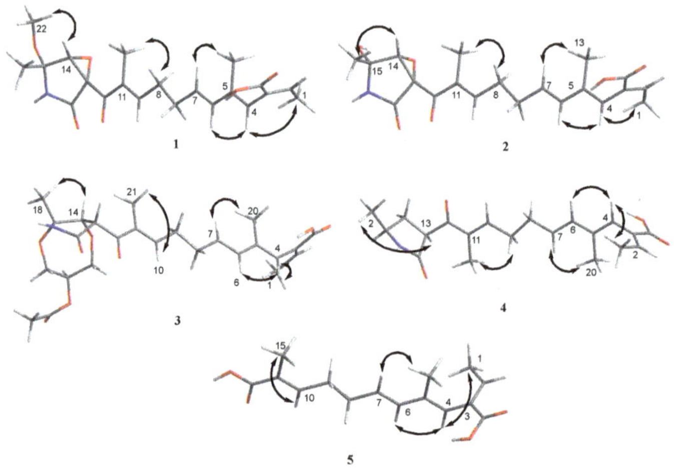

Figure 4. Key NOESY (double arrow) correlations of compounds **1**–**5**.

Table 3. ^{1}H and ^{13}C NMR spectroscopic data for **4** (700, 175 MHz, DMSO-d_6).

No.	4a		4b	
	δ_C, Type	δ_H (*J* in Hz)	δ_C, Type	δ_H (*J* in Hz)
1	15.7 CH_3	1.64 (d, 7.1)	15.7 CH_3	1.64 (d, 7.1)
2	138.4 CH	6.76 (q, 7.1)	138.4 CH	6.76 (q, 7.1)
3	131.5 C		131.5 C	
4	123.8 CH	5.95 s	123.8 CH	5.95 s
5	136.2 C		136.4 C	
6	134.7 CH	6.26 (d, 15.6)	134.7 CH	6.26 (d, 15.6)
7	128.8 CH	5.78 (dt, 15.6, 7.0)	128.8 CH	5.78 (dt, 15.6, 7.0)
8	31.2 CH_2	2.30 m	31.2 CH_2	2.30 m
9	29.0 CH_2	2.38 (q, 7.3)	29.0 CH_2	2.38 (q, 7.3)
10	145.4 CH	6.85 (td, 7.2, 0.7)	145.8 CH	6.91 (td, 7.2, 0.7)
11	136.4 C		136.6 C	
12	198.3 C		198.4 C	
13	47.8 CH	4.30 (q, 9.1)	48.9 CH	4.33 (q, 9.1)
14	33.0 CH_2	1.72 m 2.27 m	33.3 CH_2	1.80 m 2.33 m
15	47.3 CH	3.60 (q, 6.6)	48.4 CH	3.65 (q, 6.6)
16-NH		7.90 s		7.90 s
17	172.5 C		172.8 C	
18	22.1 CH_3	1.10 (d, 6.2)	22.3 CH_3	1.12 (d, 6.2)
19	168.2 C		168.2 C	
20	14.3 CH_3	1.56 s	14.3 CH_3	1.56 s
21	11.4 CH_3	1.72 s	11.5 CH_3	1.72 s

Compound **5** was obtained as a yellow oil. It showed a $[M + H]^+$ ion peak at m/z 265.1431 in the positive-ion HRESIMS (calcd. for $C_{15}H_{21}O_4^+$, 265.1434) that was appropriate for a molecular formula of $C_{15}H_{20}O_4$. The NMR data of **5** showed that it shared the same trienedioic acid skeleton as epolactaene 4a [25]. The main difference was the absence of a methyl in **5**. The large vicinal coupling constant (J = 15.6 Hz) between H-6 (δ_H 6.23) and H-7 (δ_H 5.72) suggested an *E*-configuration for the $\Delta^{6(7)}$ double bond. In addition, the NOESY correlations between H_3-1 (δ_H 1.65) and H-4 (δ_H 5.94), H-4 and H-6 and H_2-9 (δ_H 2.45) and H_3-15 (δ_H 1.74) indicated an *E*-configuration for the other double bonds (Figure 4). Therefore, the structure of **5**, named as steckfusarin F, was elucidated.

The other two known compounds were elucidated as (3*S*)-3,5-dimethyl-8-methoxy-3,4-dihydro-1*H*-isochromen-6-ol (**6**) [26] and 4-methyl-5,6-dihydro-2*H*-pyran-2-one (**7**) [27] by comparing their NMR and MS data with the data reported in the literature.

The new compounds (**1**–**5**) were evaluated for antioxidant, antibacterial, antifungal, antiviral, cytotoxic, anti-inflammatory, cholesterol-lowering and PFKFB3 and PI3K kinase inhibitory activities in vitro.

The isolated compounds (**1**–**5**) were evaluated for their cytotoxic activities against twenty human cancer cell lines (A549, MKN-45, HCT 116, HeLa, K-562, MCF7, HepG2, SF126, DU145, CAL-62, 786-O, TE-1, 5637, GBC-SD, L-02, PATU8988T, HOS, A-375, A-673 and 293T) via CCK-8 (Dojindo) method. However, none of the compounds showed obvious cytotoxicity. Likewise, none of the new compounds did exhibited any growth inhibition when tested against *Klebsiella pneumonia*, *Candida albicans Berkhout*, *Staphyloccocus aureus*, *Colletotrichum gloeosporioiles*, *Magnaporthe grisea* and *Clerotinia miyabeana Hanzawa*. In addition, none of the new compounds exhibited cholesterol-lowering or PFKFB3 and PI3K kinase inhibitory activities. Compound **1** showed weak inhibitory activity against lipopolysaccharide-inducted nitric oxide (NO) in RAW 264.7 cells at a concentration of 20 μM.

The radical scavenging activities of all the isolated compounds were tested against DPPH; compounds **1**, **2** and **4** showed radical scavenging activity against DPPH, with 37.3%, 54.3% and 45.8% inhibition at concentrations of 100 μg/mL, respectively. Due to the low mass of the active compounds, only compound **1** was measured for IC_{50}. It had an IC_{50} value of 74.5 μg/mL, whereas ascorbic acid, used as a positive control, had an IC_{50} value of 7.5 μg/mL. Furthermore, molecular docking analyses between the active compounds and superoxide dismutase (PDB ID: 7wx0) were performed to gain functional and structural insights (Figure 5). The results showed that compound **1** could interact with superoxide dismutase at the entrance of the catalytic pocket, with a calculated binding affinity of −6.3 kcal/mol. Compound **1** interacted with the residues SER-32 and GLN-153 via two hydrogen bonds, and five hydrophobic interactions were formed between compound **1** and the residues LYS-4, VAL-6, HIS-20, ALA-152 and GLN-153. The active site of **1** contained methoxy and carboxyl groups. The molecular docking results demonstrated that the negative binding free energy value between compound **2** and superoxide dismutase was −6.6 kcal/mol. Interaction modes revealed six hydrogen bonding interactions with the residues GLN-23, ARG-79, ARG-79, SER-102, LEU-103 and ILE-104. In addition, there were two hydrophobic interactions with LEU-103 and ALA-105 in the active site of superoxide dismutase. The active site of **2** contained hydroxyl and carboxyl groups and an epoxy ring. The size of compound **1** was 51 × 51 × 47, centered at x: 17.894, y: 7.9, z: 82.091. The size of compound **2** was 49 × 44 × 49, centered at x: 17.894, y: 7.9, z: 82.091. Compounds **1** and **2** had no π-π stacking and π-cation interactions. The docking studies suggested that compounds **1** and **2** could inhibit superoxide dismutase by tightly binding to catalytic amino acid residues through different types of interactions.

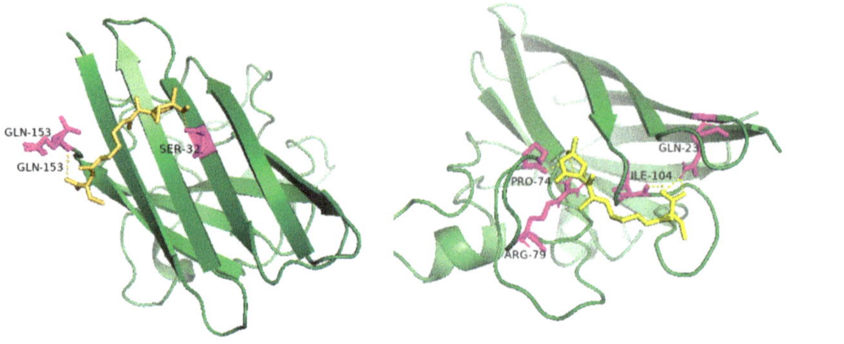

1 2

Figure 5. Low-energy binding conformations of **1** and **2** bound to superoxide dismutase (generated by molecular docking).

Although many activities of fusarin derivatives have been reported [4,8,13,17], no obvious activities were detected for compounds **1–5**. It was reported that the tetrahydropyran ring and methylation of the acid group from the sidechain were crucial to the activity of these compounds [21]. However, none of the new compounds were methylated at C-19, which may be the reason why they did not display obvious activities.

3. Materials and Methods

3.1. General Experimental Procedures

Optical rotations were recorded using a Perkine Elmer 341 polarimeter (Hertford, UK). ECD spectra were recorded with a Chirascan circular dichroism spectrometer (Applied Photophysics, Surrey, UK). UV spectra were recorded on a UV-2600 UV–vis spectrophotometer (Shimadzu, Kyoto, Japan). The 1D and 2D NMR spectra were recorded on a Bruker AC 500 and 700 NMR (Broker, Fallanden, Switzerland) spectrometer with TMS as the internal standard. HRESIMS spectra were measured with a Bruker micro TOF-QII (Bruker, Fallanden, Switzerland) mass spectrometer in positive/negative ion mode. Silica gel GF-254 (10–40 mm) was used for thin-layer chromatography (TLC) (Qingdao Marine Chemical Factory, Qingdao, China). Sephadex LH-20 (Amersham Biosciences, Uppsala, Sweden) and silica gel (200–300 mesh, 100–200 mesh) (Qingdao Marine Chemical Factory, Qingdao, China) were applied in column chromatography (CC). HPLC was carried out on a Hitachi Primaide with a YMC ODS Series column (YMC-Pack ODS-A, YMC Co. Ltd. (Kyoto, Japan), 250 × 10 mm i.d., S-5 μm, 12 nm). All solvents were analytical grade (Tianjin Fuyu Chemical and Industry Factory). The fermentation culture medium and reagents were obtained from Guangzhou Haili Aquarium Technology Company, Guangzhou, China.

3.2. Fungal Strain

The fungal strain *Penicillium steckii* SCSIO 41040 was isolated from a green algae *Botryocladia* sp. that was collected from the South China Sea. The isolated fungal strain was stored on Muller Hinton broth (MB) agar (malt extract 16 g, agar 18 g, sea-salt 30 g, water 1 L and pH 7.4–7.8) slants at 4 °C and then deposited in CAS Key Laboratory of Tropical Marine Bio-resources and Ecology, South China Sea Institute of Oceanology, Chinese Academy of Sciences, Guangzhou, PR China. The ITS1-5.8S-ITS2 sequence region (550 base pairs (bp), GenBank accession no. OP349656) of strain SCSIO 41040 was amplified via the PCR process. DNA sequencing showed that it shared significant homology (100%) with *Penicillium steckii* (GenBank accession no. NR111488).

3.3. Fermentation and Extraction

A few loops of cells of the strain SCSIO 41040 were inoculated into a 500 mL Erlenmeyer flask containing 150 mL of seed medium (malt extract 1%, yeast extract 0.4%, glucose 0.4% and pH 7.2) and cultivated on a rotary shaker at 180 rpm and 28 °C for 48 h as a seed culture. Then, a large-scale fermentation of the fungal strain SCSIO 41040 was incubated at 25 °C in 1 L conical flasks containing a solid medium (300 mL/flask) composed of 200 g rice and 220 mL 3.2% (NaCl 3.2 g/H_2O 100 mL) artificial seawater. After 32 days, the fermented material from each flask was extracted successively with EtOAc (700 mL/flask) and the combined EtOAc extract was suspended in MeOH and extracted with petroleum ether to remove rice oil. Finally, the MeOH solution was concentrated under reduced pressure to obtain a reddish-brown extract (56.0 g).

3.4. Isolation and Purification

The extract was subjected to silica gel column chromatography (CC), eluting with a gradient CH_2Cl_2-MeOH (100:0–0:100, v/v) to give 15 fractions based on TLC properties. Fr.3 was separated by semipreparative HPLC (40% MeCN/H_2O + 0.3% TFA, 2.0 mL/min) to provide **6** (10.7 mg, t_R 17 min) and **7** (9.2 mg, t_R 8 min). Fr.5 was separated into 8 subfractions (Fr.5.1–5.8) using ODS silica gel chromatography via elution with MeOH/H_2O (5–100%). Fr.5.6 was directly separated via semipreparative HPLC (56% MeOH/H_2O + 0.3% TFA, 2.7 mL/min) to offer **4** (4.5 mg, t_R 44 min). Fr.5.7 was separated by semipreparative HPLC (35% MeOH/H_2O + 0.3% TFA, 2.7 mL/min) to give **5** (8.5 mg, t_R 37 min). Fr.6 was applied to ODS silica gel chromatography to give eight fractions (Fr.6.1–6.8). Fr.6.2 and Fr.6.3 were further divided into 6 (Fr.6.2.1–Fr.6.2.6) and 11 (Fr.6.3.1–Fr.6.3.11) parts through a Sephadex LH-20 with MeOH and semipreparative HPLC (56% MeOH/H_2O + 0.3% TFA, 2.5 mL/min), respectively. Fr.6.2.2 was purified by semipreparative HPLC (58% MeOH/H_2O + 0.3% TFA, 2.7 mL/min) to yield **3** (3.0 mg, t_R 15 min). Fr.6.2.6 was separated by semipreparative HPLC (37% MeCN/H_2O + 0.3% TFA, 3.0 mL/min) to yield **1** (23.2 mg, t_R 42 min). Fr.6.3.10 was directly separated via semipreparative HPLC (34% MeCN/H_2O + 0.3% TFA, 2.5 mL/min) to provide **2** (5.8 mg, t_R 33 min).

Steckfusarin A (**1**): yellow oil; $[\alpha]_D^{25} - 14.8°$; UV (MeOH) λ_{max} (log ε) 235 (3.94) nm; ^{1}H and ^{13}C NMR data, Table 1; HRESIMS m/z 390.1912 [M + H]$^+$ (calcd for $C_{21}H_{28}NO_6^+$, 389.1911).

Steckfusarin B (**2**): yellow oil; $[\alpha]_D^{25} + 24.4°$; UV (MeOH) λ_{max} (log ε) 213 (4.27), 233 (4.27) nm; ^{1}H and ^{13}C NMR data, Table 1; HRESIMS m/z 347.1609 [M − H]$^-$ (calcd for $C_{20}H_{24}NO_6^-$, 374.1609).

Steckfusarin C (**3**): pink oil; $[\alpha]_D^{25} - 1.7°$; UV (MeOH) λ_{max} (log ε) 212 (4.03), 232 (4.01) nm; ^{1}H and ^{13}C NMR data, Tables 1 and 2; HRESIMS m/z 476.2290 [M + H]$^+$ (calcd for $C_{25}H_{34}NO_8^+$, 476.2279).

Steckfusarin D (**4**): yellow oil; $[\alpha]_D^{25} - 8.5°$; UV (MeOH) λ_{max} (log ε) 204 (4.21) nm; ^{1}H and ^{13}C NMR data, Table 2; HRESIMS m/z 346.2016 [M + H]$^+$ (calcd for $C_{20}H_{28}NO_4^+$, 346.2013).

Steckfusarin E (**5**): yellow oil; UV (MeOH) λ_{max} (log ε) 216 (4.14) nm; ^{1}H and ^{13}C NMR data, Table 1; HRESIMS m/z 265.1431 [M + H]$^+$ (calcd for $C_{15}H_{21}O_4^+$, 265.1434).

3.5. ECD Calculation

The relative configurations of the new compounds were subjected to random conformational searches using Spartan'14 and Gaussian 09 software with the Merck molecular force field (MMFF) and density functional theory (DFT)/TDDFT, respectively. The MMFF conformational search produced low-energy conformers with a Boltzmann population of more than 5%, which were geometrically optimized using the DFT method at the B3LYP/6-311G* level in MeOH using the IEFPCM model. For the stable conformers of the new compounds, the overall theoretical calculation of ECD was achieved in MeOH using time-dependent density functional theory at the B3LYP/6-311G* level. We used the 0.2–0.4 eV half bandwidth Multiwfn to generate the ECD spectra of different conformations and calculated the contribution according to the Boltzmann of each conformation after the UV correction [28].

3.6. Cell Culture and Cytotoxic Bioassay

The obtained fusarin analogues (**1–5**) were evaluated for their cytotoxic activities against twenty cell lines, A549, MKN-45, HCT 116, HeLa, K-562, MCF7, HepG2, SF126, DU145, CAL-62, 786-O, TE-1, 5637, GBC-SD, L-02, PATU8988T, HOS, A-375, A-673 and 293T (Shanghai Cell Bank, Chinese Academy of Sciences). The cytotoxic activity was determined via the CCK-8 (Dojindo) method [29].

3.7. Antimicrobial Assay

All new compounds were tested for antibacterial activity against *Klebsiella pneumonia*, *Candida albicans Berkhout*, *Staphyloccocus aureus*, *Colletotrichum gloeosporioiles*, *Magnaporthe grisea* and *Clerotinia miyabeana Hanzawa*. The tests were performed in 96-well plates using a modification of the broth microdilution method [30].

3.8. Anti-Inflammatory Assay

All new compounds were evaluated for their inhibitory activity against LPS-induced NF-κB activation in RAW264.7 cells using a luciferase reporter gene assay [31]. The RAW264.7 cells, stably transfected with the NF-κB luciferase reporter gene, were plated in triplicate into 96-well plates for all treatments and controls. The compounds (20 μM) and BAY11-7082 (an NF-κB inhibitor used as a positive control (5 μM); Sigma-Aldrich, St. Louis, MO, USA) were used to pretreat the cells for 30 min, followed by stimulation with 5 μg/mL LPS for 8 h. The cells were harvested and luciferase activity measured using a luciferase assay system (Promega, Madison, WI, USA).

3.9. Cholesterol Transport Mechanism

Pancreatic triglyceride lipase (PTL) and Niemann-Pick C1-like 1 (NPC1L1) were the crucial targets involved in cholesterol cellar uptake. The inhibitory activity of the compounds against PTL was evaluated by colorimetry. Surface plasmon resonance (SPR) was used to analyze the binding of the compounds with NPC1L1, and the activities of targeted compounds for NPC1L1 were studied [32,33].

3.10. PFKFB3 Kinase Inhibitory Activity

PFKFB3 kinase inhibitory activity was measured using an ADP-Glo Kinase Assay kit (Promega) according to a published modified protocol [34]. Compound solution (1 μL, with a final concentration of 20 μM in 1% DMSO) and 2 μL enzyme solution were added to 384-well plates, followed by incubation at room temperature (RT) for 30 min. After 2 h incubation at RT, 4 μL ADP-Glo Reagent was added and then the incubation continued for 1 h prior to the addition of 8 μL Kinase Detection Reagent and a further incubation of 1 h. The luminous signal was measured using an Envision flat panel reader (PerkinElmer, Waltham, MA, USA).

3.11. PI3K Kinase Inhibitory Activity

An amount of 0.5 μL of the compounds was preincubated with 14.5 μL of enzyme and PIP2 substrate for 10 min before addition of 5 μL of ATP to achieve a final ATP concentration of 10 μM [35]. The total reaction volume was 20 μL, and the reaction was allowed to proceed for 45 min at room temperature before the addition of the stop solution and detection mixture provided in the kit. The luminous signal was measured using an Envision flat panel reader (PerkinElmer).

3.12. Measurement of AChE Inhibition Activity

We evaluated the AChE enzyme inhibitory activity of compounds according to the slightly modified spectrophotometric method [36]. Tacrine was used as a positive control.

3.13. Molecular Docking Analysis

The crystal structure of neuraminidase (PDB ID: 7wx0) was retrieved from the Protein DataBank. We used AutoDockTools (Version 1.5.7) to conduct the molecular docking. The structures were generated in ChemBio3D Ultra 14.0 (ChemBioOffice version 14.0), followed by an MM2 calculation to minimize the conformation energy. The docking pose that had the lowest binding energy was represented as the most favorable binding conformation.

4. Conclusions

In this study, five new fusarin derivatives were obtained from the marine algicolous fungus *Penicillium steckii* SCSIO 41040, which had been fermented using rice solid medium. The new structures, including absolute configurations, were determined using spectroscopic methods coupled with the calculated ECD. Bioassays were used to screen the antioxidant, antibacterial, antifungal, antiviral, cytotoxic, anti-inflammatory, cholesterol-lowering and PFKFB3 and PI3K kinase inhibitory activities. Compound **1** exhibited antioxidant activity against DPPH, with an IC_{50} value of 74.5 μg/mL. In addition, compound **1** showed weak anti-inflammatory activity at a concentration of 20 μM. Compared with the previously reported fusarin derivatives [10,14,19,23], compounds **1–5** have no obvious antitumor activities, possibly because the new compounds lacked methylation at C-19.

Supplementary Materials: The following supporting information can be downloaded at: https://www.mdpi.com/article/10.3390/md21100532/s1. Figures S1–S47: ^{1}H, ^{13}C, DEPT135, COSY, HSQC, HMBC, NOESY, HRESIMS and UV spectra of the new compounds **1–5**. Tables S1 and S2: Energies of **1** and **2** at MMFF94 force field and B97-3c level.

Author Contributions: Y.S., J.S. and W.C.; formal analysis, J.W. (Jiamin Wang) and Y.S.; investigation, J.S. and X.Z.; project administration, Y.T. and X.P.; resources, J.W. (Junfeng Wang) and Y.L.; writing original draft, Y.S.; writing-review and editing, J.W. (Junfeng Wang). All authors have read and agreed to the published version of the manuscript.

Funding: This work was financially supported by the Guangdong Basic and Applied Basic Research Foundation (2021B1515120046, and 2021A1515011523), Guangdong Local Innovation Team Program (2019BT02Y262), the Hainan Provincial Joint Project of Sanya Yazhou Bay Science and Technology City (Grant Nos. 2021JJLH0097 and 2021CXLH0013) and Hainan Provincial Natural Science Foundation of China (823CXTD393).

Institutional Review Board Statement: Not applicable.

Data Availability Statement: Not applicable.

Acknowledgments: We are grateful to ZH Xiao, AJ Sun, XH Zheng, Y Zhang and X Ma in the analytical facility at SCSIO for recording the spectroscopic data and to WZ Xiang for providing the samples and pictures of the green algae *Botryocladia* sp.

Conflicts of Interest: The authors declare no conflict of interest.

References

1. Yao, G.; Chen, X.; Zheng, H.; Liao, D.; Yu, Z.; Wang, Z.; Chen, J. Genomic and chemical investigation of bioactive secondary metabolites from a marine-derived fungus *Penicillium steckii* P2648. *Front. Microbiol.* **2021**, *12*, 600911. [CrossRef]
2. Hu, X.; Li, X.; Wang, B.-G.; Meng, L.-H. Uncommonp polyketides from *Penicillium steckii* AS-324, a marine endozoic fungus isolated from deep-sea coral in the magellan seamount. *Int. J. Mol. Sci.* **2022**, *23*, 6332. [CrossRef] [PubMed]
3. Hu, X.; Li, X.; Wang, B.; Meng, L. Tanzawaic acid derivatives: Fungal polyketides from the deep-sea coral-derived endozoic *Penicillium steckii* AS-324. *J. Nat. Prod.* **2022**, *85*, 1398–1406. [CrossRef] [PubMed]
4. Chen, C.; Chen, W.; Pang, X.; Liao, S.; Wang, J.; Lin, X.; Yang, B.; Zhou, X.; Luo, X.; Liu, Y. Pyrrolyl 4-quinolone alkaloids from the mangrove endophytic fungus *Penicillium steckii* SCSIO 41025: Chiral resolution, configurational assignment, and enzyme inhibitory activities. *Phytochemistry* **2021**, *186*, 112730. [CrossRef] [PubMed]
5. Song, Y.; Tan, Y.; She, J.; Chen, C.; Wang, J.; Hu, Y.; Pang, X.; Wang, J.; Liu, Y. Tanzawaic acid derivatives from the marine-derived *Penicillium steckii* as inhibitors of RANKL-Induced osteoclastogenesis. *J. Nat. Prod.* **2023**, *86*, 1171–1178. [CrossRef]
6. Shin, H.; Pil, G.; Heo, S.-J.; Lee, H.-S.; Lee, J.; Lee, Y.-J.; Lee, J.; Won, H. Anti-inflammatory activity of tanzawaic acid derivatives from a marine-derived fungus *Penicillium steckii* 108YD142. *Mar. Drugs* **2016**, *14*, 14. [CrossRef]

7. Maragos, C.M.; Busman, M.; Plattner, R.D. Development of monoclonal antibodies for the fusarin mycotoxins. *Food Addit. Contam.* **2008**, *25*, 105–116. [CrossRef]
8. Nagumo, Y.; Kakeya, H.; SHOJI, M.; Hayashi, Y.; Dohmae, N.; Osada, H. Epolactaene binds human Hsp60 Cys^{442} resulting in the inhibition of chaperone activity. *Biochem. J.* **2005**, *387*, 835–840. [CrossRef]
9. Donzelli, B.G.; Krasnoff, S.B.; Churchill, A.C.; Vandenberg, J.D.; Gibson, D.M. Identification of a hybrid PKS-NRPS required for the biosynthesis of NG-391 in *Metarhizium robertsii*. *Curr. Genet.* **2010**, *56*, 151–162. [CrossRef]
10. Kakeya, H.; Onozawa, C.; Sato, M.; Arai, K.; Osada, H. Neuritogenic effect of epolactaene derivatives on human neuroblastoma cells which lack high-affinity nerve growth factor receptors. *J. Med. Chem.* **1997**, *40*, 391–394. [CrossRef]
11. Kuramochi, K.; Yukizawa, S.; Ikeda, S.; Sunoki, T.; Arai, S.; Matsui, R.; Morita, A.; Mizushina, Y.; Sakaguchi, K.; Sugawara, F.; et al. Syntheses and applications of fluorescent and biotinylated epolactaene derivatives: Epolactaene and its derivative induce disulfide formation. *Bioorg. Med. Chem.* **2008**, *16*, 5039–5049. [CrossRef] [PubMed]
12. Lu, S.J.; Ronai, Z.A.; Li, M.H.; Jeffrey, A.M. *Fusarium moniliforme* metabolites: Genotoxicity of culture extracts. *Carcinogenesis* **1988**, *9*, 1523–1527. [CrossRef] [PubMed]
13. Cheng, S.J.; Jiang, Y.Z.; Li, M.H.; Lo, H.Z.A. A mutagenic metabolite produced by *Fusarium moniliforme* isolated from Linxian county, China. *Carcinogenesis* **1985**, *6*, 903–905. [CrossRef]
14. Sondergaard, T.E.; Hansen, F.T.; Purup, S.; Nielsen, A.K.; Bonefeld-Jorgensen, E.C.; Giese, H.; Sorensen, J.L. Fusarin C acts like an estrogenic agonist and stimulates breast cancer cells in vitro. *Toxicol. Lett.* **2011**, *205*, 116–121. [CrossRef] [PubMed]
15. Kleigrewe, K.; Aydin, F.; Hogrefe, K.; Piecuch, P.; Bergander, K.; Wurthwein, E.U.; Humpf, H.U. Structure elucidation of new fusarins revealing insights in the rearrangement mechanisms of the Fusarium mycotoxin fusarin C. *J. Agric. Food Chem.* **2012**, *60*, 497–505. [CrossRef] [PubMed]
16. Li, K.; Su, Z.; Gao, Y.; Lin, X.; Pang, X.; Yang, B.; Tao, H.; Luo, X.; Liu, Y.; Zhou, X. Cytotoxic minor piericidin derivatives from the actinomycete strain *Streptomyces psammoticus* SCSIO NS126. *Mar. Drugs.* **2021**, *19*, 428. [CrossRef] [PubMed]
17. Luo, X.; Lin, X.; Salendra, L.; Pang, X.; Dai, Y.; Yang, B.; Liu, J.; Wang, J.; Zhou, X.; Liu, Y. Isobenzofuranones and isochromenones from the deep-sea derived fungus *Leptosphaeria* sp. SCSIO 41005. *Mar. Drugs* **2017**, *15*, 204. [CrossRef]
18. Kakeya, H.; Takahashi, I.; Okada, G.; Isono, K.; Osada, H. Eolactaene, a novel neuritogenic compound in human neuroblastoma cells, produced by a marine fungus. *J. Antibiot.* **1995**, *48*, 733–735. [CrossRef]
19. Kakeya, H.; Kageyama, S.; Nie, L.; Onose, R.; Okada, G.; Beppu, T.; Norbury, C.J.; Osada, H. Lucilactaene, a new cell cycle inhibitor in p53-transfected cancer cells, produced by a *Fusarium* sp. *J. Antibiot.* **2011**, *54*, 850–854. [CrossRef]
20. Yamaguchi, J.; Kakeya, H.; Uno, T.; Shoji, M.; Osada, H.; Hayashi, Y. Determination by asymmetric total synthesis of the absolute configuration of lucilactaene, a cell-cycle inhibitor in p53-transfected cancer cells. *Angew. Chem. Int. Ed. Engl.* **2005**, *44*, 3110–3115. [CrossRef]
21. Kumarihamy, M.; Rosa, L.H.; Duke, S.O.; Tekwani, B.L.; Techen, N.; Ferreira, D.; Croom, E.M.; Khan, S.; Nanayakkara, N.P.D. Antimalarials and phytotoxins from *Botryosphaeria dothidea* identified from a seed of diseased *Torreya taxifolia*. *Molecules* **2021**, *26*, 59. [CrossRef] [PubMed]
22. Kleigrewe, K.; Sohnel, A.C.; Humpf, H.U. A new high-performance liquid chromatography-tandem mass spectrometry method based on dispersive solid phase extraction for the determination of the mycotoxin fusarin C in corn ears and processed corn samples. *J. Agric. Food. Chem.* **2011**, *59*, 10470–10476. [CrossRef] [PubMed]
23. Abdelhakim, I.A.; Mahmud, F.B.; Motoyama, T.; Futamura, Y.; Takahashi, S.; Osada, H. Dihydrolucilactaene, a potent antimalarial compound from *Fusarium* sp. RK97-94. *J. Nat. Prod.* **2021**, *85*, 63–69. [CrossRef] [PubMed]
24. Gelderblom, W.C.A.; Marasas, W.F.O.; Steyn, P.S.; Thiel, P.G.; Merwe, K.J.v.d.; Rooyen, P.H.v.; Vleggaar, R.; Wesselsb, P.L. Structure Eelucidation of fusarin C, a mutagen produced by *Fusarium moniliforme*. *J. Chem. Soc. Chem. Commun.* **1984**, *2*, 122–124.
25. Kuramochi, K.; Sunoki, T.; Tsubaki, K.; Mizushina, Y.; Sakaguchi, K.; Sugawara, F.; Ikekita, M.; Kobayashi, S. Transformation of thiols to disulfides by epolactaene and its derivatives. *Bioorg. Med. Chem.* **2011**, *19*, 4162–4172. [CrossRef]
26. Smetanina, O.F.; Yurchenko, A.N.; Pivkin, M.V.; Yurchenko, E.A.; Afiyatullo, S.S. Isochromene metabolite from the facultative marine fungus *Penicillium citrinum*. *Chem. Nat. Compd.* **2011**, *47*, 118–119. [CrossRef]
27. Ying, Y.; Zhang, L.; Shan, W.; Zhan, J. Secondary metabolites of *Peyronellaea* sp. XW-12, an endophytic fungus of *Huperzia serrata*. *Chem. Nat. Compd.* **2014**, *50*, 723–725. [CrossRef]
28. Lu, T.; Chen, F. Multiwfn: A multifunctional wavefunction analyzer. *J. Comput. Chem.* **2012**, *33*, 580–592. [CrossRef]
29. Tan, Y.; Yang, B.; Lin, X.; Luo, X.; Pang, X.; Tang, L.; Liu, Y.; Li, X.; Zhou, X. Nitrobenzoyl sesquiterpenoids with cytotoxic activities from a marine-derived *Aspergillus ochraceus* fungus. *J. Nat. Prod.* **2018**, *81*, 92–97. [CrossRef]
30. Wang, J.; Cong, Z.; Huang, X.; Hou, C.; Chen, W.; Tu, Z.; Huang, D.; Liu, Y. Soliseptide A, a cyclic hexapeptide possessing piperazic acid groups from *Streptomyces solisilvae* HNM30702. *Org. Lett.* **2018**, *20*, 1371–1374. [CrossRef]
31. Tan, Y.; Deng, W.; Zhang, Y.; Ke, M.; Zou, B.; Luo, X.; Su, J.; Wang, Y.; Xu, J.; Nandakumar, K.S.; et al. A marine fungus-derived nitrobenzoyl sesquiterpenoid suppresses receptor activator of NF-kappaB ligand-induced osteoclastogenesis and inflammatory bone destruction. *Br. J. Pharmacol.* **2020**, *177*, 4242–4260. [CrossRef] [PubMed]
32. Zhang, R.; Liu, W.; Zeng, J.; Meng, J.; Jiang, H.; Wang, J.; Xing, D. Niemann-Pick C1-Like 1 inhibitors for reducing cholesterol absorption. *Eur. J. Med. Chem.* **2022**, *230*, 114111. [CrossRef] [PubMed]
33. Zhang, R.; Song, Z.; Wang, X.; Xue, J.; Xing, D. One-step modification to identify dual-inhibitors targeting both pancreatic triglyceride lipase and Niemann-Pick C1-like 1. *Eur. J. Med. Chem.* **2021**, *216*, 113358. [CrossRef]

34. Li, J.; Zhou, Y.; Eelen, G.; Zhou, Q.T.; Feng, W.B.; Labroska, V.; Ma, F.F.; Lu, H.P.; Dewerchin, M.; Carmeliet, P.; et al. A high-throughput screening campaign against PFKFB3 identified potential inhibitors with novel scaffolds. *Acta Pharmacol. Sin.* **2022**, *44*, 680–692. [CrossRef] [PubMed]
35. Wang, J.; Gong, G.Q.; Zhou, Y.; Lee, W.J.; Buchanan, C.M.; Denny, W.A.; Rewcastle, G.W.; Kendall, J.D.; Dickson, J.M.J.; Flanagan, J.U.; et al. High-throughput screening campaigns against a PI3Kalpha isoform bearing the H1047R mutation identified potential inhibitors with novel scaffolds. *Acta Pharmacol. Sin.* **2018**, *39*, 1816–1822. [CrossRef] [PubMed]
36. Ellman, G.; Courtney, K.; Andres, V.; Featherstone, R. A new and rapid colorimetric determination of acetylcholinesterase activity. *Biochem. Pharmacol.* **1961**, *7*, 88–95. [CrossRef]

marine drugs

Article

Trichoderols B-G, Six New Lipids from the Marine Algicolous Fungus *Trichoderma* sp. Z43

Zhen-Zhen Shi [1], Xiu-Li Yin [1] and Nai-Yun Ji [1,2,*]

[1] Yantai Institute of Coastal Zone Research, Chinese Academy of Sciences, Yantai 264003, China; zzshi@yic.ac.cn (Z.-Z.S.); xlyin@yic.ac.cn (X.-L.Y.)
[2] Shandong Saline-Alkaline Land Modern Agriculture Company, Dongying 257345, China
* Correspondence: nyji@yic.ac.cn; Tel.: +86-535-210-9176

Abstract: Six new lipids, trichoderols B-G (**1–6**), along with a known one, triharzianin B (**7**), were isolated from the culture of *Trichoderma* sp. Z43 obtained from the surface of the marine brown alga *Dictyopteris divaricata*. Their structures and relative configurations were identified by interpretation of 1D/2D NMR and MS data. Compounds **1–7** were assayed for inhibiting the growth of three phytopathogenic fungi (*Fusarium graminearum*, *Gaeumannomyces graminis*, and *Glomerella cingulata*), four marine phytoplankton species (*Amphidinium carterae, Heterocapsa circularisquama, Heterosigma akashiwo*, and *Prorocentrum donghaiense*), and one marine zooplankton (*Artemia salina*). Compounds **1**, **4**, and **7** exhibited weak antifungal activities against three phytopathogenic fungi tested with MIC $\geq$ 64 μg/mL. All compounds displayed moderate antimicroalgal activity with $IC_{50} \geq 15$ μg/mL and low toxicity to the brine shrimp *Artemia salina*.

Keywords: *Trichoderma*; lipid; antifungal activity; antimicroalgal activity

Citation: Shi, Z.-Z.; Yin, X.-L.; Ji, N.-Y. Trichoderols B-G, Six New Lipids from the Marine Algicolous Fungus *Trichoderma* sp. Z43. *Mar. Drugs* **2023**, *21*, 453. https://doi.org/10.3390/md21080453

Academic Editors: Bin-Gui Wang and Haofu Dai

Received: 14 July 2023
Revised: 8 August 2023
Accepted: 14 August 2023
Published: 17 August 2023

1. Introduction

Crop diseases caused by pathogenic fungus have seriously restricted healthy development of agriculture in the world. *Fusarium* head blight of wheat, take-all disease, and plant anthracnose caused by *Fusarium graminearum*, *Gaeumannomyces graminis*, and *Glomerella cingulata*, respectively, lead to huge economic losses in agriculture every year [1–5]. In addition, some marine phytoplankton species including *Amphidinium carterae, Heterocapsa circularisquama*, *Heterosigma akashiwo*, and *Prorocentrum donghaiense* can induce red tides that harm aquaculture industry [6–9]. Thus, it is imperative to search for natural antifungal and antimicroalgal drugs with high activity and safety. On the other hand, *Trichoderma* species are known for producing various metabolites with novel structures and intriguing bioactivities [10–12]. In recent years, natural products originated from marine-derived *Trichoderma* species have gained the attention of researchers [13], and a large number of new compounds with significant activities have been found, such as polyketides [14], terpenoids [15–17], and steroids [18]. Although *Trichoderma* species have proven to be a treasure-house of new natural products, only a few short-chain lipids have been isolated from this species. For example, triharzianins A-D were purified from *T. harzianum* obtained from mushroom *Tricholoma matsutake* [19], harzianumols A-H were separated from *T. harzianum* obtained from sponge *Petrospongia nigra* [20], and trichoderol A was acquired from *Trichoderma* sp. obtained from soil [21]. Moreover, these lipids displayed multifarious bioactivities such as feeding attractants and antimicrobial activities [19,21]. During our investigation of the chemical diversity and biological activity of marine-derived *Trichoderma*, the epiphytic *Trichoderma* sp. Z43 obtained from the marine brown alga *Dictyopteris divaricata* was examined. Its aerial mycelia grew rapidly. Most of them were white on the PDA medium at 25 °C, and some parts of the plate were yellow. Alternate permutation of the mycelia was observed through a common microscope. As a result, six new lipids, trichoderols B-G (**1–6**), along with a known one, triharzianin B (7), were isolated and identified. Herein, the details of isolation, structure elucidation, and bioactivities of these compounds are described.

2. Results and Discussion

The organic extracts of marine-derived *Trichoderma* sp. Z43 isolated from the marine brown alga *Dictyopteris divaricata* were subjected to a series of column chromatography processes to produce six new lipids, namely, trichoderols B-G (**1–6**), along with a known one, triharzianin B (**7**) (Figure 1). Compound **7** was unambiguously identified by comparing its NMR data and specific rotation value with those reported in the literature [19].

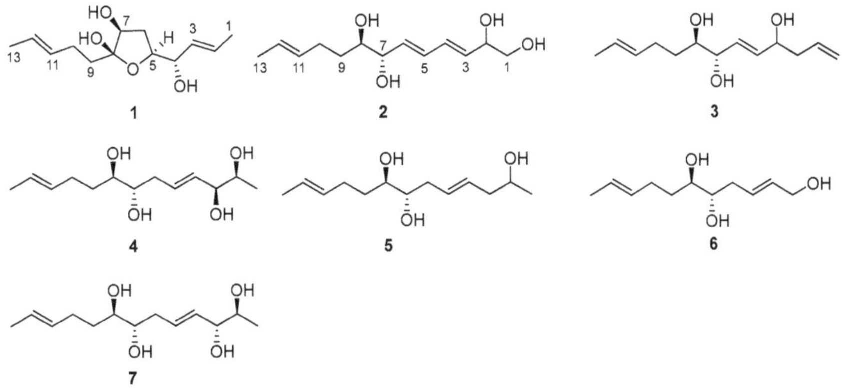

Figure 1. Chemical structures of **1–7**.

2.1. Structural Elucidation

Compound **1** was isolated as a colorless oil with a molecular formula of $C_{13}H_{22}O_4$ established by HRESIMS (m/z 265.1420 $[M + Na]^+$), implying three degrees of unsaturation. The ^{1}H NMR spectrum (Table 1) showed two methyl doublets, one double quartet, one double doublet, and two multiplets assignable to four olefinic protons and one double multiplet and two double doublets attributable to three oxymethines. The ^{13}C NMR and DEPT spectra (Table 2) displayed the presence of two methyls, three methylenes, seven methines, and one nonprotonated carbon. COSY correlations of H-1/H-2/H-3/H-4/H-5/H-6/H-7 established the linkage of C-1 to C-7 (Figure 2). A pentenyl group was confirmed by the COSY correlation of H-9 with H-10 and the HMBC correlations from H-10/H-13 to C-11 and C-12 and from H-9 to C-11, which was then elongated to C-7 via C-8 by the HMBC correlations from H-6/H-9 to C-8. To satisfy the molecular formula, an ether linkage was situated between C-5 and C-8, which was verified by the HMBC correlation from H-5 to C-8 (Figure 2). Thus, the planar structure of **1** was validated. The double bond at C-2 was attributed to be *trans* by the large coupling constant (J = 15.2) between H-2 and H-3. The chemical shifts of other two olefinic carbon atoms (C-11, δ_C 130.9; C-12, δ_C 125.5) were highly similar to those of triharzianin B (**7**) (C-11, δ_C 130.9; C-12, δ_C 124.6) [19], suggesting a *trans* configuration of the C-11 double bond, which was further verified by the IR absorption at 969 cm^{-1}. The relative configurations of H-5, OH-7, and OH-8 were confirmed by NOESY correlations of H-7 with H-5 and H-9 (Figure 3), and the relationship of H-4 and H-5 was deduced to be *threo* due to NMR data that were similar to those of (-)-(*S*)-1-[(2*S*,5*S*)-5-[2-propenyl]tetrahydrofuran-2-yl]prop-2-en-1-ol [22].

Table 1. ^{1}H NMR data for **1–6** (δ in ppm, *J* in Hz).

Pos	1 ($CDCl_3$)	2 (CD_3OD)	3 ($CDCl_3$)	4 (CD_3OD)	5 (CD_3OD)	6 (CD_3OD)
1a	1.72, d (6.6)	3.48, dd (11.1, 5.1)	5.17, brs	1.14, d (6.4)	1.14, d (6.2)	4.02, d (5.2)
1b		3.46, dd (11.2, 6.8)	5.14, brs			
2	5.81, dq (15.2, 6.6)	4.15, dt (6.3, 5.2)	5.80, m	3.65, qd (6.4, 6.1)	3.75, sext (6.2)	5.69, dt (15.4, 5.2)
3a	5.28, dd (15.2, 7.4)	5.71, dd (13.1, 6.3)	2.37, dddt (13.9, 6.6, 5.3, 1.2)	3.84, dd (7.1, 6.1)	2.18, m	5.72, dt (15.4, 6.7)
3b			2.28, dddt (13.9, 7.4, 6.6, 1.0)		2.14, m	
4a	4.20, dm (7.4)	6.32, dd (13.1, 10.6)	4.24, dt (6.1, 5.5)	5.60, ddm (15.5, 7.1)	5.54, dt (15.4, 6.4)	2.31, m
4b						2.19, m
5	4.52 dd (5.5, 3.1)	6.29, dd (13.1, 10.6)	5.84, ddd (15.5, 5.6, 0.9)	5.75, m	5.56, dt (15.4, 6.4)	3.44, m
6a	2.44, dd (13.5, 6.9)	5.73, dd (13.1, 5.7)	5.73, ddd (15.5, 6.3, 1.1)	2.33, m	2.35, dm (13.8)	3.43, m
6b	1.45, dddd (13.5, 5.5, 2.6, 1.2)			2.22, dd (14.2, 7.3)	2.12, m	
7	3.88, dd (6.9, 2.6)	3.93, dd (6.5, 5.7)	3.96, dd (6.3, 5.9)	3.47, m	3.40, m	1.52, m
8a		3.42, ddd (9.4, 5.7, 3.2)	3.49, ddd (9.3, 5.9, 3.5)	3.44, m	3.39, m	2.15, m
8b						2.04, m
9a	1.99, m	1.54, m	1.58, m	1.53, m	1.67, m	5.45, m
9b		1.38, m	1.49, m		1.41, m	
10a	2.24, m	2.16, m	2.18, m	2.16, m	2.19, m	5.45, m
10b		2.03, m	2.09, m	2.04, m	2.02, m	
11	5.49, m	5.44, m	5.44, m	5.46, m	5.46, m	1.64, d (4.7)
12	5.50, m	5.43, m	5.47, m	5.46, m	5.46, m	
13	1.64, d (4.6)	1.63, d (5.2)	1.65, d (6.0)	1.64, d (3.9)	1.64, d (3.9)	

Table 2. ^{13}C NMR data for **1–6** (δ in ppm).

Pos	1 ($CDCl_3$)	2 (CD_3OD)	3 ($CDCl_3$)	4 (CD_3OD)	5 (CD_3OD)	6 (CD_3OD)
1	18.1, CH_3	67.2, CH_2	118.9, CH_2	18.7, CH_3	22.9, CH_3	63.7, CH_2
2	131.4, CH	73.9, CH	134.0, CH	71.6, CH	68.5, CH	132.7, CH
3	126.5, CH	134.3, CH	42.0, CH_2	77.9, CH	43.6, CH_2	129.9, CH
4	80.0, CH	132.2, CH	70.8, CH	132.9, CH	130.1, CH	37.1, CH_2
5	78.2, CH	132.6, CH	135.4, CH	131.1, CH	130.7, CH	75.0, CH
6	36.1, CH_2	134.3, CH	130.0, CH	37.3, CH_2	37.3, CH_2	73.9, CH
7	74.2, CH	76.7, CH	75.4, CH	75.0, CH	75.9, CH	33.9, CH_2
8	111.3, C	75.1, CH	74.2, CH	73.9, CH	74.7, CH	30.0, CH_2
9	28.5, CH_2	33.7, CH_2	32.8, CH_2	33.9, CH_2	33.6, CH_2	132.2, CH
10	26.5, CH_2	29.9, CH_2	28.9, CH_2	30.0, CH_2	29.9, CH_2	126.0, CH
11	130.9, CH	132.2, CH	130.7, CH	132.2, CH	132.3, CH	18.1, CH_3
12	125.5, CH	126.0, CH	125.9, CH	126.0, CH	126.0, CH	
13	18.1, CH_3	18.1, CH_3	18.1, CH_3	18.1, CH_3	18.1, CH_3	

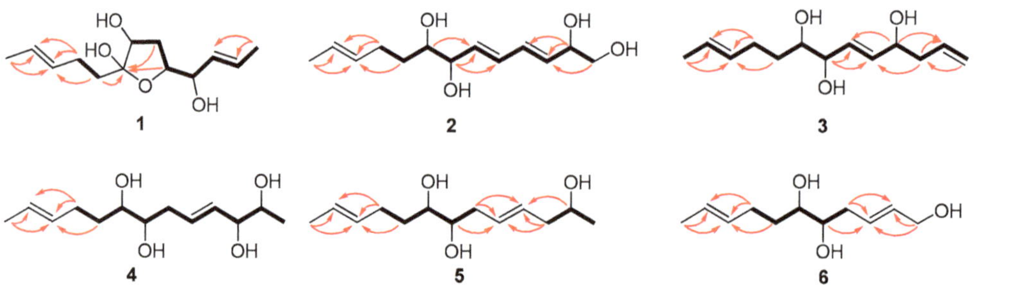

Figure 2. Key COSY and HMBC correlations of **1–6** (bold lines for COSY and arrows for HMBC).

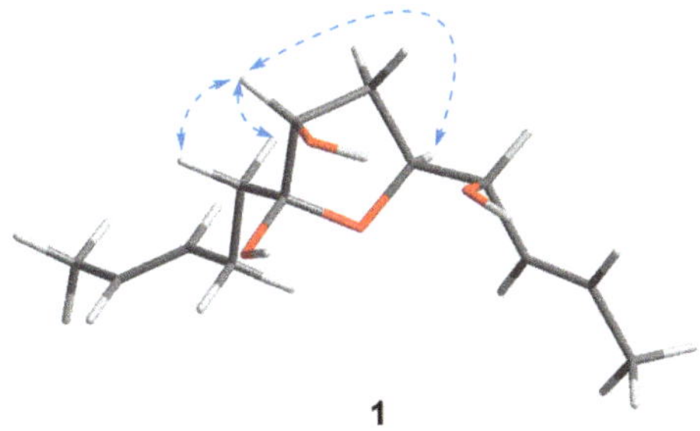

1

Figure 3. Key NOESY correlations of **1**.

Compound **2** was purified as a colorless oil and assigned a molecular formula of $C_{13}H_{22}O_4$ by interpretation of HRESIMS (m/z 265.1420 $[M + Na]^+$) data. In conjunction with HSQC data, the ^1H NMR spectrum (Table 1) revealed notable signals including one methyl doublet, one double triplet, one double doublet, and one doublet of double doublets ascribable to three oxygenated methines, two double doublets attributable to an oxygenated methylene, and four double doublets and two multiplets assignable to six olefinic protons. DEPT experiments displayed 13 resonances in the ^{13}C NMR spectrum, which were assigned to one methyl, three methylenes, and nine methines. COSY correlations of H-1/H-2/H-3/H-4/H-5/H-6/H-7/H-8/H-9/H-10 established the linkage of C-1 to C-10, which was further confirmed by the HMBC correlations from H-1 to C-3, from H-2 to C-3 and C-4, from H-7 to C-5 and C-6, and from H-8 to C-6 (Figure 2). A propenyl group was located at C-10, verified by the HMBC correlations from H-9 to C-11 and from H-10/H-13 to C-11 and C-12 (Figure 2). Thus, the planar structure of **2** was affirmed. The double bonds at C-3 and C-5 were determined to be *trans* by the large coupling constants (J = 13.1), and the identical ^{13}C NMR data of **2** (C-11, δ_C 132.2; C-12, δ_C 126.0) with those of triharzianin C (C-11, δ_C 132.2; C-12, δ_C 126.0) [19] suggested the double bond at C-11 to be *trans*. The relationship of OH-7 and OH-8 was assigned to be *erythro* by comparison of NMR data with those for triharzianin C [19]. Although the chemical shifts of C-1, C-2, and C-3 in **2** were the same as those of (*S*)-but-3-ene-1,2-diol [23], the relative configuration of OH-2 could not be confirmed due to the existence of only one chiral carbon atom (C-2) in this moiety. Thus, compound **2** was named (3*E*,5*E*,11*E*)-trideca-3,5,11-trien-1,2,7,8-tetraol, and its structural formula was $HOCH_2CH(OH)CH{=}CHCH{=}CHCH(OH)CH(OH)CH_2CH_2CH{=}CHCH_3$.

Compound **3** was acquired as a colorless oil, and its molecular formula was determined to be $C_{13}H_{22}O_3$ by HRESIMS (249.1448 $[M + Na]^+$) data. The ^1H NMR spectrum, along with HSQC data, exhibited notable signals including one methyl doublet, one double triplet, one double doublet, one doublet of double doublets assignable to three oxygenated methines, two broad singlets attributable to a terminal double bond, and three multiplets and two doublets of double doublets ascribable to five olefinic protons. The ^{13}C NMR spectrum displayed 13 resonances, sorted into one methyl, four methylenes, and eight methines. The ^1H and ^{13}C NMR data (Tables 1 and 2) revealed the presence of a similar skeleton to **2**. COSY correlations of H-1/H-2/H-3/H-4/H-5/H-6/H-7/H-8/H-9/H-10/H-11/H-12/H-13 and HMBC correlations from H-1 to C-3, from H-3 to C-2 and C-5, from H-4 to C-2 and C-6, from H-7 to C-5 and C-6, from H-10/H-13 to C-11 and C-12, and from H-9 to C-11 (Figure 2) confirmed the planar structure of **3**. The double bonds at C-5 and C-11 were deduced to be *trans* by the large coupling constant (J = 15.5) between H-5 and H-6 and by the identical NMR data of C-11 and C-12 with those of **2**. The relative configurations of OH-7 and OH-8 were determined to be the same as those of **2** on the basis of highly similar chemical shifts and coupling constants of them. Despite the similar NMR data of C-1 to C-5 in **3** with those of 12(*S*)-hydroxy-5(*Z*),8(*Z*),10(*E*),14(*E*)-eicosatetraenoic acid [24], the relative configuration of OH-4 was still unsolved. Compound **3** was named (5*E*,11*E*)-trideca-1,5,11-trien-4,7,8-triol, and its structural formula was $CH_2{=}CHCH_2CH(OH)CH{=}CHCH(OH)CH(OH)CH_2CH_2CH{=}CHCH_3$.

Compound **4** was obtained as a colorless oil and given a molecular formula of $C_{13}H_{24}O_4$ by interpretation of HRESIMS data (*m*/*z* 243.1558 $[M]^-$), requiring two degrees of unsaturation. Its NMR data (Tables 1 and 2) were highly similar to those of triharzianin B (**7**) [19], except for chemical shifts of C-2 and C-3 (C-2, δ_C 70.2; C-3, δ_C 76.6 for triharzianin B; C-2, δ_C 71.6; C-3, δ_C 77.9 for **4**), which indicated that **4** and triharzianin B possessed the same planar structure. COSY correlations of H-1/H-2/H-3/H-4/H-5/H-6/H-7/H-8/H-9/H-10 and HMBC correlations from H-9 to C-11 and from H-10/H-13 to C-11 and C-12 (Figure 2) further afforded to this structure. The geometry of two double bonds at C-2 and C-11 and the relative configurations of OH-7 and OH-8 were deduced to be the same as those of triharzianin B (**7**) [19] on the basis of their identical NMR data. The relationship of H-2 and H-3 was *threo* by analysis of NMR data with those for separacenes A and B [25]. Therefore, compound **4** was named (4*E*,11*E*)-trideca-4,11-diene-2,3,7,8-tetraol, and its structural formula was $CH_3CH(OH)CH(OH)CH{=}CHCH_2CH(OH)CH(OH)CH_2CH_2CH{=}CHCH_3$.

Compound **5** was isolated as a colorless oil. HRESIMS analysis gave the molecular formula of $C_{13}H_{24}O_3$, consistent with two degrees of unsaturation. Its NMR data (Tables 1 and 2) resembled those of **4**, except for the presence of signals for a methylene and the lack of signals for a hydroxymethine group. COSY correlations of H-1/H-2/H-3 confirmed the kinkage from C-1 to C-3, which was then elongated to C-10 by the HMBC correlations from H-2 to C-4, from H-3/H-6 to C-4 and C-5, and from H-7 to C-5 and the COSY correlations of H-6/H-7/H-8/H-9/H-10. A propenyl group was situated at C-10 by the HMBC correlations from H-9 to C-11 and from H-10/H-13 to C-11 and C-12 (Figure 2). Thus, the planar structure of **5** was confirmed. The configurations of OH-7, OH-8, and the double bond at C-11 were the same as those of **4** due to their similar NMR data, and the geometry of double bond at C-4 was deduced to be *trans* by the large coupling constant ($J = 15.4$) between H-4 and H-5. The relative configuration of OH-2 was uncertain in spite of comparing the NMR data of **5** with those of (*S*)-2-hexanol and (*R*)-octan-2-ol carefully [26]. Thus, compound **5** was named (4*E*,11*E*)-trideca-4,11-dien-2,7,8-triol, and its structural formula was $CH_3CH(OH)CH_2CH{=}CHCH_2CH(OH)CH(OH)CH_2CH_2CH{=}CHCH_3$.

Compound **6** was obtained as a colorless oil and was given a molecular formula of $C_{11}H_{20}O_3$ by analysis of HRESIMS data, requiring two degrees of unsaturation. Its NMR data exhibited high similarities to those of **4** except for the presence of signals for a hydroxymethylene group and lack of signals for a methyl and two hydroxymethine groups. COSY correlations of H-1/H-2/H-3/H-4/H-5/H-6/H-7/H-8 determined the linkage of C-1 to C-8, and a propenyl group was located at C-8, confirmed by the HMBC correlations from H-7 to C-9 and from H-8/H-11 to C-9 and C-10 (Figure 2). Other HMBC correlations further verified the structure of **6** (Figure 2). The large coupling constant ($J = 15.4$) between H-2 and H-3 demonstrated that the double bond at C-2 was *trans*. The identical NMR data of the spin system in **6** (from C-2 to C-11) and **4** (from C-4 to C-13) (Tables 1 and 2) ascertained the *trans* configuration of a double bond at C-9 and the *erythro* relationship of the vicinal diol (OH-5 and OH-6). Compound **6** was named (2*E*,9*E*)-undeca-2,9-dien-1,5,6-triol, and its structural formula was $HOCH_2CH{=}CHCH_2CH(OH)CH(OH)CH_2CH_2CH{=}CHCH_3$.

2.2. Bioactivity of Isolated Compounds

Compounds **1**–**7** were assayed for antifungal activity against *Fusarium graminearum*, *Gaeumannomyces graminis*, and *Glomerella cingulata*. The result showed that compounds **1**, **4**, and **7** displayed weak antifungal activity (Table 3). Compound **1** could inhibit the three phytopathogenic fungi tested with MIC values ranging from 64 to 256 μg/mL. Compounds **4** and **7** possessed identical inhibition against *Glomerella cingulata* and *Gaeumannomyces graminis*, with MIC values of 128 and 256 μg/mL, respectively. In addition, all the isolates were evaluated for antimicroalgal activity against *Amphidinium carterae*, *Heterocapsa circularisquama*, *Heterosigma akashiwo*, and *Prorocentrum donghaiense* (Table 4). It was worth noting that only compound **1** was active against all the phytoplankton species tested with IC_{50} values ranging from 15 to 28 μg/mL. Moreover, the antimicroalgal activity of compounds **2**–**7** ($IC_{50} \geq 30$ μg/mL) was weaker than that of compound **1**. The above results suggested that the tetrahydrofuran ring could improve antifungal and antimicroalgal activity of these lipids by analysis of their structure–activity relationships. In addition, the brine shrimp lethality of **1–7** was also estimated, with the lethal rates against *Artemia salina* of these compounds being less than 10% at 100 μg/mL (Table 4). All isolates showed low toxicity to the brine shrimp *Artermia salina*, which demonstrated the safety of their further exploitation. An in-depth study, such as chemical modification, should be conducted to promote the bioactivity of these compounds, increasing their prospective use in the development of antifungal and antimicroalgal agents.

Table 3. Inhibition of three phytopathogenic fungi by **1–7**.

Compounds	MIC (μg/mL)		
	Fusarium graminearum	*Gaeumannomyces graminis*	*Glomerella cingulata*
1	256	256	64
2	-	-	-
3	-	-	-
4	-	256	128
5	-	-	-
6	-	-	-
7	-	256	128
amphotericin B	2.0	2.0	1.0

-: no inhibition effect at 256 μg/mL.

Table 4. Inhibition of phytoplankton species by **1–7**.

Compounds	IC_{50} (μg/mL)				Lethal Rate (at 100 μg/mL)
	Amphidinium carterae	*Heterocapsa circularisquama*	*Heterosigma akashiwo*	*Prorocentrum donghaiense*	*Artemia salina*
1	15	24	28	22	9.6%
2	61	-	43	37	5.2%
3	66	78	-	-	4.6%
4	-	-	53	42	6.1%
5	50	68	-	-	8.4%
6	44	-	82	-	3.2%
7	-	-	49	30	5.4%
$K_2Cr_2O_7$	1.2	1.0	0.8	1.4	100%

-: no inhibition effect at 100 μg/mL.

3. Materials and Methods

3.1. General Experimental Producers

NMR spectra were obtained on a Bruker Avance III 500 NMR spectrometer (500 and 125 MHz for ^{1}H and ^{13}C, respectively) using tetramethylsilane (TMS) as an internal standard (Bruker Corp., Billerica, MA, USA). Chemical shifts are reported in parts per million (δ) in $CDCl_3/CD_3OD$ (δ_H reported referred to $CDCl_3/CD_3OD$ at 7.26/3.31 ppm; δ_C reported referred to $CDCl_3/CD_3OD$ at 77.16/49.00 ppm) and coupling constants (J) in Hz. Optical rotations were measured on an SGW-3 polarimeter (Shanghai Shenguang Instrument Co., Ltd., Shanghai, China). IR spectra were recorded on a Nicolet iS50 FT-IR spectrometer (Thermo Fisher Scientific, Waltham, MA, USA); peaks are reported in cm^{-1}. High-resolution ESI mass spectra were determined on a Xevo G2-XS QTof mass spectrometer (Water Crop., Milford, MA, USA). Column chromatography (CC) was carried out with silica gel (200–300 mesh, Qingdao Haiyang Chemical Co., Qingdao, China), RP-18 (AAG12S50, YMC Co., Ltd., Kyoto, Japan), and Sephadex LH-20 (GE Healthcare, Uppsala, Sweden). Thin-layer chromatography (TLC) was performed with precoated silica gel plates (GF-254, Qingdao Haiyang Chemical Co., Qingdao, China).

3.2. Fungal Material and Fermentation

The fungal strain *Trichoderma* sp. Z43 was isolated from the surface of marine brown alga *Dictyopteris divaricata* collected from Zhoushan, China, in July 2018. The species was identified according to morphological characteristics and analysis of ITS regions of its rDNA, deposited at GenBank (OR196112). Mass fermentation was performed statically at room temperature for 30 days in 200 × 1 L Erlenmeyer flasks, each containing 300 mL of media, by adding 40.0 g glucose, 10.0 g peptone, and 7.0 g yeast extract powder into 1000 mL purified water.

3.3. Extraction and Isolation

At the end of fermentation, the mycelia of cultures were obtained by filtration, which were then dried at room temperature, smashed, and extracted with CH_2Cl_2 and MeOH (1:1, *v*/*v*). After removing organic solvents under reduced pressure, the residue was partitioned between ethyl acetate (EtOAc) and H_2O to afford an EtOAc-soluble extract (25.6 g). The filtrate was extracted with EtOAc and then dried to give an extract (10.0 g). The two parts were merged based on the similarity of TLC profiles and subjected to silica gel CC with step-gradient solvent systems of petroleum ether (PE)/EtOAc and CH_2Cl_2/MeOH to afford 11 fractions (Frs. 1–11). Fr. 8 eluted with CH_2Cl_2/MeOH (20:1) and was further purified by RP-18 CC to afford Fr. 8-1 (MeOH/H_2O, 3:7), Fr. 8-2 (MeOH/H_2O, 2:3), and Fr. 8-3 (MeOH/H_2O, 9:11). Fr. 8-1 was further purified by preparative TLC (CH_2Cl_2/MeOH, 20:1) and Sephadex LH-20 CC (MeOH) to give compound **7** (2.6 mg). Compound **1** (4.2 mg) was isolated from Fr. 8-2 by preparative TLC (CH_2Cl_2/MeOH, 20:1) and Sephadex LH-20 CC (MeOH). Compounds **3** (4.0 mg), **5** (4.5 mg), and **6** (2.7 mg) were purified from Fr. 8-3 by TLC (CH_2Cl_2/MeOH, 10:1) and Sephadex LH-20 CC (MeOH). Fr. 9 eluted with CH_2Cl_2/MeOH (10:1) and was further purified by RP-18 CC (MeOH/H_2O, 1:1) and preparative TLC (CH_2Cl_2/MeOH, 10:1) as well as Sephadex LH-20 CC (MeOH) to afford compounds **2** (3.1 mg) and **4** (3.8 mg).

3.4. Spectral and Physical Data of Compounds **1–6**

Trichoderol B (**1**): colorless oil; $[\alpha]^{20}{}_D - 10.2$ (*c* 0.14, CH_3OH); IR (KBr) v_{max} 3443, 2923, 2854, 1633, 1453, 1384, 969 cm^{-1}; ^1H and ^{13}C NMR data, Tables 1 and 2; HRESIMS *m*/*z* 265.1420 $[M + Na]^+$ (calcd for $C_{13}H_{22}NaO_4$, 265.1416).

Trichoderol C (**2**): colorless oil; $[\alpha]^{20}{}_D - 4.2$ (*c* 0.10, CH_3OH); IR (KBr) v_{max} 3430, 2926, 2858, 1633, 1554, 1394, 1030 cm^{-1}; ^1H and ^{13}C NMR data, Tables 1 and 2; HRESIMS *m*/*z* 265.1420 $[M + Na]^+$ (calcd for $C_{13}H_{22}NaO_4$, 265.1416).

Trichoderol D (**3**): colorless oil; $[\alpha]^{20}{}_D + 7.9$ (*c* 0.13, CH_3OH); IR (KBr) v_{max} 3422, 2924, 2856, 1661, 1633, 1538, 1394, 1025 cm^{-1}; ^1H and ^{13}C NMR data, Tables 1 and 2; HRESIMS *m/z* 249.1448 $[M + Na]^+$ (calcd for $C_{13}H_{22}NaO_3$, 249.1467).

Trichoderol E (**4**): colorless oil; $[\alpha]^{20}{}_D - 12.2$ (*c* 0.13, CH_3OH); IR (KBr) v_{max} 3415, 2924, 1634, 1538, 1385, 1023 cm^{-1}; ^1H and ^{13}C NMR data, Tables 1 and 2; HRESIMS *m*/*z* 243.1588 $[M]^-$ (calcd for $C_{13}H_{23}O_4$, 243.1598).

Trichoderol F (**5**): colorless oil; $[\alpha]^{20}{}_D - 4.1$ (*c* 0.15, CH_3OH); IR (KBr) v_{max} 3417, 2924, 2855, 1632, 1554, 1384, 966 cm^{-1}; ^1H and ^{13}C NMR data, Tables 1 and 2; HRESIMS *m*/*z* 251.1614 $[M + Na]^+$ (calcd for $C_{13}H_{24}NaO_3$, 251.1623).

Trichoderol G (**6**): colorless oil; $[\alpha]^{20}{}_D - 5.7$ (*c* 0.09, CH_3OH); IR (KBr) v_{max} 3441, 2923, 2853, 1633, 1539, 1385, 985 cm^{-1}; ^1H and ^{13}C NMR data, Tables 1 and 2; HRESIMS *m*/*z* 223.1301 $[M + Na]^+$ (calcd for $C_{11}H_{20}NaO_3$, 223.1310).

3.5. Assay for Antifungal Activity

Antifungal activity against *Fusarium graminearum*, *Gaeumannomyces graminis*, and *Glomerella cingulata* was performed using the microdilution method in a 96-well plate, as described previously [17]. Briefly, a stock solution of each fungus tested was diluted in potato dextrose broth (PDB) to 5 × 10^5 cfu/mL. Each sample was prepared in dimethyl sulfoxide (DMSO) and was diluted to final concentrations of 5120, 2560, 1280, 640, 320, 160, 80, 40, 20,

10, and 5 μg/mL in a set of capped test tubes by two-fold serial dilution. An amount of 5 μL diluent was added into each well of a 96-well flat-bottom microtiter plate containing 95 μL fungal suspension (the final sample concentrations were 256 to 0.25 μg/mL), and the fungi were cultivated at 28 °C for 48 h. The MIC value for each sample was defined as the minimum concentration of the compound with invisible microbial growth. Amphotericin B and DMSO were chosen as positive and negative controls, respectively.

3.6. Assay for Antimicroalgal Activity

The inhibition of four marine phytoplankton species (*Amphidinium carterae, Heterocapsa circularisquama, Heterosigma akashiwo,* and *Prorocentrum donghaiense*) was assayed using our previously reported method [17]. In brief, each microalga inoculum was cultured for 5 days using the sterilized f/2 medium in an incubator (20 °C, 14:10 light–dark cycle, 2000 lx light) and reached the exponential growth phase. The microalga suspension was diluted to 4–5 $\times$ 10^4 cells/mL and then added into a 96-well flat-bottom microtiter plate with 195 μL in each well. An amount of 5 μL sample solution (in DMSO) was pipetted into each well (the final sample concentrations were 100 to 0.125 μg/mL) and mixed uniformly. After 24 h inoculation, the live cells were counted using hemocytometer, and the inhibition rate was calculated as follows. Inhibition rate (IR) = $(N_{CK} - N_T)/N_{CK} \times 100\%$ (N_{CK}: the number of live algal cells under negative control, N_T: the number of live algal cells under treatment). DMSO and $K_2Cr_2O_7$ were taken as negative and positive controls, respectively.

3.7. Assay for Brine Shrimp Lethal Activity

The inhibition of the brine shrimp *Artermia salina* was assayed according to the procedures described in our previous report [15]. In brief, brine shrimp eggs were left to hatch in sea water for 48 h at 25 °C under natural light. About 10 brine shrimp were placed in a 96-well flat-bottom microtiter plate with a volume of 195 μL sea water in each well. An amount of 5 μL sample solution (in DMSO) was added into each well (the final sample concentrations were 100 to 0.125 μg/mL) and mixed uniformly. The lethality was observed after 24 h of cultivation. DMSO and $K_2Cr_2O_7$ served as negative and positive controls, respectively.

4. Conclusions

Chemical investigation towards the marine algicolous fungus *Trichoderma* sp. Z43 resulted in the isolation of seven lipids, including six new ones (trichoderols B-G (**1–6**)) and a known one, triharzianin B (**7**). The C_{13} and C_{11} lipids are rarely found in nature, especially in *Trichoderma* species, and these new compounds greatly enrich the chemical diversity of marine-derived natural products. Finding and stimulating silent biosynthetic gene clusters may be an effective means to excavate this kind of metabolite. These isolations were evaluated for inhibition against three phytopathogenic fungi and four marine phytoplankton species. Several of them exhibited inhibition of one or more fungi/plankton species tested, and the tetrahydrofuran ring could improve antifungal and antimicroalgal activity of these lipids by analysis of their structure–activity relationships. Moreover, all isolates exhibited low toxicity to the brine shrimp *Artermia salina*, suggesting the security for their further exploitation.

Supplementary Materials: The following supporting information can be downloaded at: https://www.mdpi.com/article/10.3390/md21080453/s1, Figures S1–S42: 1D/2D NMR, HRESIMS spectra of **1–6**.

Author Contributions: Conceptualization, N.-Y.J.; data curation, Z.-Z.S.; formal analysis, Z.-Z.S.; funding acquisition, Z.-Z.S. and N.-Y.J.; investigation, Z.-Z.S.; methodology, X.-L.Y.; project administration, N.-Y.J.; supervision, N.-Y.J.; writing—original draft, Z.-Z.S.; writing—review and editing, N.-Y.J. All authors have read and agreed to the published version of the manuscript.

Funding: This research was funded by the National Natural Science Foundation of China (42206130 and 42076096), the Natural Science Foundation of Shandong Province (ZR2020QD102), the Taishan Scholar Project Special Funding (tsqn201909164), and the Science & Technology Specific Projects in Agricultural High-Tech Industrial Demonstration Area of the Yellow River Delta (2022SZX01).

Institutional Review Board Statement: Not applicable.

Data Availability Statement: Data of the compounds are available in Supplementary Materials.

Acknowledgments: The authors acknowledge Ke Li (Yantai Institute of Coastal Zone Research, Chinese Academy of Sciences) for her technical support of HRESIMS.

Conflicts of Interest: The authors declare no conflict of interest.

References

1. Phoulivong, S.; Cai, L.; Chen, H.; McKenzie, E.H.C.; Abdelsalam, K.; Chukeatirote, E.; Hyde, K.D. Colletotrichum gloeosporioides is not a common pathogen on tropical fruits. *Fungal Divers.* **2010**, *44*, 33–43. [CrossRef]
2. Wang, B.; Li, B.H.; Wang, C.X.; Zhang, Z.F. Effects of temperature, wetness duration, and moisture on the conidial germination, infection, and disease incubation period of *Glomerella cingulate*. *Plant Dis.* **2015**, *99*, 249–256. [CrossRef] [PubMed]
3. Figueroa, M.; Hammond-Kosack, K.E.; Solomon, P.S. A review of wheat diseases-a field perspective. *Mol. Plant Pathol.* **2018**, *19*, 1523–1536. [CrossRef] [PubMed]
4. Goswami, R.S.; Kistler, H.C. Heading for disaster: *Fusarium graminearum* on cereal crops. *Mol. Plant Pathol.* **2004**, *5*, 515–525. [CrossRef]
5. Freeman, J.; Ward, E. *Gaeumannomyces graminis*, the take-all fungus and its relatives. *Mol. Plant Pathol.* **2004**, *5*, 235–252. [CrossRef]
6. Murray, S.A.; Kohli, G.S.; Farrell, H.; Spiers, Z.B.; Place, A.R.; Dorantes-Aranda, J.J.; Ruszczyk, J. A fish kill associated with a bloom of *Amphidinium carterae* in a coastal lagoon in Sydney, Australia. *Harmful Algae* **2015**, *49*, 19–28. [CrossRef]
7. Rensel, J.E.J.; Haigh, N.; Tynan, T.J. Fraser river sockeye salmon marine survival decline and harmful blooms of *Heterosigma akashiwo*. *Harmful Algae* **2010**, *10*, 98–115. [CrossRef]
8. Matsuyama, Y. Impacts of the harmful dinoflagellate *Heterocapsa circularisquama* bloom on shellfish aquaculture in Japan and some experimental studies on invertebrates. *Harmful Algae* **2012**, *14*, 144–155. [CrossRef]
9. Lin, J.N.; Yan, T.; Zhang, Q.C.; Wang, Y.F.; Liu, Q.; Zhou, M.J. In situ detrimental impacts of *Prorocentrum donghaiense* blooms on zooplankton in the East China Sea. *Mar. Pollut. Bull.* **2014**, *88*, 302–310. [CrossRef]
10. Zeilinger, S.; Gruber, S.; Bansal, R.; Mukherjee, P.K. Secondary metabolites in *Trichoderma*—Chemistry meets genomics. *Fungal Biol. Rev.* **2016**, *30*, 74–90. [CrossRef]
11. Li, M.-F.; Li, G.-H.; Zhang, K.-Q. Non-volatile metabolites from *Trichoderma* spp. *Metabolites* **2019**, *9*, 58. [CrossRef] [PubMed]
12. Keswani, C.; Mishra, S.; Sarma, B.K.; Singh, S.P.; Singh, H.B. Unraveling the efficient applications of secondary metabolites of various *Trichoderma* spp. *Appl. Microbiol. Biotechnol.* **2014**, *98*, 533–544. [CrossRef] [PubMed]
13. Zhang, J.-L.; Tang, W.-L.; Huang, Q.-R.; Li, Y.-Z.; Wei, M.-L.; Jiang, L.-L.; Liu, C.; Yu, X.; Zhu, H.-W.; Chen, G.-Z.; et al. *Trichoderma*: A treasure house of structurally diverse secondary metabolites with medicinal importance. *Front. Microbiol.* **2021**, *12*, 723828. [CrossRef]
14. Song, F.; Dai, H.; Tong, Y.; Ren, B.; Chen, C.; Sun, N.; Liu, X.; Bian, J.; Liu, M.; Gao, H.; et al. Trichodermaketones A-D and 7-*O*-methylkoninginin D from the marine fungus Trichoderma koningii. *J. Nat. Prod.* **2010**, *73*, 806–810. [CrossRef]
15. Miao, F.-P.; Liang, X.-R.; Yin, X.-L.; Wang, G.; Ji, N.-Y. Absolute configurations of unique harziane diterpenes from *Trichoderma* species. *Org. Lett.* **2012**, *14*, 3815–3817. [CrossRef] [PubMed]
16. Jiang, M.; Wu, Z.; Guo, H.; Liu, L.; Chen, S. A review of terpenes from marine-derived fungi: 2015-2019. *Mar. Drugs* **2020**, *18*, 321. [CrossRef] [PubMed]
17. Shi, Z.-Z.; Liu, X.-H.; Li, X.-N.; Ji, N.-Y. Antifungal and antimicroalgal trichothecene sesquiterpenes from the marine algicolous fungus *Trichoderma brevicompactum* A-DL-9-2. *J. Agric. Food Chem.* **2020**, *68*, 15440–15448. [CrossRef]
18. Song, Y.-P.; Shi, Z.-Z.; Miao, F.-P.; Fang, S.-T.; Yin, X.-L.; Ji, N.-Y. Tricholumin A, a highly transformed ergosterol derivative from the alga-endophytic fungus trichoderma asperellum. *Org. Lett.* **2018**, *20*, 6306–6309. [CrossRef]
19. Wang, X.-Y.; Xu, T.-T.; Sun, L.-J.; Cen, R.-H.; Su, S.; Yang, X.-Q.; Yang, Y.-B.; Ding, Z.-T. The chemical diversity, the attractant, anti-acetylcholinesterase, and antifungal activities of metabolites from biocontrol *Trichoderma harzianum* uncovered by OSMAC strategy. *Bioorg. Chem.* **2021**, *114*, 105148. [CrossRef]
20. Li, B.; Huang, Q.-X.; Gao, D.; Liu, D.; Ji, Y.-B.; Liu, H.-G.; Lin, W.-H. New C_{13} lipids from the marine-derived fungus *Trichoderma harzianum*. *J. Asian Nat. Prod. Res.* **2015**, *17*, 468–474. [CrossRef]
21. Xu, L.; Zhao, Q.; Yu, H.; Wang, J.; Wang, H.; Yang, Q.; Zhu, H.; Li, Y. Absolute configuration determination of one new compound trichoderol A from *Trichoderma* sp. fungus. *Chem. J. Chin. Univ.-Chin.* **2016**, *37*, 1972–1976.
22. Hoye, T.R.; Eklov, B.M.; Jeon, J.; Khoroosi, M. Sequencing of three-component olefin metatheses: Total synthesis of either (+)-gigantecin or (+)-14-deoxy-9-oxygigantecin. *Org. Lett.* **2006**, *8*, 3383–3386. [CrossRef] [PubMed]
23. Pandit, S.; Adhikari, A.S.; Majumdar, N. Iridium-catalyzed enantioselective ring opening of alkenyl oxiranes by unactivated carboxylic acids. *Org. Lett.* **2022**, *24*, 7388–7393. [CrossRef] [PubMed]

24. Leblanc, Y.; Fitzsinnons, B.J.; Adams, J.; Perez, F.; Rokach, J. The total synthesis of 12-HETE and 12,20-DiHETE. *J. Org. Chem.* **1986**, *51*, 789–793. [CrossRef]
25. Bae, M.; Kim, H.; Shin, Y.; Kim, B.Y.; Lee, S.K.; Oh, K.-B.; Shin, J.; Oh, D.-C. Separacenes A-D, novel polyene polyols from the marine actinomycete, *Streptomyces* sp. *Mar. Drugs* **2013**, *11*, 2882–2893. [CrossRef]
26. Wang, X.; Yang, Y.; An, B.; Wu, J.; Li, Y.; Bian, Q.; Wang, M.; Zhong, J. Asymmetric synthesis of sex pheromone of the western hemlock looper, *Lambdina fiscellaria lugubrosa* (Hulst). *Tetrahedron Lett.* **2023**, *118*, 154401. [CrossRef]

marine drugs

MDPI

Article

Pentaketides and 5-*p*-Hydroxyphenyl-2-pyridone Derivative from the Culture Extract of a Marine Sponge-Associated Fungus *Hamigera avellanea* KUFA0732

Rotchana Klaram [1,2], Tida Dethoup [2,*], Fátima P. Machado [3], Luís Gales [1,4,*], Decha Kumla [5], Salar Hafez Ghoran [6], Emília Sousa [7], Sharad Mistry [8], Artur M. S. Silva [9] and Anake Kijjoa [1,3,*]

1 ICBAS-Instituto de Ciências Biomédicas Abel Salazar, Rua de Jorge Viterbo Ferreira 228, 4050-313 Porto, Portugal; mmayrotchana1993@gmail.com
2 Department of Plant Pathology, Faculty of Agriculture, Kasetsart University, Bangkok 10240, Thailand
3 Interdisciplinary Centre of Marine Environmental Research (CIIMAR), Terminal de Cruzeiros do Porto de Leixões, Av. General Norton de Matos s/n, 4450-208 Matosinhos, Portugal; maria.mfpm@hotmail.com
4 Instituto de Biologia Molecular e Celular (i3S-IBMC), Universidade do Porto, Rua de Jorge Viterbo Ferreira 228, 4050-313 Porto, Portugal
5 Faculty of Pharmaceutical Sciences, Burapha University, 169 Long Had Bangsaen Rd, Chonburi 20131, Thailand; decha1987@hotmail.com
6 HEJ Research Institute of Chemistry, International Center for Chemical and Biological Sciences, University of Karachi, Karachi 75270, Pakistan; s_hafezghoran@yahoo.com
7 Laboratório de Química Orgânica e Farmacêutica, Departamento de Ciências Químicas, Faculdade de Farmácia, Universidade do Porto, Rua de Jorge Viterbo Ferreira 228, 4050-313 Porto, Portugal; esousa@ff.up.pt
8 Department of Chemistry, University of Leicester, University Road, Leicester LE 7 RH, UK; scm11@leicester.ac.uk
9 Departamento de Química & QOPNA, Universidade de Aveiro, 3810-193 Aveiro, Portugal; artur.silva@ua.pt
* Correspondence: agrtdd@ku.ac.th (T.D.); lgales@ibmc.up.pt (L.G.); ankijjoa@icbas.up.pt (A.K.); Tel.: +66-25791026 (T.D.); +351-918945358 (L.G.); +351-962712474 (A.K.); Fax: +351-206-2232 (A.K.)

Citation: Klaram, R.; Dethoup, T.; Machado, F.P.; Gales, L.; Kumla, D.; Hafez Ghoran, S.; Sousa, E.; Mistry, S.; Silva, A.M.S.; Kijjoa, A. Pentaketides and 5-*p*-Hydroxyphenyl-2-pyridone Derivative from the Culture Extract of a Marine Sponge-Associated Fungus *Hamigera avellanea* KUFA0732. *Mar. Drugs* **2023**, *21*, 344. https://doi.org/10.3390/md21060344

Academic Editor: Marie-Lise Bourguet-Kondracki

Received: 2 May 2023
Revised: 31 May 2023
Accepted: 31 May 2023
Published: 2 June 2023

Abstract: Five undescribed pentaketide derivatives, (*R*)-6,8-dihydroxy-4,5-dimethyl-3-methylidene-3,4-dihydro-1*H*-2-benzopyran-1-one (**1**), [(3*S*,4*R*)-3,8-dihydroxy-6-methoxy-4,5-dimethyl-1-oxo-3,4-dihydro-1*H*-isochromen-3-yl]methyl acetate (**2**), (*R*)-5, 7-dimethoxy-3-((*S*)-(1-hydroxyethyl)-3,4-dimethylisobenzofuran-1(3*H*)-one (**4b**), (*S*)-7-hydroxy-3-((*S*)-1-hydroxyethyl)-5-methoxy-3,4-dimethylisobenzofuran 1(3*H*)-one (**5**), and a *p*-hydroxyphenyl-2-pyridone derivative, avellaneanone (**6**), were isolated together with the previously reported (*R*)-3-acetyl-7-hydroxy-5-methoxy-3,4-dimethylisobenzofuran-1(3*H*)-one (**3**), (*R*)-7-hydroxy-3-((*S*)-1-hydroxyethyl)-5-methoxy-3,4-dimethylisobenzofuran-1(3*H*)-one (**4a**) and isosclerone (**7**), from the ethyl acetate extract of a culture of a marine sponge-derived fungus, *Hamigera avellanea* KUFA0732. The structures of the undescribed compounds were elucidated using 1D and 2D NMR, as well as high-resolution mass spectral analyses. The absolute configurations of the stereogenic carbons in **1**, **4b**, **5**, and **6** were established by X-ray crystallographic analysis. The absolute configurations of C-3 and C-4 in **2** were determined by ROESY correlations and on the basis of their common biosynthetic origin with **1**. The crude fungal extract and the isolated compounds **1**, **3**, **4b**, **5**, **6**, and **7** were assayed for their growth inhibitory activity against various plant pathogenic fungi viz. *Alternaria brassicicola, Bipolaris oryzae, Colletotrichum capsici, C. gloeosporiodes, Curvularia oryzae, Fusarium semitectum, Lasiodiplodia theobromae, Phytophthora palmivora, Pyricularia oryzae, Rhizoctonia oryzae* and *Sclerotium rolfsii*.

Keywords: *Hamigera avellanea*; Aspergillaceae; marine sponge-associated fungus; pentaketides; dihydrochromone; 5-*p*-hydroxy-2-pyridone; anti-plant pathogenic fungal activity

1. Introduction

The fungi of the genus *Hamigera* (Family Aspergillaceae, Order Eurotiales) are widespread and common soil fungi, despite one species being associated with beetles [1]. Although many species of Eurotiales are rich sources of secondary metabolites due to the high number of secondary metabolite gene clusters present in their genomes, only a few chemical classes of secondary metabolites have been reported from the genus *Hamigera* [2].

Yamazaki et al. isolated two new cyclic pentapeptides, avellanins A and B, from *Hamigera avellanea* Stolk and Samson [3] while Breinholt et al. described the isolation of (Z, Z)-*N,N′*-[1-[(4-hydroxyphenyl)methylene]-2-[(4-methoxyphenyl)methylene]-1,2-ethanediyl]bis-formamide from an ethanol extract of the mycelium of *H. avellanea* (CBS 501.94) [4]. Breinholt's group later noticed the inhibitory activity of the culture filtrates of the same fungus on the growth of the rice blast fungus, *Pyricularia oryzae*, so they used bioassay-guided fractionation to isolate two polyketides, hamigerone, and dihydrohamigerone. Hamigerone and dihydrohamigerone were tested for the in vitro growth inhibitory activity against plant pathogenic fungi, *P. oryzae* and *Venturia inaequalis*, and the results showed that a growth inhibitory activity of hamigerone against both pathogenic fungi was comparable to the commercial fungicides Prochloraz® and Bitertanol ® whereas dihydrohamigerone exhibited only marginal activity. However, when hamigerone was tested for its ability to protect rice plants against subsequent infection with *P. oryzae*, it was found to be far inferior to commercial fungicides [5]. Isaka et al. have described the isolation of two novel cyclopropyl diketones, hamavellones A and B, and two new 14-membered macrolactones, hamigeromycins A and B, together with a resorcyclic lactone, 89-250904-F1 (radicicol analog A), a spiro-heterocyclic γ-lactam alkaloid, pseurotin A, and anthraquinone pigments emodin, ω-hydroxyemodin, and emodin bianthrones from the ethyl acetate (EtOAc) extract of the culture broth and the mycelium of *H. avellanea* BCC 17816, isolated from a soil sample collected in Chiangmai, Thailand [6]. The Isaka group later isolated from the same fungus, hamigeromycins A and C–G [7].

On the other hand, in an attempt to use biological control as an alternative to commercial synthetic fungicides used in conventional farming, a member of our group (T. Dethoup) has evaluated the antagonistic activity of crude extracts of several marine-derived fungi against many plant pathogens that are causative agents of diseases of the economic crops [8], especially those that cause rice brown spot, rice dirty panicle [9] and sheath blight diseases [10].

Due to the in vitro inhibitory activity of hamigerone, isolated from *H. avellanea* (CBS 501.94), against the rice blast fungus, *P. oryzae*, we have examined the growth inhibitory activities of the crude EtOAc extract of a marine-derived *H. avellanea* KUFA0732, isolated from the marine sponge *Mycale* sp., which was collected from the coral reef at Samaesan Island in Chonburi province, Thailand, as well as some of the isolated compounds against various plant pathogenic fungi, viz. *Bipolaris oryzae* (brown spot of rice), *Curvularia oryzae* (leaf spot of rice), *Fusarium semitectum* (dirty particle of rice), *P. oryzae* (rice blast), *Rhizoctonia oryzae* (sheath blight of rice), *Alternaria brassicicola* (black spot of Chinese kale), *Colletotrichum capsici* (anthracnose of chili), *C. gloeosporiodes* (anthracnose of mango), *Lasiodiplodia theobromae* (fruit rot of mangosteen), *Phytophthora palmivora* (root and stem rot of durian), and *Sclerotium rolfsii* (stem rot of bean).

Fractionation of the EtOAc extract of the culture *H. avellanea* KUFA0732 using silica gel column chromatography, followed by purification using preparative TLC, a Sephadex LH-20 column, and crystallization, led to the isolation of the undescribed (*R*)-6,8-dihydroxy-4,5-dimethyl-3-methylidene-3,4-dihydro-1*H*-2-benzopyran-1-one (**1**), [(3*S*,4*R*)-3,8-dihydroxy-6-methoxy-4,5-dimethyl-1-oxo-3,4-dihydro-1*H*-isochromen-3-yl]methyl acetate (**2**), (*R*)-5,7-dimethoxy-3-((*S*)-(1-hydroxyethyl)-3,4-dimethylisobenzofuran-1(3*H*)-one (**4b**), (*S*)-7-hydroxy-3-((*S*)-1-hydroxyethyl)-5-methoxy-3,4-dimethylisobenzofuran 1(3*H*)-one (**5**), and avellaneanone (**6**), together with the previously reported (*R*)-3-acetyl-7-hydroxy-5-methoxy-3,4-dimethylisobenzofuran-1(3*H*)-one (**3**) [11–13], (*R*)-7-hydroxy-3-((*S*)-1-hydroxyethyl)-5-methoxy-3,4-dimethylisobenzofuran-1(3*H*)-one (**4a**) [13] and isosclerone (**7**) [14–16]

(Figure 1). The structures of the undescribed compounds were established based on an extensive analysis of the 1D and 2D NMR spectra as well as HRMS data. In the case of **1**, **4b**, **5**, and **6**, the absolute configurations of their stereogenic carbons were established by X-ray analysis, while the absolute configurations of the stereogenic carbons in **2** were determined by ROESY correlations and comparison of its ^{1}H NMR data with those of **1**.

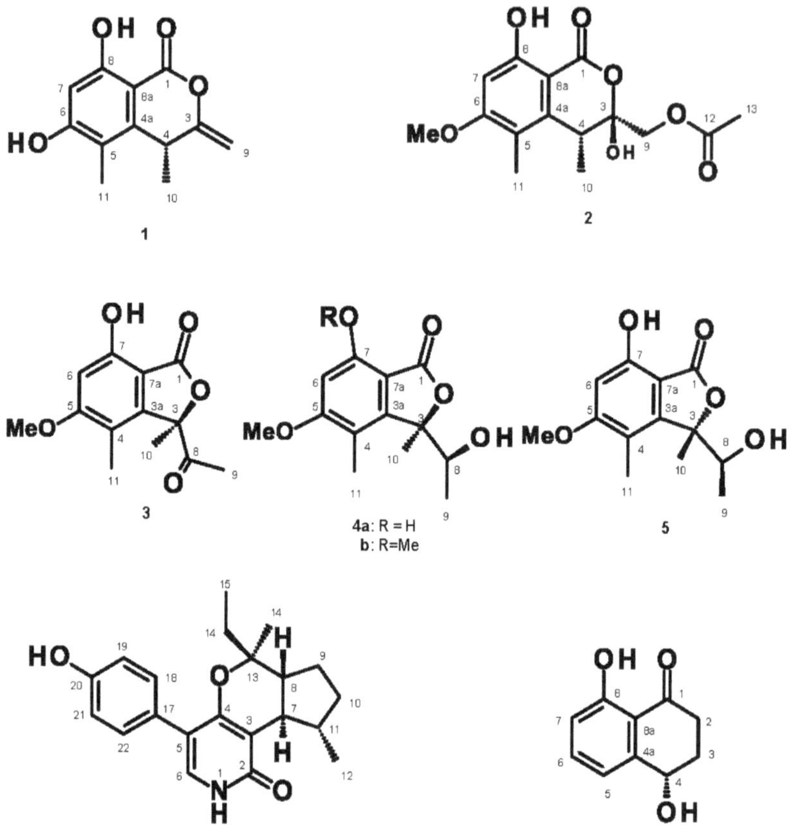

Figure 1. Structures of **1–7**.

2. Results and Discussions

The structures of (*R*)-3-acetyl-7-hydroxy-5-methoxy-3,4-dimethylisobenzofuran-1(3*H*)-one (**3**) [11–13], (*R*)-7-hydroxy-3-((*S*)-1-hydroxyethyl)-5-methoxy-3,4-dimethylisobenzofuran-1(3*H*)-one (**4a**) [13] and isosclerone (**7**) [14–16] were elucidated by analysis of their 1D and 2D NMR spectra (Figures S14–S23 and S42–S46, Tables S1, S2 and S7) and by comparison of their NMR spectral data and the sign of their optical rotations with those reported in the literature.

Compound **1** was isolated as white crystals (mp = 217–218 °C), and its molecular formula $C_{12}H_{12}O_4$ was established based on (+)-HRESIMS *m/z* 221.0814 [M + H]$^+$ (calculated for $C_{12}H_{13}O_4$, 221.0814) (Figure S6), requiring seven degrees of unsaturation. The ^{13}C NMR spectrum (Table 1, Figure S2) exhibited twelve carbon signals which, in combination with DEPT and HSQC spectra (Figure S4), can be classified as one conjugated lactone carbonyl (δ_C 166.3), six non-protonated sp^2 (δ_C 164.1, 161.8, 157.4, 143.6, 113.2, 97.7), one protonated sp^2 (δ_C 101.0), one methylene sp^2 (δ_C 96.2), one methine sp^3 (δ_C 34.2) and two methyl (δ_C 22.3 and 10.2) carbons. The ^{1}H NMR spectrum (Table 1, Figure S1) displayed a broad singlet of a hydrogen-bonded hydroxyl proton at δ_H 10.70, one aromatic singlet at δ_H 6.33, two doublets of the olefinic protons at δ_H 4.75 (J = 1.6 Hz) and 4.73 (J = 1.6 Hz), one methine quartet at δ_H 4.02 (J = 7.2 Hz), one methyl singlet at δ_H 2.02 and one methyl doublet at δ_H 1.27 (J = 7.0 Hz). The type and chemical shift values of the carbon atoms revealed that **1** is a 3, 4, 5, 6, 8-pentasubstituted chromone. That the hydroxyl group at δ_H 10.70 was on C-8, the methyl group at δ_H 2.02, s (δ_C 22.3, Me-11) was on C-5, and another hydroxyl group was on C-6 was evidenced by the chemical shift value of the hydrogen-bonded OH-8 as well as by the HMBC correlations (Table 1, Figures 2 and S5) from H_3-11 to C-5 (δ_C 113.2), C-4a (δ_C 143.6), 164.1 (C-6), and H-7 (δ_H 6.33, s) to C-5, C-6, C-8 (δ_C 161.8) and C-8a (δ_C 97.7). The COSY correlation from the methyl doublet at δ_H 1.27 (J = 7.0 Hz, Me-10) to the quartet at δ_H 4.02 (J = 7.2 Hz/δ_H 34.2) (Table 1, Figures 2 and S3) revealed that Me-10 was on C-4. This was supported by HMBC correlations from H_3-10 to C-4, C-4a, and C-3 (δ_C 157.4), and H-4 to C-4a, C-5, C-8a, and C-11. That the methylidene group was on C-3 was supported by HMBC correlations from the two doublets of the olefinic protons at δ_H 4.73 (J = 1.6 Hz/δ_C 96.2) and δ_H 4.75 (J = 1.6 Hz/δ_H 96.2) to C-3 and C-4 (Table 1, Figures 2 and S5). Taking together the HRMS and 1D and 2D NMR data, a planar structure of **1** was elucidated as 6,8-dihydroxy-4,5-dimethyl-3-methylidene-3,4-dihydro-1*H*-2-benzopyran-1-one.

Table 1. ^{1}H and ^{13}C NMR data (DMSO-d_6, 300 and 75 MHz), COSY, and HMBC for **1**.

Position	δ_C, Type	δ_H, (J in Hz)	COSY	HMBC
1	166.3, C			
3	157.4, C			
4	34.2, CH	4.02 q (7.2)	H_3-10	C-4a, 5, 8a, 11
4a	143.6, C			
5	113.2, C			
6	164.1, C			
7	101.0, CH	6.33,s	H_3-11	C-5, 6, 8, 8a
8	161.8, C			
8a	97.7, C			
9	96.2, CH_2	4.75, d (1.6) 4.73, d (1.6)		C-3, 4
10	10.2, CH_3	1.27, d (7.0)	H-4	C-3, 4, 4a
11	22.3 CH_3	2.02, s	H-7	C-4a, 5, 6
OH-8	-	10.70, brs		

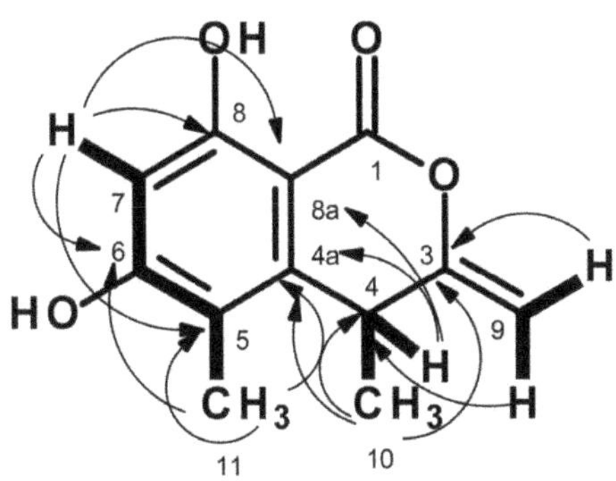

Figure 2. Key COSY (bold line) and HMBC (arrow) correlations in **1**.

Compound **1** has one stereogenic center (C-4), and it is, therefore, necessary to determine the absolute configuration of this carbon. Since **1** could be obtained as a suitable crystal for X-ray analysis using a diffractometer equipped with CuKα radiation, its stereostructure was obtained. The Ortep view of **1**, shown in Figure 3, revealed that the configuration of C-4 is *R*. Therefore, the absolute structure of **1** was elucidated as (*R*)-6,8-dihydroxy-4,5-dimethyl-3-methylidene-3,4-dihydro-1*H*-2-benzopyran-1-one.

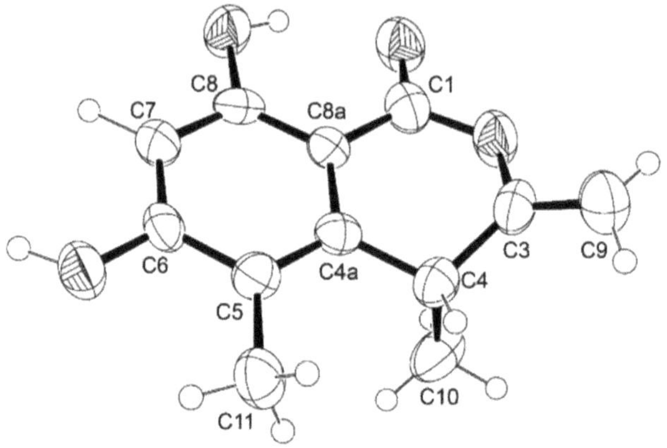

Figure 3. Ortep view of **1**.

A literature search revealed that **1** has never been previously reported; however, a structurally similar compound, but with opposite stereochemistry at C-4, i.e. (*S*)-8-hydroxy-6-methoxy-4,5-dimethyl-3-methyleneisochromen-1-one, was reported by Tayone et al. [17] from the EtOAc extract of the culture broth of *Leptosphaeria* sp. KTC 727, which was collected from woody debris in Rebun Island, Hokkaido, Japan. The absolute configuration at C-4 was determined as *S* by comparison of the calculated and experimental ECD spectra. Later, the same compound was also isolated from the mycelium and culture broth extracts of the fungus *Xylaria* sp. BL321, isolated from the leaves of the mangrove plant *Acanthus ilicifolius* L., was collected in Guangdong, China. The absolute configuration at C-4 was also determined as *S* by comparison of the calculated and experimental ECD spectra [18].

Compound **2** was isolated as a pale yellow viscous mass. The molecular formula $C_{15}H_{18}O_7$ of **2** was obtained from the (+)-HRESIMS *m/z* 311.1133 [M + H]+ (calculated for $C_{15}H_{19}O_7$, 311.1131) (Figure S13), indicating seven degrees of unsaturation. The ^{13}C NMR spectrum (Table 2, Figure S8) displayed 15 carbon signals which, in conjunction with DEPT and HSQC spectra (Figure S10), can be classified as one ester carbonyl (δ_C 170.7), one conjugated ester carbonyl (δ_C 168.2), two oxygen-bearing sp^2 (δ_C 164.9 and 163.1), four non-protonated sp^2 (δ_C 141.5, 115.5, 102.2 and 99.0), one pronated sp^2 (δ_C 97.5), one oxymethylene sp^3 (δ_C 65.9), one methoxy (δ_C 55.9), one methine sp^3 (δ_C 35.9) and three methyl (δ_C 20.8, 16.2, and 10.1) carbons. The ^{1}H NMR spectrum (Table 2, Figure S7), in combination with the HSQC spectrum (Figure S10), exhibited a singlet of a hydrogen-bonded hydroxyl proton at δ_H 11.14, one aromatic singlet at δ_H 6.37 (δ_C 97.5), two terminally coupled doublets at δ_H 4.25 (J = 11.8 Hz/δ_C 65.9) and 4.59 (J = 11.8 Hz/δ_C 65.9), one methoxyl singlet at δ_H 3.85 (δ_C 55.9), one quartet at δ_H 4.33 (J = 7.1 Hz/δ_C 35.9), two methyl singlets at δ_H 2.19 (δ_C 20.8) and 2.08 (δ_C 10.1) and one methyl doublet at δ_H 1.19 (J = 7.1 Hz/δ_C 16.2). The presence of a 4,5-dimethyl-8-hydroxy-6-methoxy-3,4-dihydro-1*H*-2-benzopyran-2-one was substantiated by HMBC correlations from the singlet at δ_H 11.14 (OH-8) to the carbons at δ_C 163.1 (C-8) and 97.5 (C-7), the singlet at $\delta_{Hz/}$6.37 (H-7) to C-8, the carbons at δ_C 115.5 (C-5) and 99.0 (C-8a), a methyl singlet at δ_H 2.08 (H$_3$-11) to the carbon at δ_C 164.9 (C-6), 141.5 (C-4a) and 115.5 (C-5), and the methyl doublet at δ_H 1.19 (H$_3$-10) to C-4a, the carbons at δ_C 102.2 (C-3) and 35.9 (C-4) (Table 2, Figures 4 and S11). The chemical shift value of C-4 is characteristic of a hemiketal carbon, and thus one of its substituents must be a hydroxyl group. Another substituent on C-3 was an acetoxymethyl group, supported by the HMBC correlations from H_2-9 (δ_H 4.25, d, J = 11.8 Hz) and 4.59, d, J = 11.8 Hz) to C-3 and the carbonyl of the acetyl group at δ_C 170.7 (C-12) and the methyl protons of the acetyl group (δ_H 2.19) to C-12 (Table 2, Figures 4 and S11).

Table 2. ^{1}H and ^{13}C NMR data (DMSO-d_6, 300 and 75 MHz), COSY, HMBC, and ROESY for **2**.

Position	δ_C, Type	δ_H, (J in Hz)	COSY	HMBC	ROESY
1	168.2, CO				
3	102.2, C				
4	35.9, CH	4.33, q (7.1)	H_3-10,	C-3, 4a, 5, 8a, 10	H_3-10, H-11
4a	141.5, C				
5	115.5, C				
6	164.9, C				
7	97.5, CH	6.37, s		C-3, 5, 8, 8a, 12	OMe-6
8	163.1, C				
8a	99.0, C				
9a b	65.9, CH_2	4.25, d (11.8) 4.59, d (11.8)	H-9b H-9a	C-12 C-3, 12	H_3-10 H_3-10
10	16.2, CH_3	1.19, d (7.1)	H-4	C-3, 4, 4a	H-9a, 9b
11	10.1, CH_3	2.08, s		C-3, 4a, 6	H-4
12	170.7, CO				
13	20.8, CH_3	2.19, s		C-12	
OMe-6	55.9, CH_3	3.85, s		C-6	H-7
OH-8	-	11.14, s		C-7, 8	

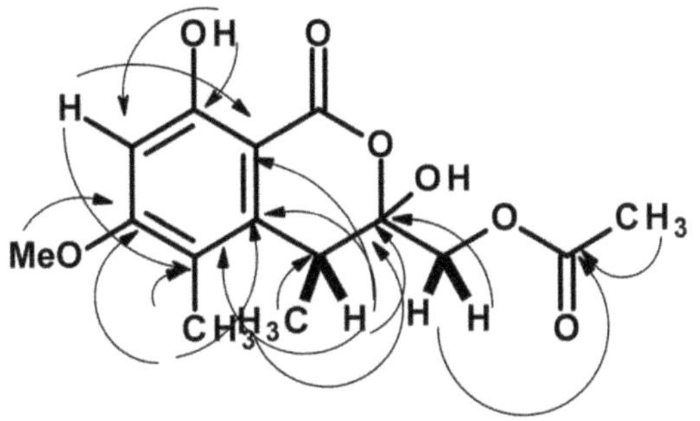

Figure 4. Key COSY (bold line) and HMBC (arrow) correlations in **2**.

Therefore, the planar structure of **2** was elucidated as (3,8-dihydroxy-6-methoxy-4,5-dimethyl-1-oxo-3,4-dihydro-1*H*-isochromen-3-yl)methyl acetate.

Compound **2** has two stereogenic carbons (C-3 and C-4); therefore, it is necessary to establish their absolute configurations. Since **2** is a viscous mass and could not be obtained as a suitable crystal for X-ray diffraction, the ROESY spectrum was obtained. The ROESY spectrum (Table 2, Figure S12) showed strong correlations from H-4 to H_3-10 and H_3-11, confirming that H-4 is β and in the equatorial position while Me-10 is α and in the axial position. Thus, the stereochemistry at C-4 in **2** is the same as in **1**. Moreover, H_3-10 also showed a strong correlation to H-9a and H-9b, while H-9a only showed a weak correlation to H-4. These ROESY correlations revealed that CH_2-9 is α and OH-3 is β. Therefore, based on the absolute configuration at C-4 of **1**, the absolute configurations at C-3 and C-4 in **2** are established as 3*S*, 4*R*.

Biosynthetically, **1** and **2** are pentaketides, and **2** could be considered to derive directly from **1** according to the proposed pathways shown in Figure 5. A Claisen condensation of one acetyl CoA (**I**) with four malonyl CoA (**II**) gives a pentaketide (**III**), which undergoes cyclization and is followed by enolization to give **IV**, and further dehydration of **IV** gives rise to **V**. Stereospecific methylation at C-3 and C-5 of **V** by *S*-adenosyl methionine (SAM) gives rise to **1**. Epoxidation of the exocyclic double bond of **1** gives an epoxide in **VI**. Enzymatic hydrolysis of the epoxide and methylation of a phenolic hydroxyl group in **VI** produces **VII**, which undergoes acetylation by acetyl CoA to give **2**. Consequently, the absolute configuration at C-4 in **2** must be the same as that of C-4 in **1**, i.e., 4*R*.

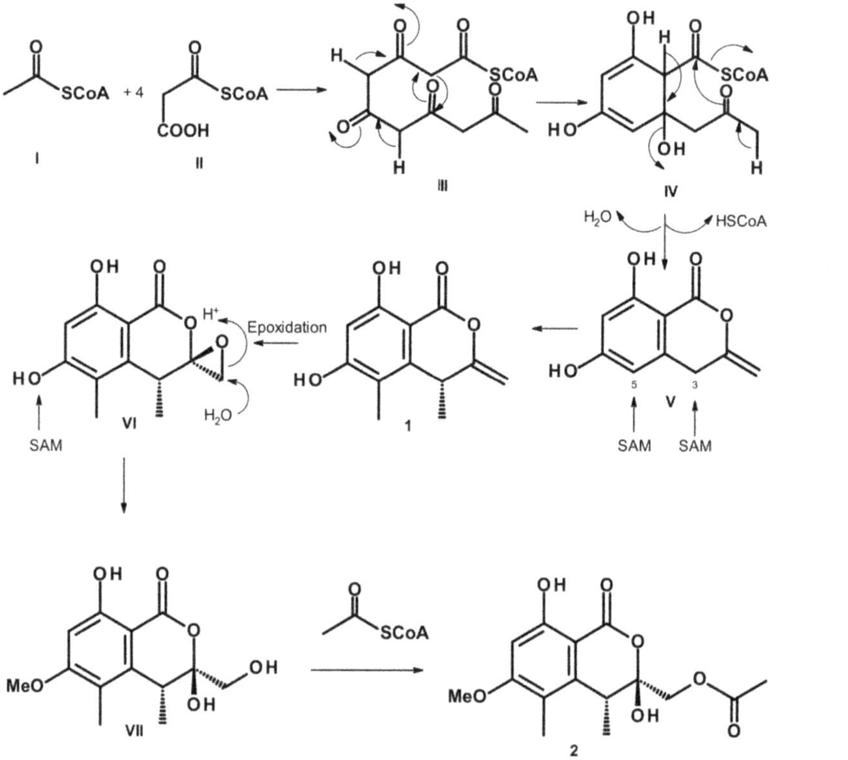

Figure 5. Common biosynthetic pathways of **1** and **2**.

Interestingly Tayone et al. [17] have described the isolation of a mixture of two diastereomers of 3,8-dihydroxy-3-hydroxymethyl-6-methoxy-4,5-dimethylisochroman-1-one, whose structures are similar to **VII**, except for the stereochemistry at C-4 which is opposite to that of **VII**, from the EtOAc extract of the culture broth of *Leptosphaeria* sp. KTC 727. These authors have detected the difference in the ^{1}H and ^{13}C NMR chemical shift values between the major component (3*R*, 4*S*) and the minor component (3*S*, 4*S*) in the mixture but were not able to isolate them individually. On the contrary, the ^{1}H and ^{13}C NMR spectra (Figures S7 and S8) of **2** displayed clearly only one set of signals, implying that **2** is not a mixture of two diastereomers.

^{1}H and ^{13}C NMR data (Table S2, Figures S19 and S20) of **4a** resemble those of (*R*)-7-hydroxy-3-((*R*)-1-hydroxyethyl)-5-methoxy-3,4-dimethylisobenzofuran-1(3*H*)-one, isolated from the culture extract of a marine mangrove-associated fungus, *Xylaria* sp. BL321 [18] as well as (*R*)-7-hydroxy-3-((*S*)-1-hydroxyethyl)-5-methoxy-3,4-dimethylisobenzofuran 1(3*H*)-one, isolated from the culture extract of the fungus, *Leptosphaeria* sp. KTC 727, collected from woody debris at Akaiwa, Rebun island in Hokkaido, Japan [13]. Table S3 compared ^{1}H and ^{13}C NMR data of **4a** with those of both (*R*)-7-hydroxy-3-((*R*)-1-hydroxyethyl)-5-methoxy-3,4-dimethylisobenzofuran 1(3*H*)-one and (*R*)-7-hydroxy-3-((*S*)-1-hydroxyethyl)-5-methoxy-3,4-dimethylisobenzofuran 1(3*H*)-one. From the ^{1}H and ^{13}C NMR chemical shift values in Table S3, it is not possible to determine if the absolute configurations at C-3 and C-8 in **4a** are 3*R*, 8*R*, or 3*R*, 8*S*. However, the two stereoisomers differ in the sign of the optical rotations. While (*R*)-7-hydroxy-3-((*R*)-1-hydroxyethyl)-5-methoxy-3,4-dimethylisobenzofuran 1(3*H*)-one is dextrorotatory ($[\alpha]^{25}_D$ + 355.4, c 0.05, MeOH), (*R*)-7-hydroxy-3-((*S*)-1-hydroxyethyl)-5-methoxy-3,4-dimethylisobenzofuran 1(3*H*)-one is levorotatory ($[\alpha]^{25}_D$ −27, c 0.02, $CDCl_3$) [13]. Since **4a** is levorotatory ($[\alpha]^{25}_D$ −73, c 0.06, MeOH), we concluded that **4a** is (*R*)-7-hydroxy-3-((*S*)-1-hydroxyethyl)-5-methoxy-3,4-dimethylisobenzofuran 1(3*H*)-one.

Compound **4b** was isolated as white crystals (mp = 174–176 °C), and its molecular formula $C_{14}H_{18}O_5$ was determined based on the (+)-HRESIMS *m/z* 267.1233 $[M + H]^+$ (calculated for $C_{14}H_{19}O_5$, 267.1232) (Figure S29), requiring six degrees of unsaturation. The ^{1}H and ^{13}C NMR spectra (Table 3, Figures S24 and S25) are quite similar to those of **4a**, except for the presence of an extra methoxyl substituent (δ_H 3.96, s/δ_C 56.1) on the aromatic ring and the lack of a broad signal of the phenolic hydroxyl group around 7 ppm. Thus, the structure of **4b** was deduced as a methoxy derivative of **4a**. The correlations observed in the COSY, HSQC, and HMBC spectra (Table 3, Figures S26–S28) allowed the establishment of the planar structure of **4b** as 5, 7-dimethoxy-3-(1-hydroxyethyl)-3,4-dimethylisobenzofuran 1(3*H*)-one.

Table 3. ^{1}H and ^{13}C NMR data (DMSO-d_6, 300 and 75 MHz), COSY, and HMBC for **4b**.

Position	δ_C, Type	δ_H, (J in Hz)	COSY	HMBC
1	168.1, CO			
3	88.7, C			
3a	152.8, C			
4	111.7, C			
5	164.5, C			
6	94.5, CH	6.04, s	OMe-5, H_3-11	C-1, 4, 5, 7, 7a
7	158.3, C			
7a	105.4, C			
8	70.9, CH	4.20, m	H_3-9	
9	17.8, CH_3	0.86, d (6.5)	H-8	C-3, 8
10	21.5, CH_3	1.74, s		C-3, 3a, 8
11	11.2, CH_3	2.10, s		C-3a, 4, 5
OMe-5	56.0, CH_3	3.91, s		C-5
OMe-7	56.1, CH_3	3.96, s		C-7

Since **4b** is also levorotatory ($[\alpha]^{25}_D$ −300, c 0.05, MeOH) similar to **4a** ($[\alpha]^{25}_D$ −73, c 0.06, MeOH), we concluded that the absolute configurations at C-3 and C-8 in **4b** are 3*R* and 8*S* also. After many attempts, we were finally able to obtain **4b** as a suitable crystal and an X-ray analysis was performed. The Ortep view of **4b** (Figure 6) confirmed the stereochemistry of C-3 and C-8 as 3*R* and 8*S*. A literature search revealed that **4b** has never been previously reported.

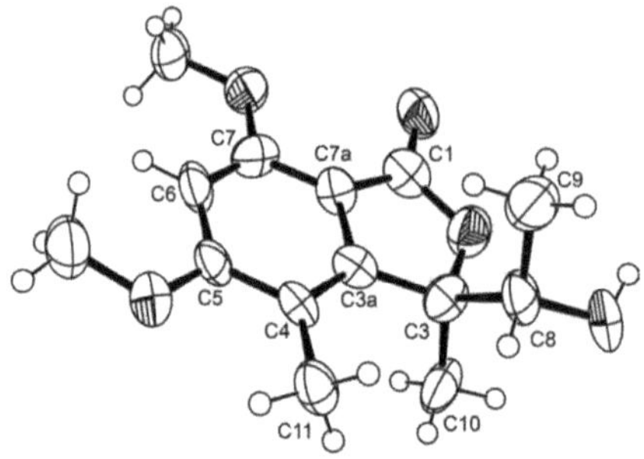

Figure 6. Ortep diagram of **4b**.

Compound **5** was isolated as a white crystal (mp = 116–118 °C). The (+)-HRESIMS spectrum showed the $[M + H]^+$ peak at m/z 253.1077 (calcd for $C_{13}H_{17}O_5$, 253.1076) (Figure S35), revealing the same molecular formula as **4a**, i.e., $C_{13}H_{16}O_5$. Therefore, **5** is an isomer of **4a**. The ^1H and ^{13}C NMR spectra (Table 4, Figures S30 and S31) of **5** resembled those of **4a**; however, the protons of the three methyl groups in **5** showed different chemical shift values from those in **4a**. Analysis of the COSY, HSQC, and HMBC spectra (Table 4, Figures S32–S34) revealed that the planar structure of **5** is the same as that of **4a**.

Table 4. ^1H and ^{13}C NMR data (DMSO-d_6, 300 and 75 MHz), COSY and HMBC for **5**.

Position	δ_C, Type	δ_H, (J in Hz)	COSY	HMBC
1	171.1, CO			
3	91.6, C			
3a	149.8, C			
4	112.4, C			
5	165.3, C			
6	98.2, CH	6.43, s	H_3-11, OMe-5	C-1, 4,7, 7a
7	156.3, C			
7a	103.7, C			
8	70.5, CH	4.26, m	H_3-9	
9	17.9, CH_3	1.40, d (6.4)		C-3, 8
10	21.1, CH_3	1.67, s		C-3, 3a, 8
11	11.1, CH_3	2.17,s	H-6	C-3a, 4, 5
OMe-5	56.3, CH_3	3.87, s	H-6	C-5
	OH-7	7.67, brs		

Curiously, the proton chemical shift value of the doublet of Me-9 (δ_H 1.40, J = 6.4 Hz) in **5** is higher than that of **4a** (δ_H 0.89, J = 6.5 Hz) around 0.5 ppm while the proton chemical shift value of its Me-10 singlet (δ_H 1.67) was lower than that of **4a** (δ_H 1.78) just around 0.1 ppm. The differences in chemical shift values of the methyl protons of Me-9 and Me-10 can be attributed to the anisotropic effect of the benzene ring on both Me-9 and Me-10. As the chemical shift value of Me-10 in **5** is higher than that of its counterpart in **4a**, while Me-10 in **5** is lower than that of its counterpart in **4a**, the absolute configuration at C-3 in **5** must be opposite to that at C-3 in **4a**, i.e., 3*S*. Since **5** could be obtained in a suitable crystal, an X-ray analysis was performed. The Ortep view of **5**, shown in Figure 7, not only confirmed the 3*S* configuration but also determined the configuration at C-8 as 8*S*. Therefore, a complete structure of **5** is (*S*)-7-hydroxy-3-((*S*)-1-hydroxyethyl)-5-methoxy-3,4-dimethylisobenzofuran 1(3*H*)-one, which is the enantiomer of (*R*)-7-hydroxy-3-((*R*)-

1-hydroxyethyl)-5-methoxy-3,4-dimethylisobenzofuran 1(3*H*)-one ([α]^{25}$_D$ +355.4, c 00.05, MeOH) [18]. This was also supported by the fact that **5** is levorotatory ([α]^{25}$_D$ −320, c 0.05, MeOH). From the Ortep view (Figure 7), the methyl protons of Me-10 of the α-axial 1-hydroxyethyl substituent should experience a paramagnetic field of the benzene ring, while the protons of the β-equatorial Me-9 should experience a slight diamagnetic effect of the benzene ring. A literature search revealed that **5** also has never been reported.

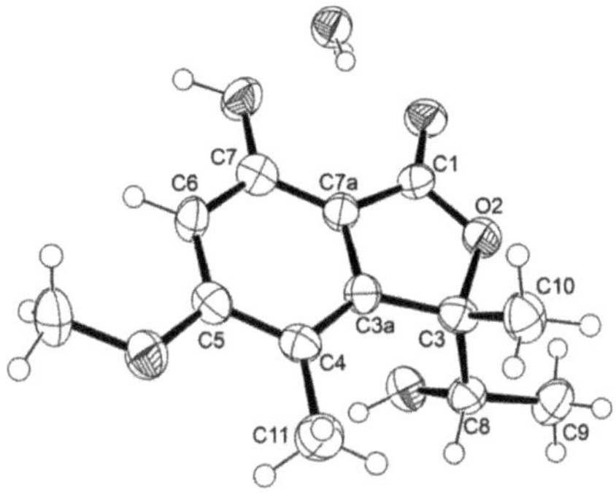

Figure 7. Ortep diagram of **5**.

Analysis of the structures of **3**, **4a**, **4b**, and **5** revealed that they all derived from a cyclization of a linear pentaketide, similar to that of **1** and **2**. Figure 8 depicts proposed biosynthetic pathways for **3**, **4a**, **4b**, and **5**.

The intermediate **VIII**, derived from a cyclization and aromatization of a linear pentaketide **III**, through **IV**, undergoes stereospecific methylation at the benzyl carbon and a methylation at the aromatic carbon and the phenolic hydroxyl group to give **IX**. Stereospecific hydroxylation of the benzyl carbon of **IX** gives rise to **X**, which undergoes a lactonization to give **3**. Stereospecific reduction of the carbonyl ketone of the side chain of **3** produces **4a**. Methylation of another phenolic hydroxyl group of **4a** gives rise to **4b**.

The route leading to the formation of **5** also begins with intermediate **VIII**. In this route, the stereospecific methylation of the benzyl carbon to form **XI** occurs on the oppo face to the methylation to form **IX**. Oxidation of the benzyl carbon and lactonization of **XI** gives rise to **XII**, which, after reduction of the ketone carbonyl of the side chain, leads to the formation of **5**.

Compound **6** was isolated as colorless crystals (mp = 209–210 °C), and its molecular formula $C_{21}H_{25}NO_3$ was established on the basis of (+)-HRESIMS *m/z* 362.1734 [M + Na]$^+$ (calculated for $C_{21}H_{25}NO_3Na$, 362.1732) (Figure S41), requiring ten degrees of unsaturation. The ^{13}C NMR spectrum (Table 5, Figure S37) displayed 19 carbon signals which, in combination with DEPT and HSQC spectra (Figure S39), can be categorized as one conjugated amide carbonyl (δ_C 162.4), two oxygen-bearing sp^2 (δ_C 160.2 and 156.6), three non-protonated sp^2 (δ_C 110.4, 113.6 and 126.7), five protonated sp^2 (δ_C 115.1 (2C), 130.6 (2C), 131.5), one oxyquaternary sp^3 (δ_C 82.2), three methine sp^3 (δ_C 51.9, 43.8, 35.3), three methylene sp^3 (δ_C 34.5, 33.8, 25.2) and three methyl (δ_C 25.1, 18.6, 7.7) carbons. The ^{1}H NMR spectrum (Table 5, Figure S36), in conjunction with the HSQC spectrum (Figure S39), showed two broad singlets at δ_H 9.38 and 10.93, two doublets of four *ortho*-coupled aromatic protons at δ_H 6.73 (J = 8.6 Hz, 2H/δ_C 115.1) and 7.15 (J = 8.6 Hz, 2H/δ_C 130.6), one olefinic singlet at δ_H 6.98 (δ_C 131.5), a double doublet at δ_H 2.06 (J = 12.0, 8.7 Hz/δ_C 43.8), a double of double doublet at δ_H 1.71 (J = 12.0, 12.0, 7.0 Hz/δ_C 51.9), a quartet at δ_H 1.55 (J = 7.1 Hz/δ_C 33.8), one methyl singlet at δ_H 1.11 (δ_C 18.6), one methyl doublet at δ_H 1.47 (J = 6.2 Hz/δ_C 25.1) and one methyl triplet at δ_H 0.85 (J = 7.3 Hz/δ_C 7.7), in addition to various overlapped multiplets. The presence of a 4-hydroxy phenyl moiety was evidenced by the COSY correlation from the doublet at δ_H 7.15 (J = 8.6 Hz/H-18, H-22) to the doublet at δ_H 6.73 (J = 8.6 Hz/H-19, H-21) (Figure S38) as well as the HMBC correlations from H-18 to C-20 (δ_C 156.6) and C-22 (δ_C 130.6), H-22 to C-18 (δ_C 130.6), and C-20, H-19 to

C-17 (δ_C 126.7), and C-21 (δ_C 115.1), and H-21 to C-17, and C-19 (Figure S40). That another portion of **6** consists of a 3,4,5-trisubstituted 2-pyridone was substantiated by the HMBC correlations from H-6 (δ_H 6.98, s) to C-2 (δ_C 162.4), C-4 (δ_C 160.2), and C-5 (δ_C 113.6). Since H-18 and H-22 showed HMBC cross peaks to C-5 while H-6 showed cross peaks to C-17, the *p*-hydroxyphenyl group was on C-5 of the 2-pyridone ring. That the last portion of the molecule was 1-ethyl-1,5-dimethyl-1,4a,5,6,7,7a-hexahydrocyclopenta[*c*]pyran was supported by the COSY correlations from H_3-16 (δ_H 0.85, t, J = 7.3 Hz/δ_C 7.7) to H-15 (δ_H 1.55, q, J = 7.1 Hz/δ_C 33.8), H_3-12 (δ_H 1.47, d, J = 6.2 Hz/δ_C 25.1) to H-11 (δ_H 2.15, m/δ_C 35.3), and a coupling system from H_3-12 through H-11/H-7 (δ_H 2.06, dd, J = 12.0, 8.7 Hz/δ_C 43.8)/H-8 (δ_H 1.71, ddd, J = 12.0, 12.0, 7.0 Hz/δ_C 51.9)/H-9 (δ_H 1.56, m/δ_C 25.2) and H-10 (δ_H 1.39, m and 1.99, m/δ_C 34.5) (Table 5, Figures 8 and S40). This was corroborated by HMBC correlations from H_3-16 to C-13 (δ_C 82.2) and C-15, H_3-14 (δ_H 1.11, s/δ_C 18.6) to C-8, C-13, C-15, H_3-12 to C-7, C-10 and C-11, H-7 to C-11 and C-12, H-8 to C-13 and C-14 (Table 5, Figures 8 and S40). That the 3,4-dihydro-2*H*-pyran is fused with the 2-pyridone ring at C-3 and C-4 was supported by the HMBC correlation from H-7 to H-3 and the chemical shift values of C-4 and C-13 (Table 5, Figures 8 and S40). Taking together the molecular formula, the COSY, and HMBC correlations, the planar structure of **6** was elucidated, as shown in Figure 9.

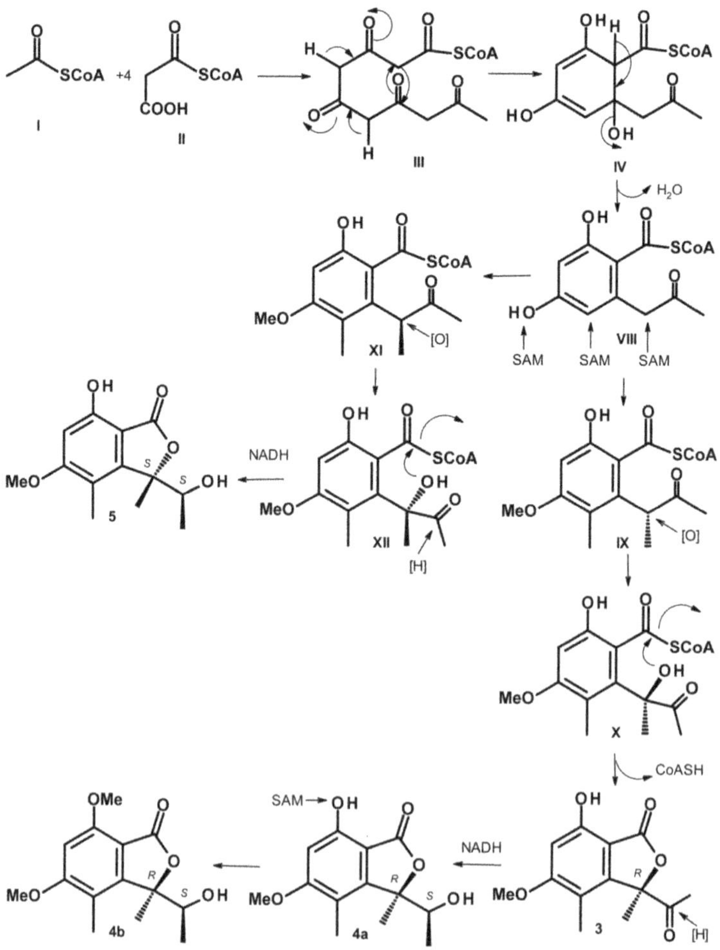

Figure 8. Proposed biosynthetic pathways for **3**, **4a**, **4b**, and **5**.

Table 5. ^{1}H and ^{13}C NMR data (DMSO-d_6, 300 and 75 MHz), COSY and HMBC for **6**.

Position	δ_C, Type	δ_H, (J in Hz)	COSY	HMBC
NH-1	-	10.93, brs		
2	162.4 CO			
3	110.4 C			
4	160.2, C			
5	113.6, C			
6	131.5, CH	6.98,s		C-2, 4.5, 17
7	43.8, CH	2.06, dd (12.0, 8.7)	H-8, 11	C-3, 11, 12
8	51.9, CH	1.71, ddd (12.0, 12.0, 7.0)	H-7, 9	C-13, 14
9	25.2 CH_2	1.56, m	H-8, 10	
10	34.5, CH_2	1.39, m 1.99, m	H-9, 11	C-9, 11
11	35.3, CH	2.15, m	H_3-12	
12	25.1, CH_3	1.47, d (6.2)	H-11	C-7, 10, 11
13	82.2, C	-		
14	18.6, CH_3	1.11, s		C-8, 13, 15
15	33.8, CH_2	1.55, q (7.1)	H_3-16	
16	7.7, CH_3	0.85, t (7.3)	H-15	C-13, 15
17	126.7, C	-		
18	130.6, CH	7.15, d (8.6)	H-19	H-5, 20, 22
19	115.1, CH	6.73, d (8.6)	H-18	H-17, 20, 21
20	156.6, C	-		
21	115.1, CH	6.73, d (8.6)	H-22	H-17, 19, 20
22	130.6, CH	7.15, d (8.6)	H-21	H-5, 18, 20
23	OH-20	9.38, brs		

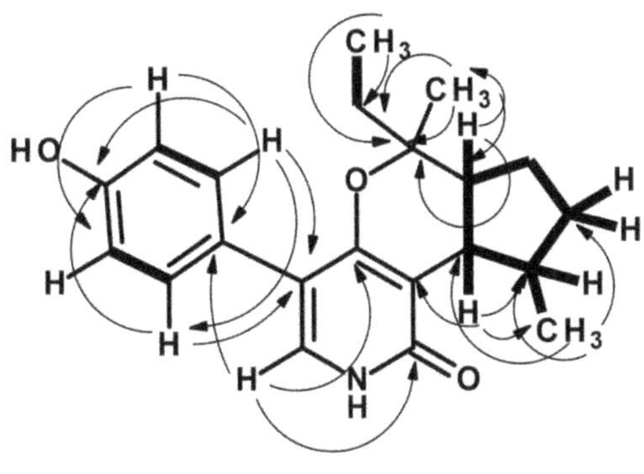

Figure 9. Key COSY (bold lines) and HMBC (arrow) correlations in **6**.

Compound **6** possesses four stereogenic centers viz. C-7, C-8, C-11 and C-13. Therefore, it is challenging to determine the absolute configurations of these carbons. Since we were able to obtain a suitable crystal of **6**, its X-ray analysis was carried out using an X-ray diffractometer equipped with CuKα radiation. The Ortep view of **6** is shown in Figure 10.

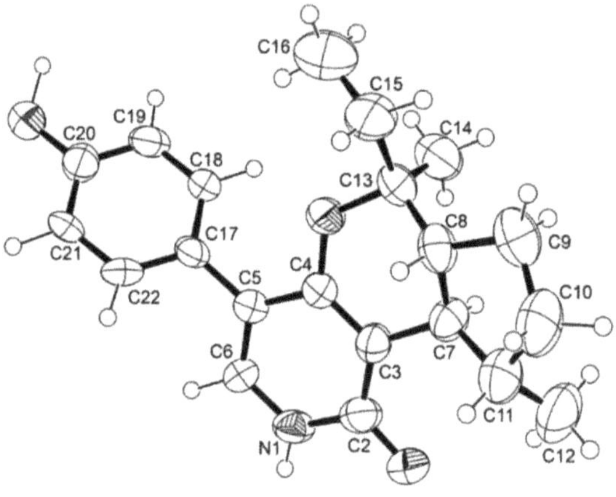

Figure 10. Ortep view of **6**.

The Ortep view showed clearly that the absolute configuration at C-7, C-8, C-12, and C-13 are 7*R*, 8*R*, 12*S*, and 13*S*. Since **6** has never been reported, we named it avellaneanone.

5-*p*-Hydroxy phenyl-2-pyridones with alkyl or alkenyl side chains on C-3 have been previously isolated from different fungal strains. Examples of these are tenellin [19] and desmethylbassianin, isolated from the entomopathogenic fungus, *Beauveria bassiana* [20], and aspyridone-A from *Aspergillus nidulans* [21]. Moreover, the alkyl and alkenyl side chains can undergo cyclization to form a tetrahydropyran ring such as sambutoxin, isolated from the wheat culture of *Fusarium sambucinum* which was obtained from a rot potato tuber [22], and a decalin ring system such as illicicolin H from the fermentation of *Cylindrocladium ilicicola* MFC 870 [23]. Additionally, the alkenyl side chain can form a pyran ring of a hexahydro-1*H*-2-benzopyran in trichodin A, isolated from a marine-derived *Trichoderma* sp. MF106, which was obtained from a Greenland Sea (Fram Strait) sample. However, only the relative configurations of the stereogenic carbons of the hexahydro-1*H*-2-benzopyran in trichodin A were established by analysis of NOESY correlations [24].

The 5-*p*-Hydroxyphenyl-2-pyridone derivatives are polyketide synthase (PKS)-non-ribosomal peptide synthetase (NRPS) metabolites, which were reported to be synthesized from ring expansion of tetrameric acid intermediates catalyzed by cytochrome P450 monooxygenases [25]. Analysis of the structure of **6** revealed that the positions of the methyl groups in its doubly methylated pentaketide are different from those of tenellin [26], suggesting that *H. avellanea* KUFA 0732 must possess a different C-methylation domain (CMeT) from other fungi. Thus we propose that the biosynthetic pathways of **6** should start from the intermediate **XIII**, which is derived from a tetramic acid derivative formed by the condensation of tyrosine with a doubly methylated pentaketide. The nucleophilic addition of the hydroxyl group of the 2-pyridone ring to C-6′ of the side chain leads to the formation of the intermediate **XIV**. Reduction of C-4′ in **XIV**, with a concomitant nucleophilic addition of C-5′ to the carbonyl group (C-1′), leads to the formation of a cyclopentane ring in **XV**. Dehydration of **XV** renders a double bond between C-1′ and C-5′ in **XVI**. Stereospecific reduction of the double bond between C-1′ and C-5′ in **XVI** by dehydrogenase gives rise to **6** (Figure 11).

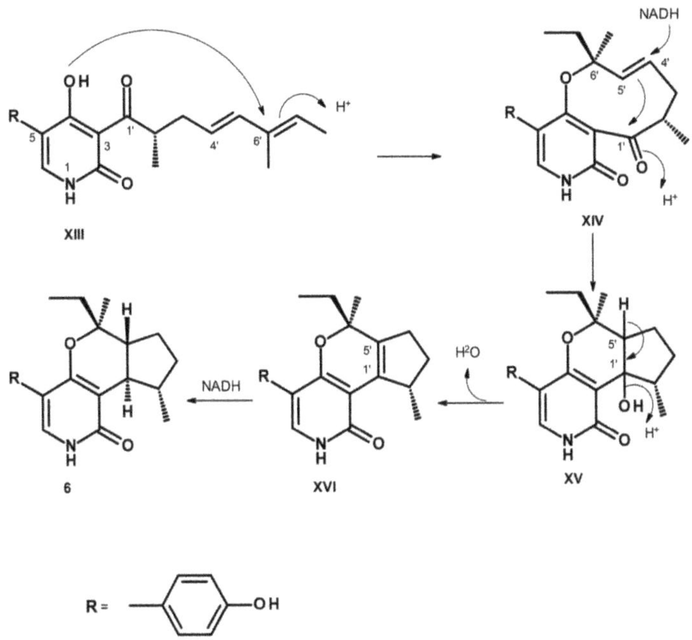

Figure 11. Proposed biosynthetic pathways of **6**.

The crude EtOAc extract and the isolated compounds **1**, **3**, **4b**, **5**, **6**, and **7** were assayed for the growth inhibitory activity against various plant pathogenic fungi that cause diseases for economic crops, viz. *Alternaria brassicicola* (black spot of Chinese Kale), *Bipolaris oryzae* (brown spot of rice), *Colletotrichum capsici* (anthracnose of chili), *C. gloeosporiodes* (anthracnose of mango), *Curvularia oryzae* (leaf spot of rice), *Fusarium semitectum* (dirty panicle of rice), *Lasiodiplodia theobromae* (fruit rot of mangosteen), *Phytophthora palmivora* (root and stem rot of durian), *Pyricularia oryzae* (rice blast), *Rhizoctonia oryzae* (sheath blight of rice) and *Sclerotium rolfsii* (stem rot of bean). The crude extract showed complete mycelial growth inhibition of all plant pathogenic fungi at 10 g/L, except for *F. semitectum*, *A. brassicicola* and *B. oryzae*. However, at a concentration of 1 g/L, the crude extract exhibited significant ($p < 0.05$) inhibition of the mycelial growth of *R. oryzae*, *C. capsici*, *C. gloeosporiodes*, and *C. oryzae*, causing more than 50% inhibition but displayed less than 50% inhibition for the rest of the pathogens tested (Table 6).

Interestingly, **3**, **5**, and **6** exhibited growth inhibition on *C. oryzae*, a causative agent of leaf spot of rice, while **1**, **2**, **4b**, and **7** were inactive (MIC > 500 µg/mL). Compounds **3** and **5** displayed strong inhibitory activity, with MIC values of 125 µg/mL, which is comparable to the positive control, mancozeb (MIC = 125 µg/Ml), while **6** only showed moderate activity, with MIC value of 250 µg/Ml. All the tested compounds were not active (MIC > 500 µg/Ml) against the rest of the plant pathogens (Table 7).

Table 6. Effects of the crude EtOAc extract of *H. avellanea* KUFA0732 on the growth of plant pathogenic fungi.

Plant Pathogen	% Mycelial Growth Inhibition at Concentrations	
	10 g/L	1 g/L
Alternaria brassicicola (black spot of Chinese Kale)	68.67 ± 2.27 [c]	47.33 ± 1.62 [d]
Bipolaris oryzae (brown spot of rice)	65.21 ± 3.12 [d]	36.09 ± 1.14 [f]
Colletotrichum capsici (anthracnose of chili)	100 ± 0 [a]	58.67 ± 2.08 [c]
C. gloeosporiodes (anthracnose of mango)	100 ± 0 [a]	62.32 ± 3.58 [b]
Curvularia oryzae (leaf spot of rice)	100 ± 0 [a]	63.67 ± 2.15 [b]
Fusarium semitectum (dirty panicle of rice)	85.34 ± 3.61 [b]	38.04 ± 1.36 [e]
Lasiodiplodia theobromae (fruit rot of mangosteen)	100 ± 0 [a]	37.42 ± 1.08 [e f]
Phytophthora palmivora (root and stem rot of durian)	100 ± 0 [a]	0 ± 0 [h]
Pyricularia oryzae (rice blast)	100 ± 0 [a]	46.25 ± 2.01 [d]
Rhizoctonia oryzae (sheath blight of rice)	100 ± 0 [a]	68.34 ± 3.20 [a]
Sclerotium rolfsii (stem rot of bean)	100 ± 0 [a]	33.47 ± 1.06 [g]

Means ± standard derivations followed by the same letter in each column are not significantly different at $p < 0.05$ when analyzed using Duncan's test of one-way ANOVA.

Table 7. Effects of **1**, **3**, **4b**, **5**, **6**, and **7** on the growth of plant pathogenic fungi.

Plant Pathogen	Plant Disease	MIC (μg/Ml)						
		1	**3**	**4b**	**5**	**6**	**7**	**Mancozeb**
Alternaria brassicicola	black spot of Chinese Kale	>500	>500	>500	>500	>500	>500	125
Bipolaris oryzae	brown spot of rice	>500	>500	>500	>500	>500	>500	125
Colletotrichum capsici	anthracnose of chili	>500	>500	>500	>500	>500	>500	125
C. gloeosporiodes	anthracnose of mango	>500	500	>500	>500	>500	>500	125
Curvularia oryzae	leaf spot of rice	>500	125	>500	125	250	>500	125
Fusarium semitectum	dirty panicle of rice	>500	>500	>500	>500	>500	>500	125
Lasiodiplodia theobromae	fruit rot of mangosteen	>500	>500	>500	>500	>500	>500	250
Phytophthora palmivora	root and stem of durian	>500	>500	>500	>500	>500	>500	125
Pyricularia oryzae	rice blast	>500	>500	>500	>500	>500	>500	125
Rhizoctonia oryzae	sheath blight of rice	>500	>500	>500	>500	>500	>500	250
Sclerotium rolfsii	stem rot of bean	>500	>500	>500	>500	>500	>500	250

3. Methods and Materials

3.1. General Experimental Procedures

The melting points were determined on a Stuart Melting Point Apparatus SMP3 (Bibby Sterilin, Stone, Staffordshire, UK) and are uncorrected. Optical rotations were measured on an ADP410 Polarimeter (Bellingham + Stanley Ltd., Tunbridge Wells, Kent, UK). The ^{1}H and ^{13}C NMR spectra were recorded at ambient temperature on a Bruker AMC instrument (Bruker Biosciences Corporation, Billerica, MA, USA) operating at 300 and 75 MHz, respectively. High-resolution mass spectra were measured with a Waters Xevo QToF mass spectrometer (Waters Corporations, Milford, MA, USA) coupled to a Waters Aquity UPLC system. A Merck (Darmstadt, Germany) silica gel GF$_{254}$ was used for preparative TLC, and Merck Si gel 60 (0.2–0.5 mm) was used for column chromatography. LiChroprep silica gel and Sephadex LH 20 were used for column chromatography.

3.2. Fungal Material

The fungus was isolated from a marine sponge *Mycale* sp., which was collected by scuba diving at a depth of 5-10 m from the coral reef at Samaesan Island (12°34′36.64″ N 100°56′59.69″ E), in the Gulf of Thailand, Chonburi Province, in May 2018. The sponge was washed with 0.01% sodium hypochlorite solution for 1 min, followed by sterilized seawater three times, and then dried on a sterile filter paper under sterile aseptic condition. The sponge was cut into small pieces (ca. 5 × 5 mm) and placed on Petri dish plates containing 15 mL potato dextrose agar (PDA) medium mixed with 300 mg/L of streptomycin sulfate and incubated at 28 °C for 7 days. The hyphal tips emerging from sponge pieces were individually transferred onto a PDA slant and maintained as pure cultures at Kasetsart University Fungal Collection, Department of Plant Pathology, Faculty of Agriculture, Kasetsart University, Bangkok, Thailand. The fungal strain KUFA 0732 was identified as *Hamigera avellanea* based on morphological characteristics such as colony growth rate and growth pattern on standard media, namely Czapek's agar, Czapek yeast autolysate agar, and malt extract agar. Microscopic characteristics, including the size, shape, and ornamentation of conidiophores and spores, were examined under a light microscope. This identification was confirmed by molecular techniques using internal transcribed spacer (ITS) primers. DNA was extracted from young mycelia following a modified Murray and Thompson method [27]. The universal primer pairs, ITS1 and ITS4, were used for ITS gene amplification [28]. PCR reactions were conducted on Thermal Cycler, and the amplification process consisted of initial denaturation at 95 °C for 5 min, 34 cycles at 95 °C for 1 min (denaturation), at 55 °C for 1 min (annealing), and at 72 °C for 1.5 min (extension), followed by the final extension at 72 °C for 10 min. PCR products were examined by agarose gel electrophoresis (1% agarose with 1 × Tris-Borate-EDTA (TBE) buffer) and visualized under UV light after staining with ethidium bromide. DNA sequencing analyses were performed using the dideoxyribonucleotide chain termination method [29] by Macrogen Inc. (Seoul, Republic of Korea). The DNA sequences were edited using FinchTV software and submitted to the BLAST program for alignment and compared with that of fungal species in the NCBI database (http://www.ncbi.nlm.nih.gov/ accessed on 1 May 2023). Its gene sequences were deposited in GenBank with the accession number OQ520878.

3.3. Extraction and Isolation

The fungus was cultured for one week at 28 °C in five Petri dishes (i.d. 90 mm) containing 20 mL of PDA per dish. The mycelial plugs (5 mm in diameter) were transferred to two 500 mL Erlenmeyer flasks containing 200 mL of potato dextrose broth and incubated on a rotary shaker at 120 rpm at 28 °C for one week. Fifty 1000 mL Erlenmeyer flasks, each containing 300 g of cooked rice, were autoclaved at 121 °C for 15 min. After cooling to room temperature, 20 mL of a mycelial suspension of the fungus was inoculated per flask and incubated at 28 °C for 30 days, after which 500 mL of EtOAc was added to each flask of the moldy rice and macerated for 7 days, and then filtered with a Whatman No. 1 filter paper. The EtOAc solutions were combined and concentrated under reduced pressure to yield 184.1 g of a crude EtOAc extract, which was dissolved in 500 mL of $CHCl_3$, washed with H_2O (3 × 500 mL), and dried with anhydrous Na_2SO_4, filtered and evaporated under reduced pressure to give 153.5 g of a crude $CHCl_3$ extract. The crude $CHCl_3$ extract (55.5 g) was applied on a silica gel column (350 g) and eluted with mixtures of petrol-$CHCl_3$ and $CHCl_3$-Me_2CO, wherein 250 mL fractions (frs) were collected as follows: frs 1–146 (petrol-$CHCl_3$, 1:1), 147–204 (petrol-$CHCl_3$, 3:7), 205–234 (petrol-$CHCl_3$, 1:9), 235–250 ($CHCl_3$), 251–322 ($CHCl_3$-Me_2CO, 9:1), 323–382 ($CHCl_3$-Me_2CO, 7:3). Frs 30–31 were combined (1.43 g) and applied over a Sephadex LH-20 column (10g) and eluted with MeOH, wherein 14 subfractions (sfrs) of 2 mL were collected. Sfrs 1–2 were combined (275.4 mg) and purified by a preparative TLC (silica gel G_{254}, $CHCl_3$: petrol: Me_2CO: HCO_2H, 90:5:5:0.1) to give 12.5 mg of **3**. Frs 39–46 were combined (397 mg) and recrystallized in $CHCl_3$ to give 24.8 mg of **1**. The mother liquor of frs 39–46 (360.2 mg) was combined with frs 47–57 (305.0 mg) and purified by preparative TLC (silica gel G_{254}, $CHCl_3$: Me_2CO: HCO_2H, 90:10:0.1) to give additional 93.6 mg of **1**. Frs 62–79 were combined (710.4 mg) and applied over a Sephadex LH-20 column (10 g), and eluted with MeOH, wherein 12 sfrs of 2 mL were collected. Sfrs 2–8 were combined (540.8 mg) and applied over another Sephadex LH-20 column (10 g) and eluted with MeOH, wherein 28 sub-subfractions (ssfrs) of 2 mL were collected. Ssfrs 11–22 were combined (281.4 mg) and purified by preparative TLC (silica gel G_{254}, $CHCl_3$:Me_2CO: HCO_2H, 90:10:0.1) to give 38.5 mg of **7** and 22.4 mg of **2**. Frs 87–107 were combined (1.23 g), and part of it (543.2 mg) was purified by preparative TLC (silica gel G_{254}, $CHCl_3$: petrol: Me_2CO: HCO_2H, 90:5:5:0.1) to give 375.3 mg of a mixture of compounds which was then applied on a Sephadex LH-20 column (10 g) and eluted with MeOH, wherein 16 sfrs were collected. Sfrs 3-16 were combined to give 80 mg of **4a**. Another part of the combined frs 87–109 (670 mg) was purified by a preparative TLC (silica gel G_{254}, $CHCl_3$: petrol: Me_2CO: HCO_2H, 90:5:5:0.1) to give 19.3 mg of **4a**. Frs 117–149 were combined (264.1 mg) and crystallized in $CHCl_3$ to give 15 mg of **5**. Frs 150–166 were combined (584.8 mg) and applied on a silica gel column (10 g), and eluted with mixtures of petrol-$CHCl_3$, $CHCl_3$, $CHCl_3$-Me_2CO, wherein 100 mL sfrs were collected as follows: sfrs 1–12 (petrol-$CHCl_3$, 1:1), 13–21 (petrol-$CHCl_3$, 3:7), 22–30 (petrol-$CHCl_3$, 1:9), 31–73 ($CHCl_3$), 74–81 ($CHCl_3$-Me_2CO, 9:1). Sfrs 35-39 were combined (143.9 mg) and purified by a preparative TLC (silica gel G_{254}, $CHCl_3$: Me_2CO: HCO_2H, 90:10:0.1) to give 18 mg of **4b** and 10 mg of **4a**. Sfrs 60–81 were combined (228.2 mg) and purified by a preparative TLC (silica gel G_{254}, $CHCl_3$: Me_2CO: HCO_2H, 90:10:0.1) to give 15.2 mg of **4b** and 5.6 mg of **4a**. Frs 326–349 were combined (1.01 g) and applied on a silica gel column (10 mg) and eluted with mixtures of petrol-$CHCl_3$, $CHCl_3$, $CHCl_3$-Me_2CO, wherein 100 mL sfrs were collected as follows: Sfrs 1–11 (petrol-$CHCl_3$, 1:1), 12–16 (petrol-$CHCl_3$, 3:7), 17–26 ($CHCl_3$), 27-54 ($CHCl_3$-Me_2CO, 9:1), 55–61 ($CHCl_3$-Me_2CO, 7:3). Sfrs 33–35 were combined (154.8 mg) and crystallized in $CHCl_3$ and Me_2CO to give 64.2 mg of **6**.

3.3.1. (*R*)-6,8-Dihydroxy-4,5-dimethyl-3-methylidene-3,4-dihydro-1*H*-2-benzopyran-1-one (**1**)

White crystal. mp 217–218 °C. $[\alpha]_D^{20}$ −140 (*c* 0.05, MeOH); For ^{1}H and ^{13}C spectroscopic data (DMSO-*d6*, 300 and 75 MHz), see Table 1; (+)-HRESIMS *m/z* 221.0814 $[M + H]^+$ (calcd for $C_{12}H_{13}O_4$, 221.0814).

3.3.2. (3*S*,4*R*)-3,8-Dihydroxy-6-methoxy-4,5-dimethyl-1-oxo-3,4-dihydro-1*H*-2-benzopyran-3-yl)methyl acetate (**2**)

Pale yellow viscous mass. $[\alpha]_D^{20}$ −104 (*c* 0.05, MeOH); For ^{1}H and ^{13}C spectroscopic data ($CDCl_3$, 300 and 75 MHz), see Table 2; (+)-HRESIMS *m/z* 311.1133 $[M + H]^+$ (calcd for $C_{15}H_{19}O_7$, 311.1131) and 333.0948 $[M + Na]^+$ (calculated for $C_{15}H_{18}O_7Na$, 333.0950).

3.3.3. (*R*)-5, 7-Dimethoxy-3-((*S*)-(1-hydroxyethyl)-3,4-dimethylisobenzofuran 1(3*H*)-one (**4b**)

White crystals. mp 174–176 °C. $[\alpha]_D^{20}$ −300 (*c* 0.05, MeOH); For ^{1}H and ^{13}C spectroscopic data ($CDCl_3$, 300 and 75 MHz), see Table 3; (+)-HRESIMS *m/z* 267.1233 $[M + H]^+$ (calcd for $C_{14}H_{19}O_5$, 267.1232).

3.3.4. (*S*)-7-Hydroxy-3-((*S*)-1-hydroxyethyl)-5-methoxy-3,4-dimethylisobenzofuran 1(3*H*)-one (**5**)

White crystal. mp 116–118 °C. $[\alpha]_D^{20}$ −320 (*c* 0.05, MeOH); For ^{1}H and ^{13}C spectroscopic data ($CDCl_3$, 300 and 75 MHz), see Table 4; (+)-HRESIMS *m/z* 253.1077 $[M + H]^+$ (calcd for $C_{13}H_{17}O_5$, 253.1076).

3.3.5. Avellaneanone (**6**)

White crystal. mp 219–210 °C. $[\alpha]_D^{20}$+ 240 (*c* = 0.05, MeOH) (*c* 0.05, MeOH); For ^{1}H and ^{13}C spectroscopic data (DMSO-*d6*, 300 and 75 MHz), see Table 5; (+)-HRESIMS *m/z* 362.1734 $[M + Na]^+$ (calcd for C_{21} H_{25} NO_3Na, 362.1732).

3.4. X-ray Crystal Structures

Single crystals were mounted on a cryo-loop using paratone. X-ray CuK$_\alpha$ radiation (λ = 1.54184 Å). The structures were resolved by direct methods using SHELXS-97 and refined with SHELXL-97 [30].

Full details of data collection and refinement and tables of atomic coordinates, bond lengths, angles, and torsion angles have been deposited with the Cambridge Crystallographic Data Centre.

3.4.1. X-ray Crystal Structure of **1**

The crystal was orthorhombic, space group P2$_1$2$_1$2$_1$, cell volume 1050.3(2) Å^3 and unit cell dimensions a = 5.2675(7) Å, b = 13.5239(16) Å, c = 14.7440(14) Å (uncertainties in parentheses). The calculated crystal density is 1.393 g.cm^{-3}. Non-hydrogen atoms were refined anisotropically. Hydrogen atoms were directly found from different Fourier maps and refined freely with isotropic displacement parameters or placed at their idealized positions using appropriate HFIX instructions in SHELXL and included in subsequent refinement cycles. The refinement converged to R (all data) = 13.26% and wR2 (all data) = 15.30%. CCDC deposition number 2256003.

3.4.2. X-ray Crystal Structure of **4b**

The crystal was triclinic, space group P1, cell volume 662.30(14) Å^3 and unit cell dimensions a = 7.3973(9) Å, b = 8.4697(12) Å, c = 11.2567(13) Å and angles α = 107.628(12)°, β = 92.120(11)° and γ = 98.481(11))° (uncertainties in parentheses). The calculated crystal densities were 1.393 g.cm^{-3} and 1.335 g.cm^{-3}. Two molecules were found in the asymmetric unit. Non-hydrogen atoms were refined anisotropically. Hydrogens were placed at their idealized positions using appropriate HFIX instructions in SHELXL and included in subsequent refinement cycles. The refinement converged to R (all data) = 19.00% and wR2 (all data) = 47.69%. CCDC deposition number 2260970.

3.4.3. X-ray Crystal Structure of **5**

The crystal was triclinic, space group P-1, cell volume 655.52(16) Å^3 and unit cell dimensions a = 7.8395(11) Å, b = 9.2206(12) Å, c = 9.8904(14) Å and angles α = 87.687(11)°, β = 85.728(11)° and γ = 66.857(13)° (uncertainties in parentheses). The calculated crystal density is 1.452 g.cm^{-3}. One water molecule was found in the asymmetric unit. Non-hydrogen atoms were refined anisotropically. Hydrogens were directly found from different Fourier maps and refined freely with isotropic displacement parameters. The refinement converged to R (all data) = 10.99% and wR2 (all data) = 19.13%. CCDC deposition number 2256016.

3.4.4. X-ray Crystal Structure of **6**

The crystal was monoclinic, space group I2, cell volume 1880.5(3) Å^3 and unit cell dimensions a = 8.9429(8) Å, b = 6.4805(6) Å, c = 32.453(3) Å and β = 91.065(8)° (uncertainties in parentheses). The calculated crystal density is 1.199 g.cm^{-3}. Non-hydrogen atoms were refined anisotropically. Hydrogen atoms were directly found from different Fourier maps and refined freely with isotropic displacement parameters or placed at their idealized positions using appropriate HFIX instructions in SHELXL and included in subsequent refinement cycles. The refinement converged to R (all data) = 7.81% and wR2 (all data) = 12.53%. CCDC deposition number 2256402.

3.5. Antifungal Activity Bioassays

3.5.1. Antifungal Activity of the Crude Extract against Plant Pathogenic Fungi

The in vitro growth inhibitory activity of the crude EtOAc extract of *H. avellanea* KUFA0732 against eleven plant pathogenic fungi was evaluated using the poison food technique described previously by Dethoup et al. [9]. The crude EtOAc extract (1 g) was dissolved in 1 mL of DMSO and serially diluted with 9 mL of H_2O to obtain stock solutions of 100 and 10 g/L. A volume of 1 mL of each stock solution was added to 9 mL of warm PDA, thoroughly mixed by a vortex mixer, and then poured into Petri dishes to give medium plates amended with crude extract with concentrations of 10g/L and 1 g/L, respectively. Each pathogen strain was cultured on a PDA for 7 days at 28 ± 2 °C, and a mycelial plug of 0.5 cm in diameter of each pathogen was placed on the center of the crude extract-containing PDA plates and incubated at room temperature for 14 days. A PDA plate void of the crude extract was used as a negative control. The inhibition levels were calculated using the formula: $[(x - y)/x] \times 100$, where x = colony radius of the plant pathogenic fungi in the negative control, and y = colony radius of the plant pathogenic fungi in the presence of the crude extract. Each treatment was performed with five replications and repeated three times independently. The mean inhibition levels and standard deviations were calculated from five replications and three repetitions. Data were subjected to analysis of variance, and subsequently, means were compared using Duncan's multiple range test ($p < 0.05$) in the SPSS version 19 statistical program (IBM Corporation, Somers, NY, USA).

3.5.2. Antifungal Activity of Isolated Compounds against Plant Pathogenic Fungi

The minimum inhibitory concentrations (MICs) for the antifungal activity of the isolated compounds on the plant pathogenic fungi were determined according to CLSI recommendation [31]. Each pathogen strain was cultured on PDA for 14 days at 28 ± 2 °C, and its spores were gently scraped by a sterile loop from the mycelium and then added to the PDA. A mass of 1 mg of each compound was dissolved in 100 µL of 10% DMSO, and then 900 µL of distilled H_2O was added to prepare a stock solution of 1000 µg/L. Two-fold serial dilutions of the stock solution by potato dextrose broth mixed with a spore suspension of each pathogen at 10^6 spores/mL provided the tested concentrations at 500, 250, and 125 µg/L. Each treatment was loaded at 200 µL per well into the 96-well U-shaped untreated polystyrene plates with five wells per treatment, and then the plate was incubated for 7 days at 25 °C. The MIC was determined as the lowest concentration of the compound that inhibited visible growth. Mancozeb (OXA-Sigma-Aldrich, St. Louis, MI, USA), at the same concentration, was used as a positive control.

4. Conclusions

The crude EtOAc extract of a solid-rice culture of a marine sponge-associated fungus, *Hamigera avellanea* KUFA0732, isolated from the marine sponge *Mycale* sp., which was collected from the Gulf of Thailand, was found to have the potential to inhibit the growth of various plant pathogenic fungi. Chemical study of the crude extract of the culture of this marine-derived fungus led to the isolation of five undescribed specialized metabolites, viz. (*R*)-6,8-dihydroxy-4,5-dimethyl-3-methylidene-3,4-dihydro-1*H*-2-benzopyran-1-one (**1**), (3*S*, 4*R*)-3,8-dihydroxy-6-methoxy-4,5-dimethyl-1-oxo-3,4-dihydro-1*H*-isochromen-3-yl]methyl acetate (**2**), (*R*)-5, 7-dimethoxy-3-((*S*)-(1-hydroxyethyl)-3,4-dimethylisobenzofuran-1(3*H*)-one (**4b**), (*S*)-7-hydroxy-3-((*S*)-1-hydroxyethyl)-5-methoxy-3,4-dimethylisobenzofuran 1(3*H*)-one (**5**), and avellaneanone (**6**), as well as three previously reported (*R*)-3-acetyl-7-hydroxy-5-methoxy-3,4-dimethylisobenzofuran-1(3*H*)-one (**3**), (*R*)-7-hydroxy-3-((*S*)-1-hydroxyethyl)-5-methoxy-3,4-dimethylisobenzofuran-1(3*H*)-one (**4a**) and isosclerone (**7**). The common biosynthetic pathways of the pentaketide derivatives **1** and **2**, as well as of **3**, **4a**, **4b**, and **5**, were determined. The biosynthesis of the PKS-NRPS hybrid product, avellaneanone (**6**), was also proposed based on the previously proposed pathways for other analogs. Finally, it is important to point out that although the crude EtOAc extract of *H. avellanea* KUFA0732 was able to completely inhibit the growth of many plant pathogenic fungi tested, only **3** and **5** displayed strong inhibitory activity (MIC = 125 µg/mL) comparable to the positive control, mancozeb, against *Curvularia oryzae*, while **6** exhibited only moderate activity (MIC = 250 µg/mL). These results give a new perspective on the potential application of marine-derived fungi and their specialized metabolites as biocontrol agents in agriculture.

Supplementary Materials: The following are available online at https://www.mdpi.com/article/10.3390/md21060344/s1, Figures S1–S5, S7–S12, S14–S28, S30–S34, S36–S40, S42–S46: 1D and 2D NMR spectra of compounds **1–7**. Figures S6, S13, S29, S35, S41: HRMS data for compounds **1**, **2**, **4b**, **5**, **6**. Table S1: ^{1}H and ^{13}C NMR data ($CDCl_3$, 300 and 75 MHz), COSY and HMBC for **3**. Table S2: ^{1}H and ^{13}C NMR (300 and 75 MHz, $CDCl_3$), COSY, and HMBC of **4a**. Table S3: Comparison of ^{1}H and ^{13}C NMR data of **4a** (300 and 75 MHz, $CDCl_3$) with those of (*R*)-7-hydroxy-3-((*R*)-1-hydroxyethyl)-5-methoxy-3,4-dimethylisobenzofuran-1(3*H*)-one (400 and 100 MHz, $CDCl_3$) and (*R*)-7-hydroxy-3-((*S*)-1-hydroxyethyl)-5-methoxy-3,4-dimethylisobenzofuran-1(3*H*)-one (500 and 125 MHz, $CDCl_3$). Table S4: ^{1}H and ^{13}C NMR data (DMSO-d_6, 300 and 75 MHz), COSY, and HMBC for **7**.

Author Contributions: A.K. and T.D. conceived and designed the experiment and elaborated the manuscript; R.K. performed the isolation and purification of the compounds; F.P.M. and D.K. assisted in the isolation and purification of some compounds; S.H.G. assisted in the structure elucidation of some compounds; T.D. collected, isolated, identified, cultured the fungus and performed bioassays of the extract and compounds; L.G. performed X-ray analysis; S.M. provided HRMS; A.M.S.S. provided NMR spectra; E.S. assisted in structure elucidation and preparation of the manuscript. All authors have read and agreed to the published version of the manuscript.

Funding: This work is partially supported by the national infrastructure PT-OPENSCREEN (NORTE-01-0145-FEDER-085468) and the national funds through the FCT—Foundation for Science and Technology with the scope of UIDB/04423/2020 and UIDP/04423/2020, and by the Office of the Ministry of Higher Education, Science, Research and Innovation; and the Thailand Science Research and Innovation.

Institutional Review Board Statement: Not applicable.

Data Availability Statement: The data presented in this study are available on request from the corresponding author.

Acknowledgments: R.K. thanks the Office of the Ministry of Higher Education, Science, Research and Innovation; and the Thailand Science Research and Innovation through the Kasetsart University Reinventing University Program 2021 for the scholarship.

Conflicts of Interest: The authors declare no conflict of interest.

References

1. Peterson, S.W.; Jurjevic, Z.; Bills, G.F.; Stchigel, A.M.; Guarro, J.; Vega, F.E. Genus *Hamigera*, six new species and multilocus DNA sequence based phylogeny. *Mycologia* **2010**, *102*, 847–864. [CrossRef]
2. Igarashia, Y.; Hanafusa, T.; Gohda, F.; Peterson, S.; Bills, G. Species-level assessment of secondary metabolite diversity among *Hamigera* species and a taxonomic note on the genus. *Mycology* **2014**, *5*, 102–109. [CrossRef]
3. Yamazaki, M.; Horie, Y.; Bae, K.; Maebayashi, Y.; Jisai, Y.; Fujimoto, H. New fungal metabolites avellanins A and B from *Hamigera avellanea*, with pressure effect. *Chem. Pharm. Bull.* **1987**, *35*, 2122–2124. [CrossRef] [PubMed]
4. Breinholt, J.; Kjaer, A.; Olsen, C.E.; Rassing, B.R. A bis-formamidodiphenylbutadiene from the fungus *Hamigera avellanea*. *Acta Chem. Scand.* **1996**, *50*, 643–645. [CrossRef]
5. Breinholt, J.; Kjaer, A.; Olsen, C.E.; Rassing, B.R.; Rosendahl, C.N. Hamigerone and dihydrohamigerone: Two acetate-derived, antifungal metabolites from *Hamigera avellanea*. *Acta Chem. Scand.* **1997**, *51*, 1241–1244. [CrossRef]
6. Isaka, M.; Chinthanom, P.; Veeranondha, S.; Supothina, S.; Luangsa-ard, J.J. Novel cyclopropyl diketones and 14-membered macrol ides from the soil fungus *Hamigera avellanea* BCC 17816. *Tetrahedron* **2008**, *64*, 11028–11033. [CrossRef]
7. Isaka, M.; Chinthanom, P.; Kongthong, S.; Supothina, S.; Ittiworapong, S. Hamigeromycins C–G, 14-membered macrolides from the fungus *Hamigera avellanea* BCC 17816. *Tetrahedron* **2010**, *66*, 955–961. [CrossRef]
8. Chalearmsrimuang, T.; Ismail, S.I.; Mazlan, N.; Suasaard, S.; Dethoup, T. Marine-derived fungi: A promising source of halotolerant biological control agents against plant pathogenic fungi. *J. Pure Appl. Microbiol.* **2019**, *13*, 209–223. [CrossRef]
9. Dethoup, T.; Kaewsalong, N.; Songkumorn, P.; Jantasorn, A. Potential application of a marine-derived fungus, *Talaromyces tratensis* KUFA 0091 against rice diseases. *Biol. Control* **2018**, *119*, 1–6. [CrossRef]
10. Klaram, R.; Jantasorn, A.; Dethoup, T. Efficacy of marine antagonist, *Trichoderma* spp. as halo-tolerant biofungicide in controlling rice diseases and yield improvement. *Biol. Control* **2022**, *172*, 104985. [CrossRef]
11. Chinworrungsee, M.; Kittakoop, P.; Isaka, M.; Chanphen, R.; Tanticharoen, M.; Thebtaranonth, Y. Halorosellins A and B, unique isocoumarin glucosides from the marine fungus *Halorosellinia oceanica*. *J. Chem. Soc. Perkin Trans. 1* **2002**, 2473–2476. [CrossRef]
12. Bi, Y.M.; Bi, X.B.; Fang, A.; Zhao, Q.R. Metabolites from the fungus *Cephalosporium* sp. AL031. *Arch. Pharm. Res.* **2007**, *30*, 267–269. [CrossRef] [PubMed]

13. Tayone, W.C.; Honma, M.; Kanamaru, S.; Noguchi, S.; Tanaka, K.; Nehira, T.; Hashimoto, M. Stereochemical investigations of isochromenones and isobenzofuranones isolated from *Leptosphaeria* sp. KTC 727. *J. Nat. Prod.* **2011**, *74*, 425–429. [CrossRef] [PubMed]
14. Morita, T.; Aoki, R. Isosclerone, a new metabolite of *Sclerotinia sclerotiorum* (LIB.) DE BARY. *Agric. Biol. Chem.* **1974**, *38*, 1501–1505. [CrossRef]
15. Evidente, A.; Sparapano, L.; Anna Andolfil, A.; Bruno, G. Two naphthalenone pentaketides from liquid cultures of *Phaeoacremonium aleophilum*, a fungus associated with esca of grapevine. *Phytopathol. Mediterr.* **2000**, *39*, 162–168.
16. Husain, S.M.; Müller, M. Fungal dihydroxynaphthalene-melanin: Diversity-Oriented biosynthesis through enzymatic and non-enzymatic transformations. *Synlett* **2017**, *28*, 2360–2372. [CrossRef]
17. Tayone, W.C.; Kanamaru, S.; Honma, M.; Tanaka, K.; Nehira, T.; Hashimoto, M. Absolute stereochemistry of novel isochromanone derivatives from *Leptosphaeria* sp. KTC 727. *Biosci. Biotechnol Biochem.* **2011**, *75*, 2390–2393. [CrossRef]
18. Song, Y.X.; Wang, J.; Li, S.W.; Cheng, B.; Li, L.; Chen, B.; Liu, L.; Lin, Y.C.; Gu, Y.C. Metabolites of the mangrove fungus *Xylaria* sp. BL321 from the South China Sea. *Planta Med.* **2012**, *78*, 172–176. [CrossRef]
19. Wat, C.-K.; Mcinnes, A.G.; Smith, D.G.; Wright, J.L.C.; Vining, L.C. The yellow pigments of *Beauveria* species. Structures of tenellin and bassianin. *Can. J. Chem.* **1977**, *55*, 4090–4098. [CrossRef]
20. Heneghan, M.N.; Yakasai, A.A.; Williams, K.; Kadir, K.A.; Wasil, Z.; Bakeer, W.; Fisch, K.M.; Bailey, A.M.; Simpson, T.J.; Cox, R.J.; et al. The programming role of *trans*-acting enoyl reductases during the biosynthesis of highly reduced fungal polyketides. *Chem. Sci.* **2011**, *2*, 972–979. [CrossRef]
21. Bergmann, S.; Schümann, J.; Scherlach, K.; Lange, C.; Brakhage, A.A.; Hertweck, C. Genomics-driven discovery of PKS-NRPS hybrid metabolites from *Aspergillus Nidulans. Nat. Chem. Biol.* **2007**, *3*, 213–217. [CrossRef] [PubMed]
22. Kim, J.-C.; Lee, Y.-W.; Tamura, H.; Yoshizawa, T. Sambutoxin: A new mycotoxin isolated from *Fusarium sambucinum. Tetrahedron Lett.* **1995**, *36*, 1047–1050. [CrossRef]
23. Matsumoto, M.; Minato, H. Structure of ilicicolin H, an antifungal antibiotic. *Tetrahedron Lett.* **1976**, *42*, 3827–3830. [CrossRef]
24. Wu, B.; Oesker, V.; Wiese, J.; Schmaljohann, R.; Imhoff, J.F. Two new antibiotic pyridones produced by a marine fungus, *Trichoderma* sp. Strain MF106. *Mar. Drugs* **2014**, *12*, 1208–1219. [CrossRef]
25. Guo, Y.; Contesini, F.J.; Wang, X.; Ghidinelli, S.; Tornby, D.S.; Andersen, T.E.; Mortensen, U.F.; Larsen, T.O. Biosynthesis of calipyridone A represents a fungal 2-pyridone formation without ring expansion in *Aspergillus californicus. Org. Lett.* **2022**, *24*, 804–808. [CrossRef]
26. Fisch, K.M.; Bakeer, W.; Yakasai, A.A.; Song, Z.; Pedrick, J.; Wasil, Z.; Bailey, A.M.; Lazarus, C.M.; Simpson, T.J.; Cox, R.J. Rational domain swaps decipher programming in fungal highly reducing polyketide synthases and resurrect an extinct metabolite. *J. Am. Chem. Soc.* **2011**, *133*, 16635–16641. [CrossRef]
27. Murray, M.G.; Thompson, W.F. Rapid isolation of high molecular weight plant DNA. *Nucleic Acids Res.* **1980**, *8*, 4321–4325. [CrossRef]
28. White, T.J.; Bruns, T.; Lee, S.; Taylor, J. Amplification and direct sequencing of fungal ribosomal RNA genes for phylogenetics. In *PCR Protocols: A Guide to Methods and Applications*; Innis, M.A., Gelfand, D.H., Sninsky, J.J., White, T.J., Eds.; Academic Press: New York, NY, USA, 1990; pp. 315–322.
29. Sanger, F.; Nicklen, S.; Coulson, A.R. DNA sequencing with chain-terminating inhibitors. *Proc. Natl. Acad. Sci. USA* **1977**, *72*, 5463–5467. [CrossRef]
30. Sheldrick, G.M. A short history of SHELX. *Acta Cryst.* **2008**, *A64*, 112–122. [CrossRef]
31. CLSI. *Methods for Determining Bactericidal Activity of Antimicrobial Agents*; Approved Guideline; CLSI Document M26-A; Clinical and Laboratory Standards Institute: Wayne, PA, USA, 1999.

marine drugs

MDPI

Article

Bioactive Polyketides and Benzene Derivatives from Two Mangrove Sediment-Derived Fungi in the Beibu Gulf

Bo Peng [1,†], Jian Cai [2,†], Zimin Xiao [3], Manli Liu [4], Xinlong Li [2], Bin Yang [2], Wei Fang [4], Yi-You Huang [5], Chunmei Chen [2,*], Xuefeng Zhou [2,*] and Huaming Tao [3,*]

1 Institute for Environmental and Climate Research, Jinan University, Guangzhou 511443, China; pengbo@jnu.edu.cn
2 CAS Key Laboratory of Tropical Marine Bio-Resources and Ecology, Guangdong Key Laboratory of Marine Materia Medica, South China Sea Institute of Oceanology, Chinese Academy of Sciences, Guangzhou 510301, China; caijian19@mails.ucas.ac.cn (J.C.); 13409379936@163.com (X.L.); yangbin@scsio.ac.cn (B.Y.)
3 Guangdong Provincial Key Laboratory of Chinese Medicine Pharmaceutics, School of Traditional Chinese Medicine, Southern Medical University, Guangzhou 510515, China; 15917491112@163.com
4 Hubei Biopesticide Engineering Research Center, Hubei Academy of Agricultural Science, Wuhan 430064, China; manli.liu@nberc.com (M.L.); wei.fang@nberc.com (W.F.)
5 Key Laboratory of Tropical Biological Resources of Ministry of Education, School of Pharmaceutical Sciences, Hainan University, Haikou 570228, China; hyyou@hainanu.edu.cn
* Correspondence: chenchunmei18@mails.ucas.ac.cn (C.C.); xfzhou@scsio.ac.cn (X.Z.); taohm@smu.edu.cn (H.T.)
† These authors contributed equally to this work.

Abstract: To discover bioactive natural products from mangrove sediment-derived microbes, a chemical investigation of the two Beibu Gulf-derived fungi strains, *Talaromyces* sp. SCSIO 41050 and *Penicillium* sp. SCSIO 41411, led to the isolation of 23 natural products. Five of them were identified as new ones, including two polyketide derivatives with unusual acid anhydride moieties named cordyanhydride A ethyl ester (**1**) and maleicanhydridane (**4**), and three hydroxyphenylacetic acid derivatives named stachylines H–J (**10–12**). Their structures were determined by detailed nuclear magnetic resonance (NMR) and mass spectroscopic (MS) analyses, while the absolute configurations were established by theoretical electronic circular dichroism (ECD) calculation. A variety of bioactive screens revealed three polyketide derivatives (**1–3**) with obvious antifungal activities, and **4** displayed moderate cytotoxicity against cell lines A549 and WPMY-1. Compounds **1** and **6** at 10 μM exhibited obvious inhibition against phosphodiesterase 4 (PDE4) with inhibitory ratios of 49.7% and 39.6%, respectively, while **5**, **10**, and **11** showed the potential of inhibiting acetylcholinesterase (AChE) by an enzyme activity test, as well as in silico docking analysis.

Keywords: mangrove sediment-derived fungi; polyketide; stachylines; biological activity

Citation: Peng, B.; Cai, J.; Xiao, Z.; Liu, M.; Li, X.; Yang, B.; Fang, W.; Huang, Y.-Y.; Chen, C.; Zhou, X.; et al. Bioactive Polyketides and Benzene Derivatives from Two Mangrove Sediment-Derived Fungi in the Beibu Gulf. *Mar. Drugs* **2023**, *21*, 327. https://doi.org/10.3390/md21060327

Academic Editors: Bin-Gui Wang and Haofu Dai

Received: 9 May 2023
Revised: 21 May 2023
Accepted: 24 May 2023
Published: 26 May 2023

1. Introduction

The mangrove wetland ecosystems, located at tropical and subtropical intertidal estuarine zones, possess rich biodiversity and include an enormous diversity of microorganisms [1]. From these, a vast range of fungi species in mangrove sediment play a vital role in biogeochemical cycles to sustain the mangrove wetland ecosystems [2]. Mangrove sediment-derived fungi are widely considered to be a pivotal and prolific reservoir of structurally unique and biologically active secondary metabolites with promising medicinal, agricultural, or industrial applications [2,3].

The genus *Talaromyces* is widely distributed in marine environments, soil, plants, and foods. The extreme living conditions have led the fungi to develop more specific metabolic patterns, and marine-derived *Talaromyces* spp. can produce a number of structurally diverse substances with a wide range of bioactivities, such as anti-inflammatory meroterpenoids, thioester containing benzoate derivatives that exhibit α-glucosidase inhibitory activity, and oxaphenalenone dimers with broad antibacterial activity [4].

Penicillium species are among the most widespread fungal organisms on earth and contains more than 350 species. Many *Penicillium* species can produce plentiful secondary metabolites, such as alkaloids [5], polyketides [6] and terpenoids [7], that can ascribe specific structural characteristics and significant biological activities.

As part of our research on discovering structurally novel and bioactive natural products from mangrove sediment-derived fungi [5,8,9], two fungi strains of *Talaromyces* sp. SCSIO 41050 and *Penicillium* sp. SCSIO 41411, isolated from a mangrove sediment sample, collected from Gaoqiao mangrove wetland in the Zhanjiang coastline of the northern part of Beibu Gulf, attracted our attention for the characterization of their HPLC-DAD profiles. Further chemical investigations of their crude extract led to the isolation of 23 natural products, including four maleic anhydride polyketides (**1–4**), three austdiol polyketides derivates (**5–7**), a 2*H*-pyran-2-one derivate (**8**), a benzoic acid derivate (**9**), seven stachyline derivates (**10–16**), and seven carboxylic acids and ester derivatives (**17–23**) (Figure 1). Five of them were identified as new ones: two polyketides named cordyanhydride A ethyl ester (**1**) and maleicanhydridane (**4**) from *Talaromyces* sp. SCSIO 41050, and three benzene derivatives called stachylines H–J (**10–12**) from *Penicillium* sp. SCSIO 41411. Herein, details of the isolation, structure elucidation, and biological activities of these compounds are described.

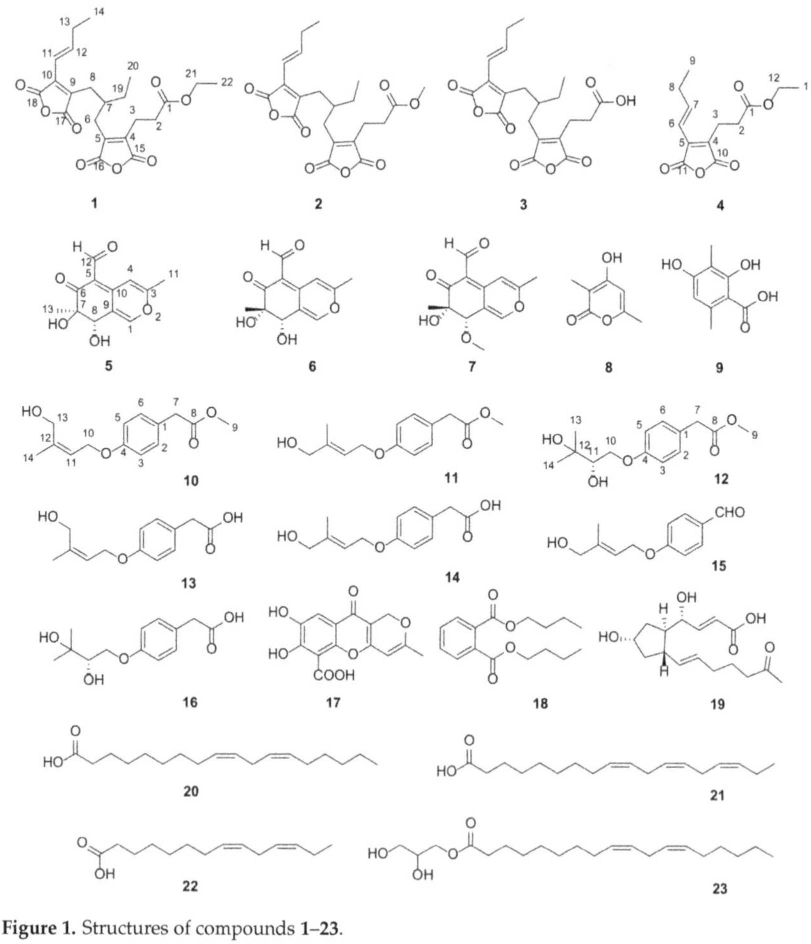

Figure 1. Structures of compounds **1–23**.

2. Results and Discussion

2.1. Structural Determination

Compound **1** was obtained as a yellow oil, and its molecular formula was established as $C_{22}H_{26}O_8$ by HRESIMS ion peak at m/z 419.1701 $[M + H]^+$ (calcd for $C_{22}H_{27}O_8^+$, 419.1700). A detailed analysis of ^1H NMR data (Table 1) of **1** exhibited the presence of two olefinic protons at δ_H 7.08 (dt, J = 15.5, 6.5 Hz, H-12) and 6.43 (dt, J = 15.5, 1.5 Hz, H-11); one methine at δ_H 2.05 (m, H-7); seven methylenes (including one oxygenated) at δ_H 4.05 (q, J = 7.0 Hz, H_2-21), 2.68 (t, J = 8.0 Hz, H_2-3), 2.58 (dd, J = 9.5, 7.0 Hz, H_2-2), 2.47/2.40 (H_2-6), 2.47/2.05 (H_2-7), 2.28 (m, H_2-13), and 1.30 (m, H_2-19); and three methyls at δ_H 1.17 (t, J = 7.0 Hz, H_3-19), 1.05 (t, J = 7.5 Hz, H_3-14), and (d, J = 0.88 Hz, H_3-20). The ^{13}C NMR data and HSQC spectrum displayed 22 carbon signals including five ester carbons at δ_C 171.5 (C-1), 166.2 (C-17), 166.0 (C-16), 165.5 (C-15), and 164.5 (C-18); four olefinic tertiary carbons at δ_C 144.1 (C-4), 143.0 (C-5), 138.1 (C-9), and 137.7 (C-10); two olefinic methine carbons at δ_C 147.7 (C-12) and 117.0 (C-11); one methine carbon at δ_C 37.6 (C-7); seven methylene carbons (including one oxygenated) at δ_C 60.3 (C-21), 30.8 (C-2), 27.9 (C-8), 27.4 (C-6), 26.6 (C-13), 25.7 (C-19), and 19.4 (C-3); and three methyl carbons at δ_C 14.0 (C-22), 12.5 (C-14), and 10.7 (C-20). The ^1H-^1H COSY correlations (Figure 2) of H_2-2/H_2-3, H_2-6/H-7/H_2-8, H-7/H_2-19/H_3-20, and H-11/H-12/H_2-13/H_3-14 revealed partial structure of CH_2-2/CH_2-3, CH_2-6/CH-7(CH_2-19/CH_3-20)/CH_2-8, and CH-11/CH-12/CH_2-13/CH_3-14. The above NMR data indicated that the structural skeleton of **1** was similar to that of the co-isolated cordyanhydride A methyl ester (**2**) [10], with two acid anhydride moieties. The main distinction was the presence of an ethyl ester group at C-1 of **1** instead of the methyl ester group in **2**, which was supported by key ^1H-^1H COSY correlation of H_2-21/H_3-22 and HMBC correlations (Figure 2) from H_2-21 to C-1 and C-22. Thus, the planar structure of **1** was defined as shown in Figure 1, and the other HMBC correlations supported the deduction. The configurations of the Δ^{11} double bonds were all deduced as E based on the large coupling constant $J_{H\text{-}11/H\text{-}12}$ = 15.5 Hz. Due to the weak Cotton effect and the failed single crystal cultivation experiment, the absolute configuration of C-7 was unsolved. Finally, compound **1** was identified to be a new maleic anhydride derivative named cordyanhydride A ethyl ester (**1**).

Compound **4** was isolated as a yellow oil. Its molecular formula was established as $C_{13}H_{17}O_5$ by HRESIMS ion peak at m/z 253.1073 $[M + H]^+$ (calculated for $C_{13}H_{17}O_5^+$, 253.1071). Comparison of NMR spectroscopic data of **4** (Table 1) with **1** revealed that **4** shared part of the structure of **1**, with one acid anhydride moiety. By further analysis of the chemical shift, the coupling constant and the molecular formula, **4** was determined as shown in Figure 1. The structure had been reported as a synthetic product in a patent without NMR data and trivial name [11]. Herein, it was discovered as a new natural product and named maleicanhydridane (**4**).

Compounds bearing acid anhydride moieties are rare in nature. Cordyanhydrides A and B, bearing two and three acid anhydride moieties, were originally described from the insect pathogen fungus *Cordyceps pseudomilitaris* [12] and Amazonian endophytic *Talaromyces* fungi [10], with the absolute configuration unsolved. To the best of our knowledge, this study is the first example to obtain cordyanhydride derivatives from marine-derived microbes. In addition, **4** should be a precursor compound for the biosynthesis of **1**–**3**.

Compound **5** has been reported as a fungal metabolite [13] and was isolated as an epimer at C-7 of 7-epiaustdiol (**6**) [14] with almost identical NMR data. The absolute configurations of 7-epiaustdiol (**6**) and 8-O-methylepiaustdiol (**7**) were determined as shown in Figure 1, because they share the matched experimental ECD curve (Figure 3) and have similar OR values ($[\alpha]_D^{25}$ +172 (c 0.05, CH_3OH) and $[\alpha]_D^{25}$ +160 (c 0.05, CH_3OH)), similar to reports in the literature. In order to determine the absolute configurations of C-7/C-8 of **5**, the NOESY analysis and ECD calculation methods were used. The NOESY correlations of OH-7/H-8 and OH-8/H_3-13 (Figure 2) supported the *trans* configuration of OH-7 and OH-8. The Boltzmann-weighted ECD curves of 7R,8S-**5** and 7S,8R-**5** were calculated and compared with the experimental ECD curve (Tables S1 and S2), which led to the determination of the 7*R*,8*S* absolute configuration of **5** (Figure 3). Thus, **5** was assigned as (7*R*, 8*S*)-austdiol (**5**).

Table 1. The NMR data of **1** and **4** (500 and 125 MHz, δ in ppm, DMSO-d_6).

Pos.	1		4	
	δ_C Type	δ_H (*J* in Hz)	δ_C Type	δ_H (*J* in Hz)
1	171.5, C		171.4, C	
2	30.8, CH_2	2.58 (dd, 9.5, 7.0)	31.0, CH_2	2.59 (t, 7.5)
3	19.4, CH_2	2.68 (t, 8.0)	18.9, CH_2	2.75 (t, 7.5)
4	144.1, C		137.9, C	
5	143.0, C		136.8, C	
6	27.4, CH_2	2.40 (m) 2.47 (overlapped)	117.0, CH	6.49 (dt, 16.0, 1.5)
7	37.6, CH	2.05 (m)	147.5, CH	7.05 (dt, 16.0, 6.5)
8	27.9, CH_2	2.47 (overlapped)	26.6, CH_2	2.29 (td, 7.1, 1.6)
9	138.1, C		12.5, CH_3	1.05 (t, 7.5)
10	137.7, C		165.8, C	
11	117.0, CH	6.43 (dt, 15.5, 1.5)	164.6, C	
12	147.7, CH	7.08 (dt, 15.5, 6.5)	60.2, CH_2	4.05 (q, 7.0)
13	26.6, CH_2	2.28 (m)	14.0, CH_3	1.17 (t, 7.0)
14	12.5, CH_3	1.05 (t, 7.5)		
15	165.5, C			
16	166.0, C			
17	166.2, C			
18	164.5, C			
19	25.7, CH_2	1.30 (m)		
20	10.7, CH_3	0.88 (t, 7.5)		
21	60.3, CH_2	4.05 (q, 7.0)		
22	14.0, CH_3	1.17 (t, 7.0)		
7-OH				
8-OH				

Compound **10** was isolated as a brown oil, and its molecular formula was determined as $C_{14}H_{18}O_4$ by HRESIMS ion peak at *m*/*z* 273.1097 $[M + Na]^+$ (calculated for $C_{14}H_{18}NaO_4^+$, 273.1097), corresponding to six indices of hydrogen deficiency. The ^{1}H NMR data (Table 2) of **10** displayed the presence of a characteristic for *para*-substituted aromatic ring [δ_H 7.15 (d, *J* = 8.5 Hz, H-2 and H-6) and 6.86 (d, *J* = 8.6 Hz, H-3 and H-5)], three methylene groups [δ_H 4.57 (d, *J* = 6.6 Hz, H_2-10), 4.00 (s, H_2-13), and 3.58 (s, H_2-7)], a trisubstituted double bond [δ_H 5.43 (t, *J* = 6.3 Hz, H-11)], and two methyl groups [δ_H 3.59 (s, H_3-9) and 1.75 (s, H_3-14)]. The ^{13}C NMR data (Table 2) of **10** displayed 14 carbon resonances consistent with four non-protonated carbons (including three sp^2 carbons and a carbonyl), five sp^2 methine groups, three methylene groups, and two methyl groups (one of them oxygenated). The ^{1}H-^{1}H COSY spectrum (Figure 2) indicated the presence of three independent spin systems of H-2/H-3, H-5/H-6, and H_2-10/H-11. Comparison of its NMR data with those of stachyline G from *Mortierella* sp. revealed closed similarities except for the presence of an additional oxygenated methyl signal [15]. Extensive analysis of its HMBC spectrum revealed key signals from H_3-9 to C-8, indicating that the methoxy was linked to C-8 and formed a methyl ester group. The geometric configuration of the $\Delta^{10,11}$ double bond was determined to be in *Z* configuration by the NOESY correlations (Figure 2) of H_2-13 with H_2-10, and H_3-14 and H-11. The gross structure was constructed as shown in Figure 1 and named stachyline H (**10**).

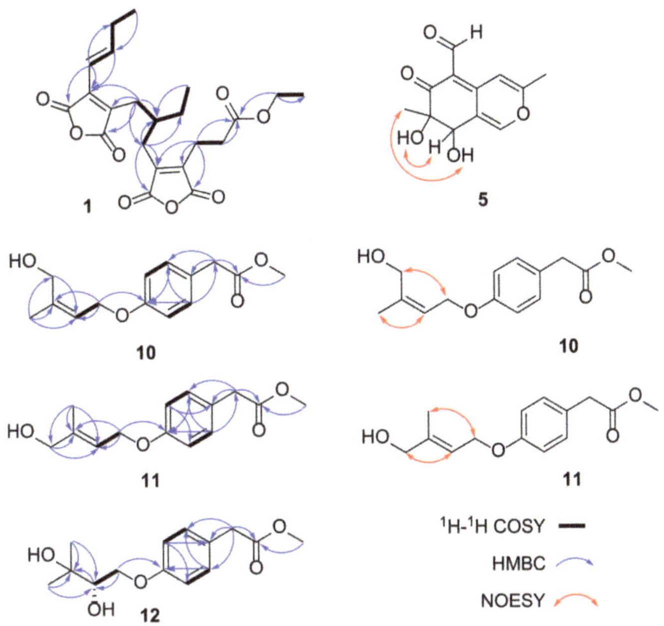

Figure 2. Key ^{1}H-^{1}H COSY, HMBC, and NOESY correlations of **1**, **5**, and **10–12**.

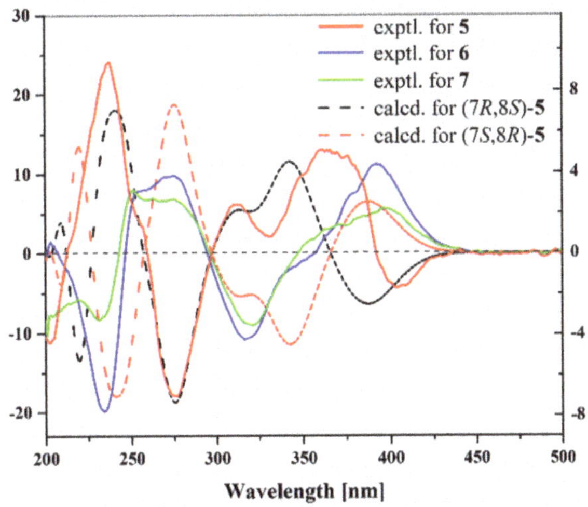

Figure 3. Experimental ECD spectra of **5–7** and calculational ECD spectrum of **5**.

Compound **11** was isolated as a brown oil, and its molecular formula was designated as $C_{14}H_{18}O_4$ by HRESIMS ion peak at m/z 268.1540 [M + NH_4]$^+$ (calculated for $C_{14}H_{22}NO_4^+$, 268.1543). Comparison of NMR spectroscopic data of **11** with **10** indicated that they shared the same planar structures, supported by the HMBC and COSY correlations (Figure 2). Upon, detailed interpretation of its ^{13}C NMR data, we found that the C-13 resonance was shielded from δ_C 21.0 in **10** to δ_C 13.8 in **11**, while the C-14 resonance was deshielded from δ_C 59.8 in **10** to δ_C 65.4 in **11**. The above obvious differences in the chemical shifts at C-13 and C-14 hinted that the geometric configuration of the $\Delta^{10,11}$ double bond is different in both compounds. The NOESY spectrum (Figure 2) of **11** revealed key signals of H_3-14 with H_2-10, and H_2-13 and H-11, indicating the double bond $\Delta^{10,11}$ was in *E* configuration. Accordingly, it was elucidated and named stachyline I (**11**).

Table 2. The NMR data of **10–12** (500 and 125 MHz, δ in ppm, DMSO-d_6).

Pos.	10		11		12	
	δ_C Type	δ_H (*J* in Hz)	δ_C Type	δ_H (*J* in Hz)	δ_C Type	δ_H (*J* in Hz)
1	126.2, C		126.2, C		126.1, C	
2	130.3, CH	7.15 (d, 8.5)	130.3, CH	7.16 (d, 8.6)	130.4, CH	7.15 (d, 8.5)
3	114.5, CH	6.86 (d, 8.6)	114.5, CH	6.87 (d, 8.6)	114.4, CH	6.88 (d, 8.6)
4	157.3, C		157.3, C		157.9, C	
5	114.5, CH	6.86 (d, 8.6)	114.5, CH	6.87 (d, 8.6)	114.4, CH	6.88 (d, 8.6)
6	130.3, CH	7.15 (d, 8.5)	130.3, CH	7.16 (d, 8.6)	130.4, CH	7.15 (d, 8.5)
7	51.6, CH_2	3.58 (s)	51.6, CH_2	3.59 (overlapped)	39.3, CH_2	3.58 (s)
8	171.9, C		171.9, C		172.0, C	
9	39.8, CH_3	3.59 (s)	39.2, CH_3	3.59 (overlapped)	51.6, CH_3	3.59 (s)
10	63.6, CH_2	4.57 (d, 6.6)	64.0, CH_2	4.57 (d, 6.5)	69.8, CH_2	4.19 (dd, 10.1, 2.3) 3.76 (dd, 10.0, 8.0)
11	121.1, CH	5.43 (t, 6.3)	118.1, CH	5.63 (t, 6.3)	75.8, CH	3.52 (d, 7.9)
12	140.4, C		140.5, C		70.9, C	
13	59.8, CH_2	4.00 (s)	65.4, CH_2	3.84 (s)	27.4, CH_3	1.13 (s)
14	21.0, CH_3	1.75 (s)	13.8, CH_3	1.65 (s)	24.3, CH_3	1.07 (s)
13-OH				4.87 (s)		
12-OH						4.40 (s)
11-OH						4.98 (s)

Compound **12** was obtained as a colorless oil, and its molecular formula of $C_{14}H_{20}O_5$ was deduced from the positive HRESIMS ion peak at m/z 286.1651 $[M + NH_4]^+$ (calculated for $C_{14}H_{24}NO_5^+$, 286.1649), implying five degrees of hydrogen deficiency. The ^{1}H NMR data (Table 2) of **12** displayed the series of typical proton signals responsive for a *para*-substituted aromatic ring [δ_H 7.15 (d, J = 8.5 Hz, H-2 and H-6) and 6.88 (d, J = 8.6 Hz, H-3 and H-5)], two methylene groups [δ_H 3.58 (s, H_2-7), 4.19 (dd, J = 10.1, 2.3 Hz, H-10a), and 3.76 (dd, J = 10.0, 8.0 Hz, H-10b)], a oxygenated methine [δ_H 3.52 (d, J = 7.9 Hz, H-11)], and three methyl groups [δ_H 3.59 (s, H_3-9), 1.13 (s, H_3-13), and 1.07 (s, H_3-14)]. The ^{13}C NMR data (Table 2) of **12** displayed 14 carbon resonances consistent with four non-protonated carbons (including one sp^3 carbon, two sp^2 carbons and a carbonyl), five sp^2 methine groups (including one oxygenated and four sp^2 hybridized), two methylene groups (one of them oxygenated), and three methyl groups (one of them oxygenated). Analysis of its NMR data revealed that the structure of **12** closely resembled that of **10**. The difference was the replacement of signals for the 4-hydroxy-2-en-3-methylbutoxy unit substituted at the C-4 position in **10** with those for the 2,3-dihydroxy-3-methylbutoxy moiety in **12**. The ^{1}H-^{1}H COSY correlation of H_2-10/H-11 and the HMBC correlations of H_2-10 with C-4 and C-11, H-11 with C-12, and H_3-13/H_3-14 with C-11 and C-12 support the above deduction. In addition, the planar structure of **12** was similar to the known compound stachyline E except for the presence of an additional oxygenated methyl [15]. The recorded optical rotation for **12** was $[\alpha]_D^{25} -4$ (c 0.1, CH_3OH), which has the same angle as that of stachyline E, $[\alpha]_D^{23} -6$ (c 0.1, CH_3OH), suggesting that the configuration of C-11 in **12** is the same as that of stachyline E. Consequently, the structure of **12** was determined and assigned stachyline J (**12**). Compounds **10–12** may be separation artifacts.

By comparing their physicochemical properties and spectroscopic data with the reported literature values, other known compounds were determined. Compounds present in SCSIO 41050 were cordyanhydride A methyl ester (**2**) [10], cordyanhydride A (**3**) [12], 7-epiaustdiol (**6**) [14], 8-*O*-methylepiaustdiol (**7**) [14], 4-hydroxy-3,6-dimethyl-2*H*-pyran-2-one (**8**) [16], and 2,4-dihydroxy-3,6-dimethylbenzoic acid (**9**) [17]. Compounds present in SCSIO 41411 were stachylines G (**13**) and F (**14**) [15], (*E*)-4-(4-hydroxy-3-methylbut-2-enyloxy)benzaldehyde (**15**) [18], stachyline E (**16**) [15], penialidins C (**17**) [19], dibutylphthalate (**18**) [20], brefeldin G (**19**) [21], 9,12-octadecadieonic acid (**20**) [22], α-linolenic acid (**21**) [23], linoleic acid (**22**) [24], and glycerol monlinoleate (**23**) [25].

2.2. Bioactive Assay

All the isolated compounds (**1–23**) were evaluated for cytotoxicity against four cancer cell lines (PC-3, 22Rv1, A549, WPMY-1), antibacterial activities against five bacteria (*Erysipelothrix rhusiopathiae* WH13013, *Streptococcus suis* SC19, *Escherichia coli* ATCC 25922, *Pseudomonas aeruginosa* ATCC 27853, *Staphylococcus aureus* ATCC 25923), and antifungal activities against five strains (*Botrytis cinerea*, *Verticillium dahlia* kieb., *Fusarium graminearum* schw., *Fusarium oxysporum* f.sp. *niveum*, *Rhizoctonia solani*). As shown in Table 3, maleicanhydridane (**4**), with one acid anhydride moiety, showed moderate cytotoxicity against cell lines A549 and WPMY-1, with IC_{50} values of 15.5 and 22.9 μM, respectively, while other compounds, including the cordyanhydride derivatives (**1–3**) with two acid anhydride moieties, were inactive (IC_{50} > 50 μM). In the antibacterial assay, only **9**, **21**, and **22** showed weak antibacterial activity against *S. suis*, with minimal inhibitory concentration (MIC) values of 50, 100, and 100 μg/mL, respectively. It is worth noting that three cordyanhydride derivatives (**1–3**) showed obvious antifungal activities, especially against *F. graminearum*, *F. oxysporum*, and *R. solani*, with MIC values of 6.25–12.5 μg/mL.

Table 3. Cytotoxic, antibacterial and antifungal activities of isolated compounds.

	Tested Compounds					
Cells (IC_{50}, μM)	**4**	**Others**				**Positive**
PC-3	>50	>50				0.12 [a]
22Rv1	>50	>50				0.03 [a]
A549	15.5	>50				29.95 [a]
WPMY-1	22.9	>50				0.51 [a]
Bacteria (MIC, μg/mL)	**9**	**21**	**22**	**Others**		**Positive**
E. rhusiopathiae	>100	>100	100	>100		12.5 [b]
S. suis	50	100	100	>100		12.5 [b]
E.coli	>100	>100	>100	>100		50 [c]
P. aeruginosa	>100	>100	>100	>100		12.5 [b]
S. aureus	100	>100	>100	>100		12.5 [b]
Fungi (MIC, μg/mL)	**1**	**2**	**3**	**9**	**Others**	
B. cinerea	12.5	25	>100	>100	>100	12.5 [d]
V. dahlia	100	>100	>100	50	>100	12.5 [d]
F. graminearum	6.25	12.5	12.5	50	>100	12.5 [d]
F. oxysporum	6.25	6.25	12.5	100	>100	25 [d]
R. solani	6.25	6.25	12.5	>100	>100	25 [d]

[a] docetaxel; [b] penicillin; [c] streptomycin; [d] cycloheximide.

The obtained compounds were screened at 20 μM for their inhibitory activities against LPS-induced NF-κB activation in RAW264.7 cells, and no compounds showed obvious activities. The enzyme inhibitory activity assay was also conducted for acetylcholine esterase (AChE) and phosphodiesterase 4 (PDE4). PDE4 was involved in the regulation of proinflammatory cytokines via the degradation of cyclic adenosine monophosphate [26]. As a result, compounds **1**, **2**, **4**, and **6** at 10 μM displayed weak or moderate inhibition against PDE4 with inhibitory ratios of 49.7%, 27.5%, 11.2% and 39.6%, respectively. Meanwhile, **5**, **10**, and **11** at 50 μM displayed weak inhibition against AChE with inhibitory ratios of 21.3%, 22.3%, and 19.9%, respectively. Although the activities were weak, this is the first report of the AChE inhibitory activities of austdiol polyketides and stachyline derivates.

In order to further understand the interaction between the compounds and AChE protein, so as to improve the activity by structure optimization in the future, docking studies were carried out for **5**, **10**, and **11** in the active site of AChE (PDB: 1UT6) to gain insights into their molecular interactions. As a result, these ligands were favorably accommodated within the binding cleft with analogous anchoring conformations, exhibiting binding free energies (designated as S value) spanning from −8.7 to −8.4 kcal/mol. Compounds **5**, **10**, and **11** interacted with the AChE active site mainly through hydrogen bonds, π-π stacking contacts, and hydrophobic interactions (Figure 4). Compound **5** formed hydrogen bonds with amino acid residues TYR121 and SER122 within the target protein at distances of 3.6 Å and 3.0 Å, respectively. It also exhibited π-π stacking contacts with TRP84 and PHE330, as well as hydrophobic interactions with TRP84, PHE330, ILE439, and TYR442. Compound **10** established hydrogen bonds with TRP84, GLY118, GLY119, SER200, TRP432, and HIS440 at distances of 2.9 Å, 3.5Å, 3.0Å, 2.8Å, 3.2Å, and 3.9Å, respectively. It also showed π-π stacking contacts with TRP84 and PHE330, in addition to hydrophobic interactions with TRP84, TYR121, PHE290, PHE330, PHE331, ILE439, and TYR442. Compound **11** formed hydrogen bonds with TRP84, SER122, GLY123, and TRP43 at distances of 2.9 Å, 3.5Å, 3.3Å, and 3.3Å, respectively. It also exhibited π-π stacking contacts with TRP84 and PHE330, as well as hydrophobic interactions with ASP72, TRP84, TYR121, PHE330, ILE439, and TYR442. The binding of these compounds to the enzyme was stabilized through these interactions.

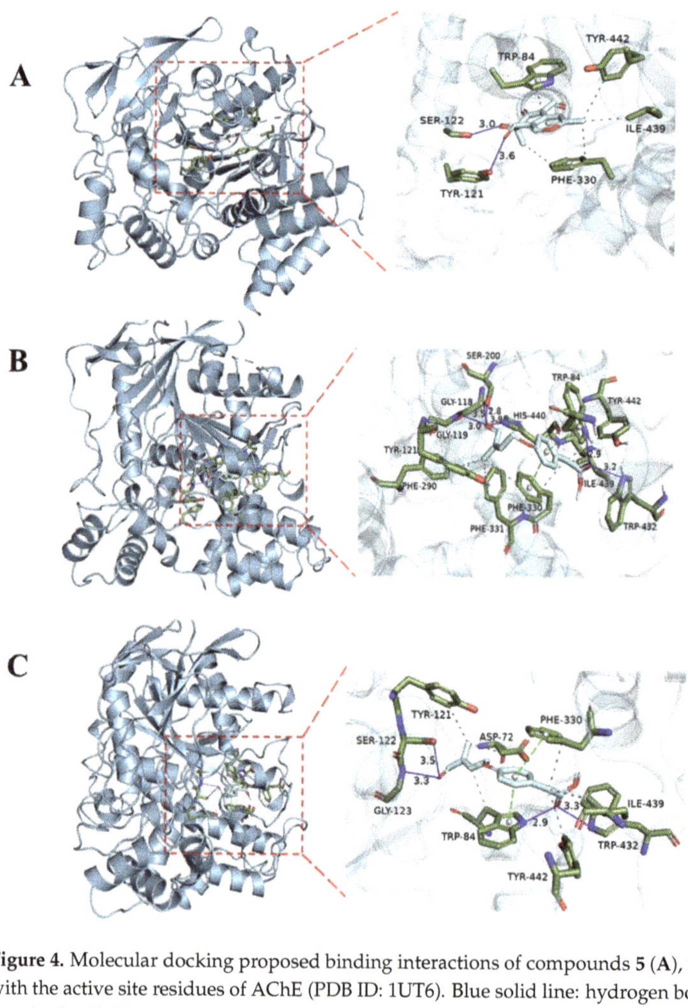

Figure 4. Molecular docking proposed binding interactions of compounds **5** (**A**), **10** (**B**), and **11** (**C**) with the active site residues of AChE (PDB ID: 1UT6). Blue solid line: hydrogen bond; black dotted line: hydrophobic interaction; green dotted line: π-π stacking interaction.

3. Materials and Methods

3.1. General Experimental Procedures

The UV spectrum was recorded on a Shimadzu UV-2600 PC spectrometer (Shimadzu, Beijing, China). The IR spectrum was obtained using an IR Affinity-1 spectrometer (Shimadzu). Optical rotations were determined with an Anton Paar MPC 500 polarimeter. HRESIMS spectra were recorded with a Bruker maXis Q-TOF mass spectrometer. The NMR spectra were recorded on a Bruker Avance-500 spectrometer (Bruker BioSpin International AG, Fällanden, Switzerland), and chemical shifts were recorded as δ-values. Semipreparative high-performance liquid chromatography (HPLC) was performed on the Hitachi Primaide with a DAD detector, using an ODS column (YMC-pack ODS-A, 10 × 250 mm, 5 μm). Thin-layer chromatography analysis (TLC) and column chromatography (CC) were carried out on plates precoated with silica gel GF254 (10–40 μm) and over silica gel (200–300 mesh) (Qingdao Marine Chemical Factory, Qingdao, China) and Sephadex LH-20 (Amersham Biosciences, Uppsala, Sweden), respectively. Spots were detected on TLC (Qingdao Marine Chemical Factory) under 254 nm UV light. All solvents employed were of analytical grade (Tianjin Fuyu Chemical and Industry Factory, Tianjin, China).

3.2. Fungal Material

The fungal strains SCSIO 41050 and SCSIO 41411 were isolated from a mangrove sediment sample, collected from Gaoqiao mangrove wetland (21.573° N, 109.767° E) in Zhanjiang, coastline of the northern part of Beibu Gulf, China. The strains were stored on MB agar (malt extract 15 g, sea salt 10 g, agar 16 g, H_2O 1 L, pH 7.4–7.8) slants in liquefied petrolatum and deposited at Key Laboratory of Tropical Marine Bio-resources and Ecology, Chinese Academy of Sciences. The strains SCSIO 41050 and SCSIO 41411 were designated as *Talaromyces* sp. and *Penicillium* sp., due to their ITS sequences (GenBank accession No. OQ867300 and OQ052995) homology with those of *Talaromyces* sp. KT240143.1 and *Penicillium* sp. NR138263.1, respectively.

3.3. Fermentation and Extraction

The fungal strains were cultured in 200 mL seed medium (15 g malt extract, 10g sea salt, 1 L H_2O) in 500 mL Erlenmeyer flasks at 28 °C for 3 days on a rotary shaker (180 rpm). Large-scale fermentations of SCSIO 41050 and SCSIO 41411 were incubated statically at 25 °C for 30 days using a rice medium (200 g rice, 2.5% sea salt, 230 mL H_2O) in the 1 L flask (×60 and ×45, respectively). The fermented culture was extracted three times with EtOAc, yielding a reddish extract (130 g) and a brown extract (53.2 g), respectively.

3.4. Isolation and Purification

The SCSIO 41050 organic extract was subjected to silica gel CC using step gradient elution with petroleum ether/CH_2Cl_2 (0–100%, v/v) and CH_2Cl_2/CH_3OH (0–100%, v/v) to obtain eight fractions (Frs. 1–8) based on TLC properties. Fraction 1 was subjected to semipreparative HPLC eluting with 88% CH_3CN/H_2O (0.4% TFA, 3 mL/min) to afford compound **4** (3.7 mg, t_R = 8.6 min). Fraction 2 was divided into four subfractions (Frs. 2-1–2-4) by semipreparative HPLC using step gradient elution with CH_3CN/H_2O (0.4% TFA, 65–85%, v/v, 0–30 min). Subfraction 2-1 was further purified by semipreparative HPLC (68% CH_3CN/H_2O (0.4% TFA), 2 mL/min) to afford **3** (18.5 mg, t_R = 18.2 min). Subfraction 2-3 was further purified by semipreparative HPLC (70% CH_3CN/H_2O (0.4% TFA), 2 mL/min) to afford **2** (40.5 mg, t_R = 20.5 min). **1** (11.6 mg, t_R = 26.2 min) was obtained from subfraction 2–4 by semipreparative HPLC, eluting with 70% CH_3CN/H_2O (0.4% TFA, 2 mL/min). Fraction 5 was divided into six subfractions (Frs. 5-1–5-4) by MPLC using step gradient elution with CH_3OH/H_2O (10–100%, v/v). Subfraction 5-1 was further purified by semipreparative HPLC (40% CH_3OH/H_2O (0.4% TFA), 2 mL/min) to afford **6** (92.6 mg, t_R = 9.7 min). Subfraction 5-3 was further divided by semipreparative HPLC (38% CH_3OH/H_2O (0.4% TFA), 2 mL/min) to afford **7** (3.4mg, t_R = 12.9 min), **5** (6.1 mg, t_R = 14.0 min), and **8** (40.8 mg, t_R = 15.7 min). **9** (26.6 mg, t_R = 17.5 min) was obtained from subfraction 5-4 by semipreparative HPLC, eluting with 53% CH_3OH/H_2O (0.4% TFA, 2 mL/min).

The SCSIO 41411 crude extract was chromatographed over an ODS RP-18 CC eluted with CH_3OH/H_2O (10–100%, v/v) to obtain fifteen fractions (Frs. 1–15). Fraction 2 was chromatographed over an ODS RP-18 CC eluted with CH_3OH/H_2O (10–100%, v/v) to obtain nine fractions (Fr. 2-1–Fr. 2-9). Fraction 2-8 was subjected to semipreparative HPLC, eluting with 62% CH_3OH/H_2O (2.5 mL/min) to afford **10** (4.0 mg, t_R = 15.6 min), **11** (5.5 mg, t_R = 13.8 min), and **17** (4.2 mg, t_R = 21.0 min). Fraction 2-5 was subjected to semipreparative HPLC, eluting with 31% CH_3CN/H_2O (3 mL/min) to afford **14** (24.0 mg, t_R = 9.5 min), **15** (1.3 mg, t_R = 14.0 min), and Fraction 2-5-2. Fraction 2-5-2 was further purified by semipreparative HPLC (55% CH_3OH/H_2O, 2 mL/min) to afford **12** (3.7 mg, t_R = 12.0 min) and **13** (1.6 mg, t_R = 9.7 min). Fraction 2-3 was subjected to semipreparative HPLC, eluting with 20% CH_3CN/H_2O (3 mL/min) to provide **16** (8.9 mg, t_R = 13.0 min) and **19** (11.4 mg, t_R = 16.5 min). Fraction 10 was subjected to semipreparative HPLC, eluting with 75% CH_3CN/H_2O (3 mL/min) to afford **18** (2.9 mg, t_R = 15.8 min). Fraction 11 was subjected to semipreparative HPLC, eluting with 90% CH_3OH/H_2O (3 mL/min) to give **20** (4.0 mg, t_R = 18.1 min) and Fraction 11-2. Further purification of Fraction 11-2 by HPLC (85% CH3CN/H_2O, 3 mL/min) yielded **21** (23.6 mg, t_R = 14.1 min), **22** (106.4 mg, t_R = 20.0 min), and **23** (85.6 mg, t_R = 15.4 min).

3.5. Spectroscopic Data of New Compounds

Cordyanhydride A ethyl ester (**1**): yellow oil; $[\alpha]_D^{25}$ +2 (c 0.05, CH_3OH); UV (CH_3OH) λ_{max} (log ε) 205 (4.24), 250 (3.84), 320 (3.45) nm; IR (film) ν_{max} 2963, 2928, 2361, 1764, 1732, 1271, 1184, 1024, 974, 920 cm^{-1}; ^1H and ^{13}C NMR data, Table 1; HRESIMS m/z 419.1701 $[M + H]^+$ (calculated for $C_{22}H_{27}O_8^+$, 419.1700), 441.1521 $[M + Na]^+$ (calculated for $C_{22}H_{26}NaO_8^+$, 441.1521).

Maleicanhydridane (**4**): yellow oil; ^1H and ^{13}C NMR data as shown in Table 1; HRESIMS m/z 253.1073 $[M + H]^+$ (calculated for $C_{13}H_{17}O_5^+$, 253.1071).

Stachyline H (**10**): brown oil; UV (CH_3OH) λ_{max} (log ε) 202 (3.81), 226 (3.51) nm; IR (film) ν_{max} 3446, 2949, 2879, 1732, 1611, 1510, 1435, 1223, 1159, 1003, 818 cm^{-1}; ^1H and ^{13}C NMR data as shown in Table 2; HRESIMS m/z 273.1097 $[M + Na]^+$ (calculated for $C_{14}H_{18}NaO_4^+$, 273.1097).

Stachyline I (**11**): brown oil; UV (CH_3OH) λ_{max} (log ε) 202 (3.88), 226 (3.61) nm; IR (film) ν_{max} 3443, 2951, 2918, 1732, 1510, 1225, 1157, 1001, 820, cm^{-1}; ^1H and ^{13}C NMR data as shown in Table 2; HRESIMS m/z 268.1540 $[M + NH_4]^+$ (calculated for $C_{14}H_{22}NO_4^+$, 268.1543) and 273.1091 $[M + Na]^+$ (calculated for $C_{14}H_{18}NaO_4^+$, 273.1097).

Stachyline J (**12**): colorless oil; $[\alpha]_D^{25}$ −4 (c 0.1, CH_3OH); UV (CH_3OH) λ_{max} (log ε) 200 (4.36), 226 (3.90) nm; IR (film) ν_{max} 3446, 2976, 1734, 1514, 1246, 1163, 1032, 831, 806, 536 cm^{-1}; ^1H and ^{13}C NMR data as shown in Table 2; HRESIMS m/z 286.1651 $[M + NH_4]^+$ (calculated for $C_{14}H_{24}NO_5^+$, 286.1649) and 291.1202 $[M + Na]^+$ (calculated for $C_{14}H_{20}NaO_5^+$, 291.1203).

3.6. ECD Calculation of 5

Conformational analyses were carried out via Monte Carlo searching by means of Spartan'14 software (v1.1.4, Wavefunction, Irvine, CA, USA) using a Molecular Merck force field. The results showed ten lowest energy conformers within an energy window of 14 Kcal/mol. Then, these conformers were further re-optimized by the TD-DFT method at the B3LYP/6-31G(d) level in methanol using the Gaussian 16 program (A.03, Guassian, Pittsburgh, PA, USA) [27]. ECD calculations were further carried out at the B3LYP/6-311+G (d, p) level in methanol by adopting 50 excited states. The ECD spectra were generated based on Boltzmann distribution theory by SpecDis (1.70.1, SpecDis, Berlin, Germany) under a half band width of 0.3 eV and shifted by −25 nm to facilitate comparison to the experimental data.

3.7. Antibacterial and Antifungal Activity Assay

The antimicrobial activities against five bacteria (*Erysipelothrix rhusiopathiae* WH13013, *Streptococcus suis* SC19, *Escherichia coli* ATCC 25922, *Pseudomonas aeruginosa* ATCC 27853, *Staphylococcus aureus* ATCC 25923) and five fungi (*Botrytis cinerea*, *Verticillium dahlia* kieb., *Fusarium graminearum* schw., *Fusarium oxysporum* f.sp. *niveum*, *Rhizoctonia solani*) were evaluated using the methods described previously [9,28]. Streptomycin and penicillin were used as positive controls against bacteria, and cycloheximide was used against fungi.

3.8. Cytotoxicity Bioassay

Cytotoxicities against PC-3 (human prostate cancer cell line), 22Rv1 (human prostate cancer cell line), WPMY-1 (human prostatic stromal myofibroblast cell line), and A549 (human lung cancer cell), purchased from Shanghai Cell Bank, Chinese Academy of Sciences, were evaluated. Cell viability was analyzed by 3-(4,5)-dimethylthiahiazo (-z-yl)-3,5-diphenytetrazoliumromide (MTT) assay as previously described [29]. In brief, cells were seeded in a 96-well plate at a density of 5×10^3 per well overnight and treated with compounds for the required time. OD_{570} values were detected using a Hybrid Multi-Mode Reader (Synergy H1, BioTek, Santa Clara, CA, USA). The experiment was independently repeated three times.

3.9. NF-κB Bioassay

The suppression of LPS-induced NF-κB activation in RAW264.7 cells was assessed using a luciferase reporter gene assay as detailed previously [30].

3.10. Enzyme Inhibitory Activities Assay

The protocols for expression, purification, and enzymatic assays of PDE4D2 were similar to those we described previously [26]. The inhibitory activity of AChE was assessed in vitro following a modified Ellman method [31].

3.11. Molecular Docking

The molecular docking simulation was implemented by utilizing the software AutoDock Tools (ADT 1.5.6) [32]. The crystal structure of AChE from *Tetronarce californica* (PDB ID: 1UT6) [33] was acquired from the Protein Data Bank (http://www.rcsb.org, accessed on 21 April 2005). The structures of ligands were generated in ChemBioOffice 18.0 (PerkinElmer Informatics, Waltham, MA, USA), followed by an MM2 calculation to minimize the conformation energy. The size of the grid box was $2.3 \times 62.7 \times 55.6$, centered at x: 31.1, y: 27.7, z: 50.7. The other docking parameters, settings, and calculations were default, and the docking results were analyzed using the software PyMOL 2.4.0 (Schrödinger, New York, NY, USA).

4. Conclusions

In conclusion, the chemical investigation of the two mangrove-sediment fungal stains *Talaromyces* sp. SCSIO 41050 and *Penicillium* sp. SCSIO 41411 afforded 23 different compounds. Among these, cordyanhydride A ethyl ester (**1**) and stachylines H–J (**10–12**) were identified as new compounds, and maleicanhydridane (**4**) as a new natural product. Although some natural maleic anhydrides have been reported from *Talaromyces* species [34,35], the discovery of three cordyanhydride derivatives (**1–3**) with obvious antifungal activities was impressive. Other active natural products were also revealed, such as **4** with moderate cytotoxicity against cell lines A549 and WPMY-1, **1** and **6** with PDE4 inhibitory activities, and **5**, **10**, and **11** with potential for inhibiting AChE. The obtained results highlight the immense potential of the mangrove wetland ecosystem to yield novel natural products as well as bioactive compounds.

Supplementary Materials: The following supporting information can be downloaded at: https://www.mdpi.com/article/10.3390/md21060327/s1, Figures S1–S39: NMR, HRESIMS, UV, IR and CD spectra of compounds **1**, **4**, and **10–12**; Tables S1 and S2: ECD calculation details for **5**; The spectroscopic data of **2–3**, **5–9**, and **13–23**; ITS sequence data of the strains.

Author Contributions: Funding acquisition, X.Z. and H.T.; Investigation, B.P., J.C., Z.X., M.L., X.L., B.Y., W.F., Y.-Y.H. and C.C.; Methodology, B.P. and J.C.; Supervision, C.C., X.Z. and H.T.; Writing—original draft, B.P., J.C. and C.C.; Writing—review & editing, X.Z. and H.T. All authors have read and agreed to the published version of the manuscript.

Funding: This research was funded by the Marine Economy Development Project of Guangdong Province (GDNRC [2023]37, GDNRC [2022]35), National Key Research and Development Program of China (2021YFD1800402), National Natural Science Foundation of China (U20A20101, 81973235), K. C. Wong Education Foundation (GJTD-2020-12), Natural Science Foundation of Guangdong Province (2021A1515011711), and Science and Technology Major Program of Hubei Province (2021ABA005).

Institutional Review Board Statement: Not applicable.

Informed Consent Statement: Not applicable.

Data Availability Statement: The data presented in this study are available on request from the corresponding author.

Acknowledgments: We are grateful to thank Zhihui Xiao, Xiaohong Zheng, Aijun Sun, Yun Zhang, and Xuan Ma in the analytical facility at SCSIO for recording spectroscopic data.

Conflicts of Interest: The authors declare no conflict of interest.

References

1. Xu, J. Bioactive natural products derived from mangrove-associated microbes. *RSC Adv.* **2015**, *5*, 841–892. [CrossRef]
2. Li, K.; Chen, S.; Pang, X.; Cai, J.; Zhang, X.; Liu, Y.; Zhu, Y.; Zhou, X. Natural products from mangrove sediments-derived microbes: Structural diversity, bioactivities, biosynthesis, and total synthesis. *Eur. J. Med. Chem.* **2022**, *230*, 114117. [CrossRef] [PubMed]
3. Zhou, X. Mangrove soil-derived Streptomyces: An important resource of pharmaceutical active natural products. *J. Holist. Integr. Pharm.* **2022**, *3*, 300–314.
4. Lei, L.; Gong, L.; Jin, M.; Wang, R.; Liu, R.; Gao, J.; Liu, M.; Huang, L.; Wang, G.; Wang, D.; et al. Research advances in the structures and biological activities of secondary metabolites from *Talaromyces*. *Front. Microbiol.* **2022**, *13*, 984801. [CrossRef]
5. Chen, C.; Chen, W.; Pang, X.; Liao, S.; Wang, J.; Lin, X.; Yang, B.; Zhou, X.; Luo, X.; Liu, Y.H. Pyrrolyl 4-quinolone alkaloids from the mangrove endophytic fungus *Penicillium steckii* SCSIO 41025: Chiral resolution, configurational assignment, and enzyme inhibitory activities. *Phytochemistry* **2021**, *186*, 112730. [CrossRef]
6. Chen, C.; Chen, W.; Tao, H.; Yang, B.; Zhou, X.; Luo, X.W.; Liu, Y.H. Diversified polyketides and nitrogenous compounds from the mangrove endophytic fungus *Penicillium steckii* SCSIO41025. *Chin. J. Chem.* **2021**, *39*, 2132–2140. [CrossRef]
7. Bai, M.; Zheng, C.J.; Chen, G.Y. Austins-type meroterpenoids from a mangrove-derived *Penicillium* sp. *J. Nat. Prod.* **2021**, *84*, 2104–2110. [CrossRef]
8. Cai, J.; Chen, C.; Tan, Y.; Chen, W.; Luo, X.; Luo, L.; Yang, B.; Liu, Y.; Zhou, X. Bioactive polyketide and diketopiperazine derivatives from the mangrove-sediment-derived fungus *Aspergillus* sp. SCSIO41407. *Molecules* **2021**, *26*, 4851. [CrossRef]
9. Cai, J.; Wang, X.; Gan, X.; Zhou, Q.; Luo, X.; Yang, B.; Liu, Y.; Ratnasekera, D.; Zhou, X. New chlorinated metabolites and antiproliferative polyketone from the mangrove sediments-derived fungus *Mollisia* sp. SCSIO41409. *Mar. Drugs* **2023**, *21*, 32. [CrossRef]
10. Da Silva, P.; de Souza, M.; Bianco, E.; da Silva, S.; Soares, L.; Costa, E.; da Silva, F.; Barison, A.; Forim, M.; Cass, Q.; et al. Antifungal polyketides and other compounds from amazonian endophytic *Talaromyces* fungi. *J. Brazil. Chem. Soc.* **2018**, *29*, 622–630. [CrossRef]
11. Omura, S.; Shiomi, K.; Asami, Y.; Kashima, Y.; Nonaka, K.; Hanaki, H. Novel Metallo-β-Lactamase Inhibitor Containing 2,5-Furandione Derivative, and Antiinfective Composition Containing the Same with β-Lactam. JP2016179964A, 13 October 2016.
12. Isaka, M.; Tanticharoen, M.; Thebtaranonth, Y. Cordyanhydrides A and B. Two unique anhydrides from the insect pathogenic fungus *Cordyceps pseudomilitaris* BCC 1620. *Tetrahedron. Lett.* **2000**, *41*, 1657–1660. [CrossRef]
13. Andrioli, W.J.; Conti, R.; Araujo, M.J.; Zanasi, R.; Cavalcanti, B.C.; Manfrim, V.; Toledo, J.S.; Tedesco, D.; de Moraes, M.O.; Pessoa, C.; et al. Mycoleptones A-C and polyketides from the endophyte *Mycoleptodiscus indicus*. *J. Nat. Prod.* **2014**, *77*, 70–78. [CrossRef] [PubMed]
14. Liu, F.; Cai, X.L.; Yang, H.; Xia, X.K.; Guo, Z.Y.; Yuan, J.; Li, M.F.; She, Z.G.; Lin, Y.C. The bioactive metabolites of the mangrove endophytic fungus *Talaromyces* sp. ZH-154 isolated from *Kandelia candel* (L.) Druce. *Planta Med.* **2010**, *76*, 185–189. [CrossRef] [PubMed]

15. Huang, Z.; Wang, W.; Han, X.; Yang, X. Three new hydroxyphenylacetic acid derivatives and a new alkaloid from endophytic fungus *Mortierella* sp. in *Epimedium acuminatum Franch.* and their antibacterial activity. *Chem. Biodivers.* **2021**, *18*, e2100741. [CrossRef]
16. Liu, S.; Tang, X.; He, F.; Jia, J.; Hu, H.; Xie, B.; Li, M.; Qiu, Y. Two new compounds from a mangrove sediment-derived fungus *Penicillium polonicum* H175. *Nat. Prod. Res.* **2022**, *36*, 2370–2378. [CrossRef]
17. Xiang, Y.; Xi, Y.; Luo, G.; Long, Y.; Yang, W. Synthesis of barbacic acid. *J. Asian Nat. Prod. Res.* **2022**, *24*, 1150–1156.
18. Chen, J.; Lin, Y.; Day, S.; Hwang, T.; Chen, I. New benzenoids and anti-inflammatory constituents from *Zanthoxylum nitidum*. *Food Chem.* **2011**, *125*, 282–287. [CrossRef]
19. Jouda, J.; Kusari, S.; Lamshöft, M.; Mouafo Talontsi, F.; Douala Meli, C.; Wandji, J.; Spiteller, M. Penialidins A–C with strong antibacterial activities from *Penicillium* sp., an endophytic fungus harboring leaves of *Garcinia nobilis*. *Fitoterapia* **2014**, *98*, 209–214. [CrossRef]
20. Wu, H.; Su, Z.; Aisa, H.; Yili, A.; Hang, B. Components of *Cichorium glandulosum* seeds. *Chem. Nat. Compd.* **2007**, *43*, 472–473. [CrossRef]
21. Zeng, F.; Chen, C.; Ali, A.A.C.; Zhou, Q.; Tong, Q.; Wang, W.; Zang, Y.; Gong, J.; Wu, Z.; Liu, J. Dibrefeldins A and B, a pair of epimers representing the first brefeldin A dimers with cytotoxic activities from *Penicillium janthinellum*. *Bioorg. Chem.* **2019**, *86*, 176–182. [CrossRef]
22. Sugimoto, T.; Miyase, T.; Kuroyanagi, M.; Ueno, A. Limonoids and quinolone alkaloids from *evodia rutaecarpa* benth. *Chem. Pharm. Bull.* **2011**, *36*, 4453–4461. [CrossRef]
23. Alamsjah, M.A.; Hirao, S.; Ishibashi, F.; Fujita, Y. Isolation and structure determination of algicidal compounds from *Ulva fasciata*. *Biosci. Biotechnol. Biochem.* **2005**, *69*, 2186–2192. [CrossRef] [PubMed]
24. Lee, J.H.; Lee, K.R. Phytochemical constituents of *Cirsium nipponicum* (max.) makino. *Korean J. Pharmacogn.* **2005**, *36*, 145–150.
25. Wang, M.; Chen, Y.; Sun, Z.; Tan, G.; Li, H.; Liu, H.; Yan, H.; Guo, X.; Zhang, W. Study on cytotoxic secondary metabolites of endophytic fungus *Diaporthe longicolla* A616 from *Pogostemon cablin*. *Chin. J. Chin. Mater. Med.* **2016**, *41*, 2112–2117.
26. Lin, T.; Huang, Y.; Tang, G.; Cheng, Z.; Liu, X.; Luo, H.; Yin, S. Prenylated coumarins: Natural phosphodiesterase-4 inhibitors from *Toddalia asiatica*. *J. Nat. Prod.* **2014**, *25*, 955–962. [CrossRef]
27. Luo, X.; Lin, X.; Tao, H.; Wang, J.; Li, J.; Yang, B.; Zhou, X.; Liu, Y. Isochromophilones A–F, cytotoxic chloroazaphilones from the marine mangrove endophytic fungus *Diaporthe* sp. SCSIO 41011. *J. Nat. Prod.* **2018**, *81*, 934–941. [CrossRef]
28. Fang, W.; Wang, J.; Wang, J.; Shi, L.; Li, K.; Lin, X.; Min, Y.; Yang, B.; Tang, L.; Liu, Y.; et al. Cytotoxic and antibacterial eremophilane sesquiterpenes from the marine-derived fungus *Cochliobolus lunatus* SCSIO41401. *J. Nat. Prod.* **2018**, *81*, 1405–1410. [CrossRef]
29. Wang, X.; Zhu, J.; Yan, H.; Shi, M.; Zheng, Q.; Wang, Y.; Zhu, Y.; Miao, L.; Gao, X. Kaempferol inhibits benign prostatic hyperplasia by resisting the action of androgen. *Eur. J. Pharmacol.* **2021**, *907*, 174251. [CrossRef]
30. Tan, Y.; Deng, W.; Zhang, Y.; Ke, M.; Zou, B.; Luo, X.; Su, J.; Wang, Y.; Xu, J.; Nandakumar, K.S.; et al. A marine fungus-derived nitrobenzoyl sesquiterpenoid suppresses receptor activator of NF-kappaB ligand-induced osteoclastogenesis and inflammatory bone destruction. *British J. Pharmacol.* **2020**, *177*, 4242–4260. [CrossRef]
31. Ellman, G.; Courtney, K.; Andres, V.; Featherstone, R. A new and rapid colorimetric determination of acetylcholinesterase activitiy. *Biochem. Pharmacol.* **1961**, *7*, 88–90. [CrossRef]
32. Trott, O.; Olson, A.J. Software news and update autodock vina: Improving the speed and accuracy of docking with a new scoring function, efficient optimization, and multithreading. *J. Comput. Chem.* **2010**, *31*, 455–461. [PubMed]
33. Rydberg, E.H.; Brumshtein, B.; Greenblatt, H.M.; Wong, D.M.; Shaya, D.; Williams, L.D.; Carlier, P.R.; Pang, Y.P.; Silman, I.; Sussman, J.L. Complexes of alkylene-linked tacrine dimers with torpedo californica acetylcholinesterase: Binding of bis5-tacrine produces a dramatic rearrangement in the active-site gorge. *J. Med. Chem.* **2006**, *49*, 5491–5500. [CrossRef] [PubMed]
34. Zhao, J.; Wang, X.; Liu, Z.; Meng, F.; Sun, S.; Ye, F.; Liu, Y. Nonadride and spirocyclic anhydride derivatives from the plant endophytic fungus *Talaromyces purpurogenus*. *J. Nat. Prod.* **2019**, *82*, 2953–2962. [CrossRef] [PubMed]
35. Zhao, Y.; Sun, C.; Huang, L.; Zhang, X.; Zhang, G.; Che, Q.; Li, D.; Zhu, T. Talarodrides A-F, nonadrides from the antarctic sponge-derived fungus *Talaromyces* sp. HDN1820200. *J. Nat. Prod.* **2021**, *84*, 3011–3019. [CrossRef] [PubMed]

marine drugs

MDPI

Article

Sclerotioloids A–C: Three New Alkaloids from the Marine-Derived Fungus *Aspergillus sclerotiorum* ST0501

Jun-Qiu Mao [1,2,3], Yao-Yao Zheng [1,2,3], Chang-Yun Wang [1,3], Yang Liu [4,5,*] and Guang-Shan Yao [2,*]

1 Key Laboratory of Marine Drugs, The Ministry of Education of China, Institute of Evolution & Marine Biodiversity, School of Medicine and Pharmacy, Ocean University of China, Qingdao 266003, China
2 Fujian Key Laboratory on Conservation and Sustainable Utilization of Marine Biodiversity, Institute of Oceanography, Minjiang University, Fuzhou 350108, China
3 Laboratory for Marine Drugs and Bioproducts, Qingdao National Laboratory for Marine Science and Technology, Qingdao 266237, China
4 Institute for Insect Biotechnology, Justus-Liebig-University of Giessen, 35392 Giessen, Germany
5 Department of Bioresources of the Fraunhofer Institute for Molecular Biology and Applied Ecology (IME), 35392 Giessen, Germany
* Correspondence: liu.yang@agrar.uni-giessen.de (Y.L.); ygshan@126.com (G.-S.Y.); Tel.: +86-0591-8378-9471 (G.-S.Y.)

Abstract: Alkaloids, as one of the largest classes of natural products with diverse structures, are an important source of innovative medicines. Filamentous fungi, especially those derived from the marine environment, are one of the major producers of alkaloids. In this study, three new alkaloids, sclerotioloids A–C (**1**–**3**), along with six known analogs (**4**–**9**), were obtained under the guidance of the MS/MS-based molecular networking from the marine-derived fungus, *Aspergillus sclerotiorum* ST0501, collected from the South China Sea. Their chemical structures were elucidated by comprehensive analysis of the spectroscopic data, including 1D and 2D NMR and HRESIMS. Additionally, the configuration of compound **2** was unambiguously determined by X-ray single crystal diffraction, and that of compound **3** was determined by the TDDFT-ECD approach. Sclerotioloid A (**1**) represents the first example of 2,5-diketopiperazine alkaloid with a rare terminal alkyne. Sclerotioloid B (**2**) showed the inhibition of NO production induced by lipopolysaccharide (LPS), with an inhibition rate of 28.92% higher than that of dexamethasone (25.87%). These results expanded the library of fungal-derived alkaloids and further prove the potential of marine fungi in the generation of alkaloids with new scaffolds.

Keywords: marine-derived fungus; *Aspergillus sclerotiorum*; secondary metabolite; alkaloid; sclerotioloids

Citation: Mao, J.-Q.; Zheng, Y.-Y.; Wang, C.-Y.; Liu, Y.; Yao, G.-S. Sclerotioloids A–C: Three New Alkaloids from the Marine-Derived Fungus *Aspergillus sclerotiorum* ST0501. *Mar. Drugs* **2023**, *21*, 219. https://doi.org/10.3390/md21040219

Academic Editors: Bin-Gui Wang and Haofu Dai

Received: 7 March 2023
Revised: 25 March 2023
Accepted: 28 March 2023
Published: 29 March 2023

1. Introduction

In the past few years, natural products isolated from marine-derived fungi have aroused great interest due to their unique structures, interesting the pharmacological and biological properties [1]. Among the marine-derived fungi, *Aspergillus* is the largest source of fungal natural products. The growth of marine-derived fungi research rose from 2015 to 2019, with natural products from marine-derived fungi sources already accounting for nearly half (47%) of the total number of new marine natural products reported by 2019 [2]. Additionally, a large part of the compounds isolated from *Aspergillus* show anti-bacterial, anti-cancer, anti-inflammatory, anti-tuberculosis, and cytotoxic activities [3]. Metabolites of marine-derived fungi, such as polyketides, alkaloids, terpenes, lactones and peptides, are rich sources of biologically active natural products [4]. It was reported that alkaloids represent about 18% of all the marine natural products in 2019 [2]. Recent studies of marine fungal metabolites looking for bioactive compounds suggest that they have the potential to become a source of new drugs [5,6]. For example, plinabulin is a 2,5-diketopiperazine alkaloid isolated from the marine-derived fungus *Aspergillus* sp., which was first used in clinical trials for the treatment of non-small-cell lung cancer (NSCLC) [7]. Ecteinascidin-743 (ET743) is a novel anti-tumor drug on the market [8]. *Aspergillus sclerotiorum* have potential to biosynthesize alkaloids, though alkaloids are the most bioactive metabolites that possess marked antimicrobial and cytotoxic activities [9]. Sclerotiamides C and F, which are isolated from the *A. sclerotiorum* LZDX-33-4, a fungus derived from a gorgonian coral (LZDX-33), showed a significant inhibitory effect against a panel of tumor cell lines with IC_{50} values ranging from 1.6 to 7.9 μM [10]. A new cytotoxic indole-3-ethenamide isolated from *A. sclerotium* PT06-1 showed moderate (3.0 mM) and weak (27 mM) cytotoxicity toward A-549 and HL-60 cells, respectively [11].

During our ongoing research for new bioactive secondary metabolites from marine-derived fungi in the South China Sea [12–15], plenty of marine-derived alkaloids have been isolated. For example, emestrins L and M were obtained from the marine-derived fungus *A. terreus* RA2905. Emestrin M displayed antibacterial activity against *Pseudomonas aeruginosa* ATCC 27,853, with a minimum inhibitory concentration (MIC) value of 64 μg/mL [16]. Three new indole-diketopiperazine alkaloids, spirotryprostatin G and cyclotryprostatins F and G, were obtained from fungal strain HBU-136, and among them, spirotryprostatin G exhibited cytotoxicity against the HL-60 cell line, with an IC_{50} value of 6.0 μM, while cyclotryprostatins F and G exhibited cytotoxicity against the MCF-7 cell line, with IC_{50} values of 7.6 and 10.8 μM, respectively [17].

The fungal strain *A. sclerotiorum* ST0501 was isolated from the inner part of an unidentified sponge, GDST-2013-05, collected from the South China Sea. In the previous work, three new 2,5-diketopiperazine alkaloids, speramide C, 3,21-*epi*-taichunamide F and 2-*epi*-amoenamide C, were obtained from the same fungus strain. Speramide C represents the first prenylated indole alkaloid with an ethylene oxide ring on the isopentenyl side chain [18]. However, in this paper, we discovered three new alkaloids (**1–3**) and six known analogs (**4–9**) from ethyl acetate extract of *A. sclerotiorum* ST0501 based on the MS/MS molecular network. Herein, we report the isolation, structure elucidation and biological activities of these isolated compounds.

2. Results and Discussion

2.1. Elucidation of Chemical Structures

In order to fully explore the alkaloids of ST0501, molecular networking analysis was performed on the crude ethyl acetate extract (EtOAc) of solid rice and wheat cultures of *A. sclerotiorum* ST0501. The results show that there are signals associated with alkaloids. The relationship between compound **2** and compound **6** is close. Based on the indication of molecular networking (Figure 1), three unreported (**1–3**) and six previously described alkaloids (**4–9**) (Figure 2) were isolated by using a combination of column chromatography including silica gel, octadecylsilyl, Sephadex LH-20 columns and semi-preparative HPLC. The structures of gartryprostatin C (**4**) [19], stephacidin A (**5**) [20], sclerotiamide (**6**) [21], notoamide B (**7**) [22], speramide C (**8**) [18] and stephacidin B (**9**) [23] were elucidated by comparison of their spectroscopic data with those previously reported in the literature.

Sclerotioloid A (**1**) was isolated as a yellow powder. Its molecular formula $C_{14}H_{12}N_2O_2$ was deduced from its HRESIMS m/z 263.0794 $[M+Na]^+$ (calcd for $C_{14}H_{12}N_2O_2Na$, 263.0797) and 1D NMR data analysis, suggesting ten degrees of unsaturation. The IR spectrum of **1** featured typical absorption bands for alkyne (2361 cm^{-1}) and conjugated ketone (1698 cm^{-1}). The ^1H NMR spectrum of **1** (Table 1) exhibited one amine proton (δ_H 8.42, s, 1-NH), five aromatic protons (δ_H 7.35–7.40, m, 1-substituted benzene ring), one olefinic proton (δ_H 7.10, s, H-10), two methylene protons (δ_H 4.13 d, J = 2.4 Hz, H-7; δ_H 4.01 d, J = 2.3 Hz, H-6) and one acetylenic proton (δ_H 3.10, t, J = 2.4 Hz, H-9). The ^{13}C NMR spectrum, in combination with HSQC spectra, revealed the presence of fourteen carbons, including two amide carbonyl (δ_C 165.7, C-5 and δ_C 163.4, C-2), six aromatic carbons (δ_C 133.4, C-11; δ_C 129.4, C-13 and C-15; δ_C 128.8, C-14; δ_C 128.6, C-12 and C-16), two olefinic carbons (δ_C 129.9, C-3 and δ_C 120.8, C-10), two alkynyl carbons (δ_C 78.2, C-8 and δ_C 74.5, C-9) and two methylene sp^3 (δ_C 44.5, C-6 and δ_C 32.9, C-7). The correlations between H-1 and H-6 in the ^1H-^1H COSY spectrum and the HMBC correlations between NH-1 and C-3 and between H-6 and C-2/C-5 revealed the existence of a 2,5-diketopiperazine ring. Furthermore, the HMBC correlations between H-10 and C-2/C-3/C-16 indicated that the benzene ring was connected to C-3 of the 2,5-diketopiperazine ring through C-10. Based on the HMBC correlations between H-7 and C-3/C-5/C-9, the alkynyl group was connected to N-4 of the 2,5-diketopiperazine ring (Figure 3). According to the (*Z*)-vinyl proton chemical shifts of H-7 (δ_H 7.03) in nocazine C [24] and H-7 (δ_H 6.54) in (3*S*,6*E*)-3-benzyl-6-benzylidenepiperazine-2,5-dione [25], the $\Delta^{3,10}$ double bond in **1** was deduced to have a *Z* configuration due to the relative downfield shift of H-10 (δ_H 7.10), where the (*Z*)-vinyl proton had a larger downfield shift than that of the (*E*)-vinyl proton because of the deshielding effect of the carbonyl in 2,5-diketopiperazine [24–27]. Therefore, the structure of **1** was elucidated.

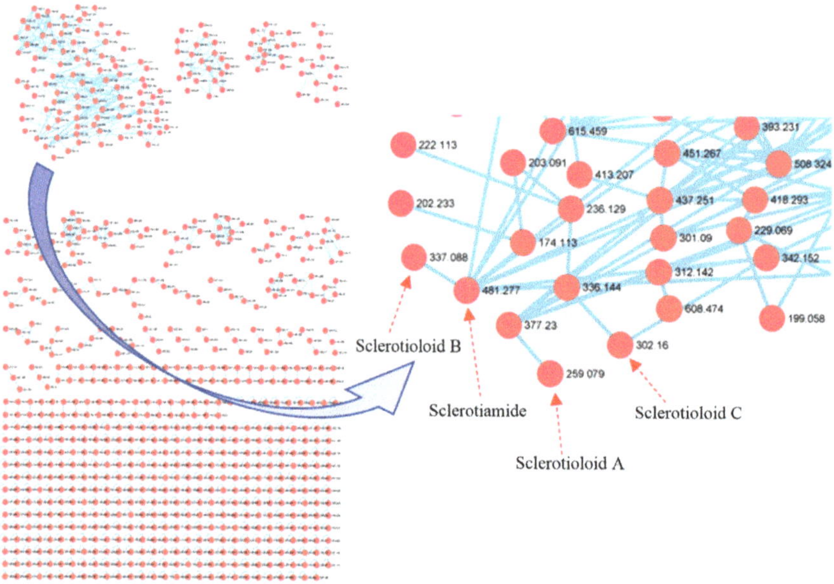

Figure 1. LC-MS/MS-based molecular networking of *Aspergillus sclerotiorum* ST0501.

Table 1. NMR data for compound **1** (DMSO-d_6, 400 MHz/100 MHz).

Position	δ_C, mult.	Δ_H, mult. (J in Hz)	COSY	HMBC
1	NH	8.42 (s)	H-6	C-3
2	163.4, C			C-6, C-10
3	129.9, C			1-NH, C-7
4	N			
5	165.7, C			C-6, C-7
6	44.5, CH_2	4.01, d, (2.4)	1-NH	C-2, C-5
7	32.9, CH_2	4.13, d, (2.4)		C-3, C-5, C-9
8	78.2, C			
9	74.5, CH	3.10, t, (2.4)		C-7
10	120.8, CH	7.10 (s)		C-2, C-16
11	133.4, C			C-16
12	128.6, CH	7.35–7.40 (m)	H-13	
13	129.4, CH	7.35–7.40 (m)	H-12, H-14	
14	128.8, CH	7.35–7.40 (m)	H-13, H-15	
15	129.4, CH	7.35–7.40 (m)	H-14, H-16	
16	128.6, CH	7.35–7.40 (m)	H-15	C-10, C-11

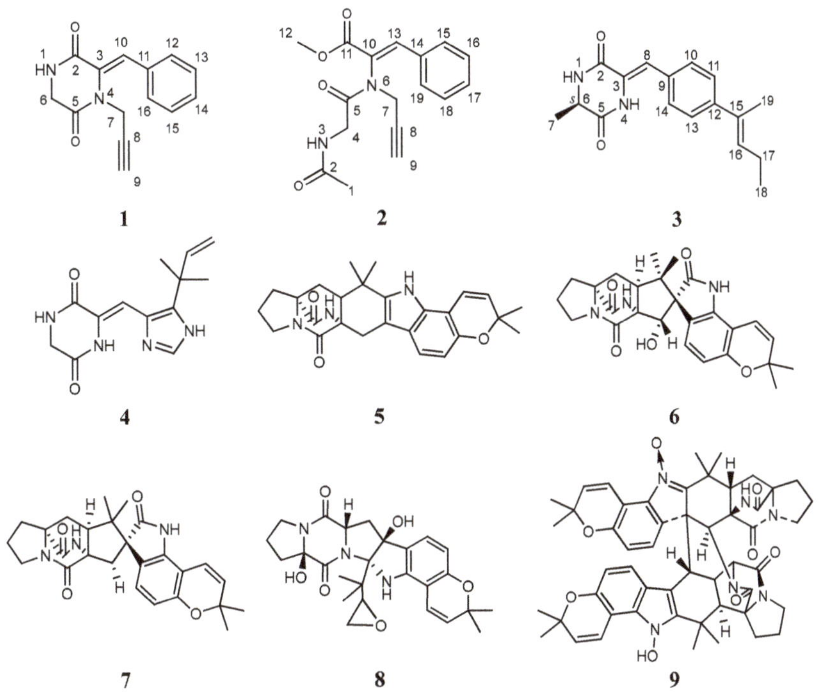

Figure 2. The structure of **1–9**.

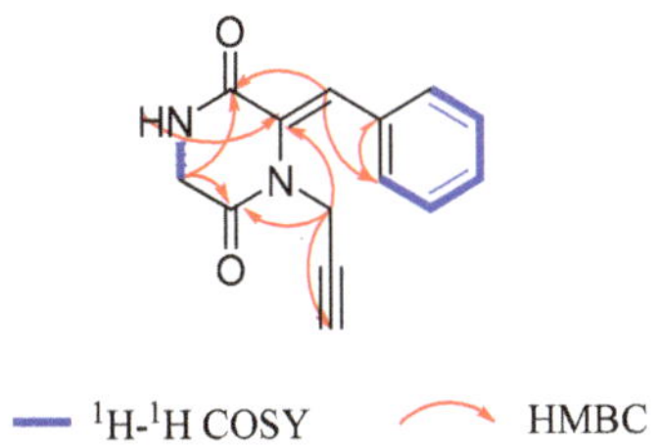

Figure 3. The key ^1H-^1H COSY and HMBC correlations of compound **1**.

Sclerotioloid B (**2**) was also isolated as a yellow powder. The HRESIMS spectrum of **2** indicates a molecular formula of $C_{17}H_{18}N_2O_4$ based on the prominent peak $[M+Na]^+$ at m/z 337.1160 (calcd for $C_{17}H_{18}N_2O_4Na$, 337.1164), as well as 1D NMR data analysis, requiring ten degrees of unsaturation. The IR spectrum of **2** featured typical absorption bands for alkyne (2361 cm^{-1}) and conjugated ketone (1557 cm^{-1}). The ^1H NMR and ^{13}C NMR spectroscopic data of **2** (Table 2) showed high similarity with **1**, except for the appearance of one methoxyl protons resonating at δ_H 3.81 (H-12) and one methyl protons resonating at δ_H 1.77 (H-1), as well as the additional carbonyl carbon resonating at δ_C 169.0 (C-2). The existence of an acetamide group was suggested by the HMBC correlations between NH-3 and C-2, between H-1 and C-2 and between H-4 and C-5, as well as the COSY correlation between NH-3 and H-4. The additional HMBC correlations between H-12 and C-11 and between H-13 and C-11/C-15 proved that the benzene ring is connected to the methyl acrylate portion at C-13 (Figure 4). Finally, a single crystal of **2** was obtained after one week of slow crystallization in 95% MeOH (H_2O) at 4 °C by optimizing the conditions. Therefore, the structure of **2** was confirmed by X-ray crystal diffraction analysis, and this further proved the *Z* configuration of the $\Delta^{10,13}$ double bond.

Table 2. NMR data for compound **2** (DMSO-d_6, 600 MHz/150 MHz).

Position	δ_C, mult.	Δ_H, mult. (J in Hz)	COSY	HMBC
1	21.9, CH_3	1.77 (s)		C-2
2	169.0, C			1-NH
3	NH	8.06 (s)	H-4	C-2
4	40.4, CH_2	3.56, dd, (17.0, 5.7) 3.66, dd, (17.1, 5.7)	1-NH	C-5
5	168.4, C			C-4
6	N			
7	35.5, CH_2	4.24, dd, (17.6, 2.5) 4.38, dd, (17.5, 2.5)		C-8, C-9
8	77.5, C			C-7
9	75.8, CH	3.15 (s)		C-7
10	126.6, C			
11	164.5, C			C-12, C-13
12	52.5, OCH_3	3.81 (s)		C-11
13	139.5 CH	7.84 (s)		C-11, C-14 C-15
14	131.5, C			C-13
15	130.2, CH	7.75 (s)	H-16	C-13
16	128.8, CH	7.46, d, (7.6)	H-15, H-17	
17	131.0, CH	7.49, d, (6.9)	H-16, H-18	
18	128.8, CH	7.46, d, (7.6)	H-17, H-19	
19	130.2, CH	7.73 (s)	H-18	

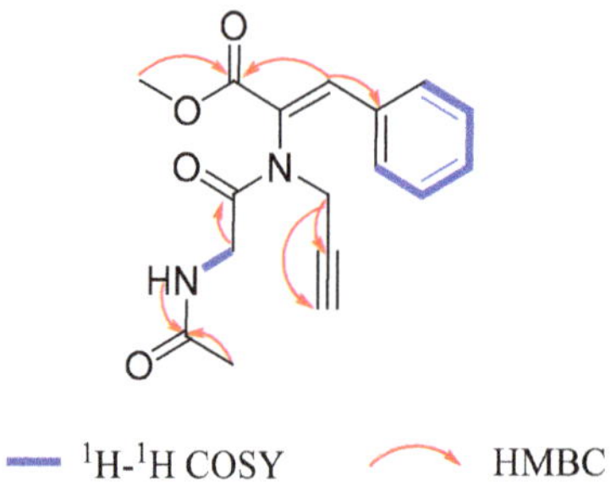

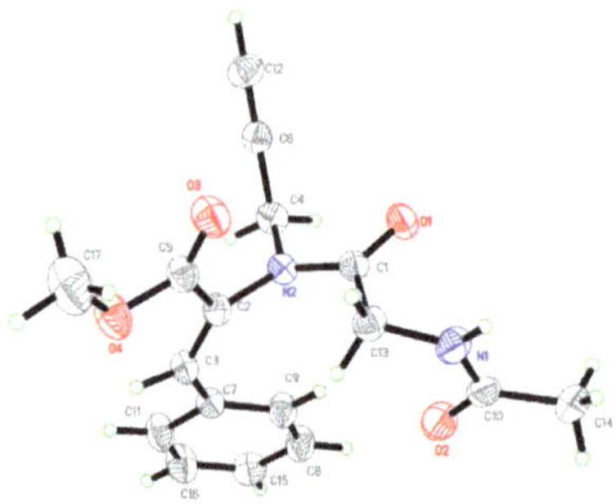

Figure 4. The key ^{1}H-^{1}H COSY and HMBC correlations of **2** (left) and the ORTEP diagram of **2** (right).

Sclerotioloid C (**3**) was isolated as a white powder. Its molecular formula, $C_{17}H_{20}N_2O_2$, is based on its HRESIMS m/z 323.1370 $[M+K]^+$ (calcd for $C_{17}H_{20}N_2O_2K$, 323.1162) and 1D NMR data analysis, suggesting nine degrees of unsaturation. The IR spectrum of **3** featured typical absorption bands for olefin (2923 cm^{-1}) and conjugated ketone (1512 cm^{-1}). The ^{1}H NMR spectra of **3** (Table 3) displayed two amine protons (δ_H 9.79, s, 4-NH; 8.35-8.37, m, 1-NH), four aromatic protons (δ_H 7.44, d, J = 8.8 Hz, H-10 and H-14; δ_H 6.95, d, J = 8.8 Hz, H-11 and H-13), two olefinic protons (δ_H 6.64, s, H-8; δ_H 5.43, tt, J = 6.7, 1.4 Hz, H-16), one methylene (δ_H 4.55, d, J = 6.7 Hz, H-17), one methine (δ_H 4.11, d, J = 6.8 Hz, H-6), and the protons of three methyl groups (δ_H 1.74, s, H-19; δ_H 1.71, d, J = 1.3 Hz, H-18; δ_H 1.33, d, J = 7.0 Hz, H-7). ^{13}C NMR and HSQC data revealed the presence of seventeen carbons, including two amide carbonyl (δ_C 167.6, C-2 and δ_C 160.7, C-5), ten sp^2 resonated between δ_C 158.2 and δ_C 114.3, one methylene sp^3 (δ_C 64.3, C-17), one methine sp^3 (δ_C 50.3, C-6) and three methyl (δ_C 25.4, C-19; δ_C 19.2, C-7; δ_C 18.0, C-18) carbons. The COSY correlations between NH-1 and H-6 and between H-6 and H-7 and the HMBC correlations between NH-4 and C-2/C-5/C-6, between H-6 and C-2/C-5/C-7 and between H-7 and C-2/C-6 revealed the existence of the 2,5-diketopiperazine ring. The COSY correlations between H-6 and H-7 and the HMBC correlations between H-7 and C-6 revealed that the methyl group is connected to C-6 of the 2,5-diketopiperazine ring. The COSY correlations between H-13 and

H-14 and the HMBC correlations between H-12 and C-13/C-14 and between H-14 and C-10 revealed the existence of 1,4-disubstituted benzene ring. The HMBC correlations between H-8 and C-2/C-10 and between H-10 and C-3 indicate that the 2,5-diketopiperazine ring is connected to C-9 of the benzene ring through C-8. The COSY correlations between H-16 and H-17 and between H-and to H-18 and the HMBC correlations between H-17 and C-12/C-16, between H-18 and C-15/C-16 and between H-19 and C-15/C-16, indicate that the *cis*-2-pentene is connected to C-12 of the benzene ring (Figure 5). Therefore, the planar structure of **3** was elucidated as a new diketopiperazine derivative. The configuration of the double bond in **3** was established by the analysis of its NOESY spectrum (Figure 5). The NOESY spectrum did not show any correlation between H-16 and H-19, confirming the *E* configuration of the double bond between C-15 and C-16. In addition, there was no NOESY correlation between H-8 and 4-NH, suggesting *Z* configuration of the double bond between C-3 and C-8. The absolute configuration of C-6 in **3** was determined by the TDDFT-ECD approach. The experimental ECD spectrum of **3** matched well with the calculated ECD spectrum of *S*-**3** (Figure 6) and further supported its absolute configuration.

Table 3. NMR data for compound **3** (DMSO-d_6, 500 MHz/125 MHz).

Position	δ_C, mult.	Δ_H, mult. (J in Hz)	COSY	HMBC	NOESY
1	NH	8.35–8.37, m	H-6	C-3, C-5	H-6, H-7
2	160.7, C			4-NH, C-6, C-8	
3	125.3, C			1-NH, C-10, C-14	
4	NH	9.79, s		C-2, C-5, C-6	H-12, H-13
5	167.6, C			1-NH, 4-NH, C-6, C-7	
6	50.3, CH	4.11, d, (6.8)	1-NH, H-7	C-2, C-5, 4-NH, C-7	1-NH, H-7
7	19.2, CH_3	1.33, d, (7.0)	H-6	C-5, C-6	1-NH, H-6
8	114.3, CH	6.64 (s)		C-2, C-10	
9	125.7, C				
10	130.8, CH	7.44, d, (8.8)	H-14	C-8, C-12, C-14	
11	114.8, CH	6.95, d, (8.8)	H-13	C-3, C-12, C-13	H-17
12	158.2, C			C-10, C-11, C-13, C-14, C-17	4-NH
13	114.8, CH	6.95, d, (8.8)	H-11	C-3, C-11, C-12	H-17
14	130.8, CH	7.44, d, (8.8)	H-10	C-10, C-12	
15	137.3, C			C-17, C-18, C-19	
16	119.8, CH	5.43, tt, (6.7, 1.4)	H-17	C-17, C-18, C-19	H-17, H-18
17	64.3, CH_2	4.55, d, (6.7)	H-16, H-18	C-12, C-15, C-16	H-11, H-13, H-16
18	18.0, CH_3	1.71, d, (1.3)	H-17	C-15, C-16	H-16
19	25.4, CH_3	1.74 (s)		C-15, C-16	

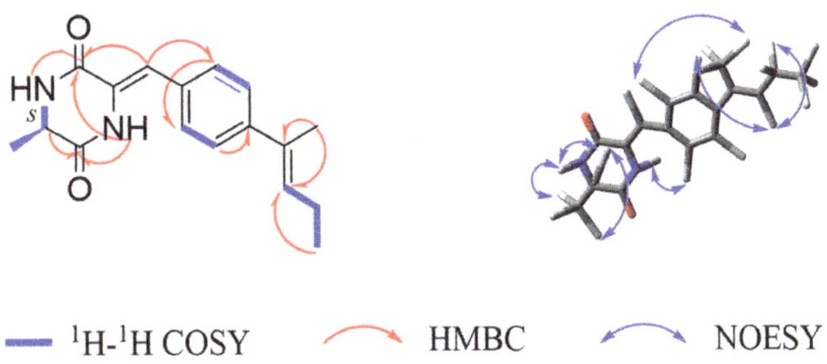

Figure 5. Key ^{1}H-^{1}H COSY, HMBC and NOESY correlations of **3**.

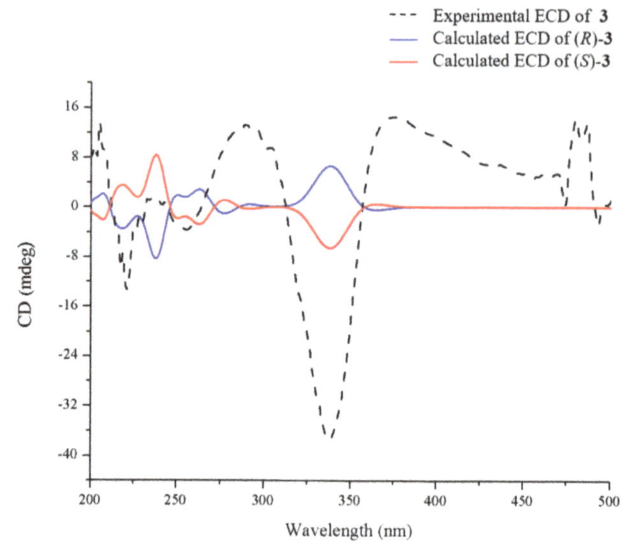

Figure 6. Experimental and calculated ECD spectra of **3**.

2.2. *Bioassays of Compounds*

2.2.1. Cytotoxicity Assay

Compound **3** was screened for inhibitory activity on 20 human tumor cell lines at a concentration of 20 μM. However, it showed no observable activities against the abovementioned tumor cell lines (Table S1).

2.2.2. Anti-Microbial Activity Assay

Compounds **1–9** were tested for the anti-microbial activity against 20 pathogenic bacteria, including *Enterobacter cloacae*, *E. hormaechei*, *Aeromonas salmonicida*, *Escherichia coli*, Pseudomonas fulva, Photobacterium halotolerans, P. aeruginosa, Staphylococcus aureus, *Bacillus subtilis*, *Vibrio Parahemolyticus* HUB183, *V. alginolyticus* HUB184, *V. anguillarum* HUB185, *Staphylococcus aureus* ATCC 43300, *Enterococcus faecalis* ATCC 51299, *E. faecium* ATCC 35667, S. aureus ATCC 33591, S. aureus ATCC 29213, S. aureus ATCC 25923, *V. vulnificus* MCCC E1758, *V. campbellii* MCCC E333 and two pathogenic fungi, *Candida albicans* ATCC 24,433 and C. parapsilosis ATCC 22019. The initial screening concentration was 50 μg/mL, and none of them showed observable activities against abovementioned pathogens.

2.2.3. Anti-Oxidant Activity Assays

To test the anti-oxidant activity, 1,1-diphenyl-2-picrylhydrazyl (DPPH) assay was used. The initial screening concentration was 50 μg/mL, but none of them showed anti-oxidant activity.

2.2.4. Anti-Inflammatory Activity Assays

Compounds **1–3** were screened for NO production inhibitory activity. Compound **2** showed inhibition of NO production induced by lipopolysaccharide (LPS), with an inhibition rate of 28.92 ± 3.49% (Table S2). Compound **2** has a higher inhibition rate than dexamethasone does, which makes it promising as a good anti-inflammatory drug.

Compared with other alkaloid compounds, the activity of compounds **1–3** is not good, and other activity models need to be further screened.

3. Materials and Methods

3.1. General Experimental Procedure

Optical rotations were measured by using a JASCO P-1020 digital polarimeter (589 nm, 20 °C) (JASCO Ltd., Tokyo, Japan). UV spectra were recorded using an HITACHI UH 5300 UV spectrophotometer (Hitachi, Tokyo, Japan). IR spectra were acquired using a Nicolet-Nexus-470 spectrometer in KBr discs (400–4000 cm^{-1}) (PerkinElmer Ltd., Boston, MA, USA). NMR spectra were recorded using a JEOL JEM-ECP NMR spectrometer (JEOL Ltd., Tokyo, Japan) with 500 MHz for ^{1}H NMR and 125 MHz for ^{13}C NMR and using TMS as an internal standard. HRESIMS were analyzed using a Thermo MAT95XP high resolution mass spectrometer (Thermo Fisher Scientific, Bremen, Germany), and ESIMS spectra were analyzed using a Thermo DSQ EImass spectrometer (Thermo Fisher Scientific, Bremen, Germany). Single-crystal X-ray crystallographic analysis was performed using a Bruker D8 venture X-ray single crystal diffractometer (Bruker, Karlsruhe, Germany). HPLC analysis was performed using an HITACHI L-2000 HPLC system coupled with a L-2455 photodiode array detector and using a semi-prepared C18 column (Kromasil 250 mm × 10 mm, 5 μm). For HPLC analysis, the mobile phase consisted of methanol (A) and water (B) with a flowrate of 2 mL/min with 10 μL injection volume (10 mg/mL) and monitored at UV length of 190–400 nm. The elution gradient was as follows: 0–40 min, 40–100% A; 40–45 min, 100% A; 45–50 min, 100–40% A; 50–55 min, 40% A. Sephadex LH-20 (Amersham Biosciences Co., Buckinghamshire, UK) and octadecylsilyl silica gel (Unicorn; 45–60 μm) were used for column chromatography (CC). Precoated silica gel plates (Yan Tai Zi Fu Chemical Group Co.; G60, F-254) were used for thin-layer chromatography analysis.

3.2. Fungal Material

The fungal strain, *A. sclerotiorum* ST0501, was isolated from the inner part of an unidentified sponge, GDST-2013-05, collected from the South China Sea (Guangdong, China) in May 2013. The fungal identification was performed by analysis of its morphological characteristics and ITS region of the rDNA [19]. The sequence data were submitted to Genbank with accession number MT534582. The strain was deposited in a −80 °C refrigerator in the laboratory.

3.3. Fermentation, Extraction and Isolation

The fungal strain was cultivated on a solid rice and wheat medium (3.6 g of natural sea salt from Yangkou saltern, China; 80 g of rice; 20 g wheat; 100 mL of H_2O in 1000-mL Erlenmeyer flask) for 30 d at room temperature. Total fermentation of 45 flasks was extracted repeatedly with equal volume of ethyl acetate (EtOAc) two times, and the organic solvent was evaporated to dryness under vacuum to give a crude extract of 24.0 g. The crude extract was first fractionated by silica gel column chromatography (CC) using a step gradient elution with EtOAc-petroleum ether (0–100%), then with methanol-EtOAc (0–100%) to provide eleven fractions (Fr.1–Fr.11). Fr.3 was separated into three subfractions (M1–M3) by silica gel CC eluted with a step gradient of petroleum ether-EtOAc (from 8/1 to 1/5, *v*/*v*). M1 was placed in a Sephadex LH-20 column and eluted with dichloromethane-methanol (1/1, *v*/*v*) to generate fourteen subfractions (N1–N14). N14 was purified by HPLC (50% methanol-H_2O) to yield **1** (1.0 mg) and **2** (6.4 mg). Fr.6 was separated into eight subfractions (O1–O8) by silica gel CC eluting with a step gradient of methanol–dichloromethane (0–100%). O4 was further separated by the silica gel CC eluted with a step gradient of dichloromethane–methanol (from 100/1 to 5/1, *v*/*v*) to produce **3** (8.8 mg). Fr.7 was separated into eighteen subfractions (P1–P18) in an ODS column eluted with 30–100% methanol–H_2O to give **4** (4.5 mg). P6 was separated by the silica gel CC eluted with dichloromethane–methanol (30/1, *v*/*v*) to produce **9** (4.8 mg). P12 was further separated by the silica gel CC eluted with dichloromethane-methanol (60/1, *v*/*v*) to generate **5** (2.3 mg). P13 was separated by the silica gel CC eluted with dichloromethane-methanol (20/1, *v*/*v*) to produce **8** (1 mg). Fr.8 was separated into five subfractions (Q1–Q5) by the silica gel CC eluted with a step gradient of dichloromethane-methanol (60/1 to 5/1, *v*/*v*) to produce **6** (116.5 mg). Q4 was separated into five subfractions

(R1–R5) in the Sephadex LH-20 column and eluted with dichloromethane–methanol (1/1, *v*/*v*). R2 was further purified by HPLC (80% methanol–H_2O) to produce **7** (1.9 mg).

3.4. LC-MS/MS and Molecular Networking Analysis

LC-MS/MS was performed using a Waters series 2695 HPLC instrument coupled with an amaZon SL ion trap Mass spectrometer (Bruker, Karlsruhe, Germany), with a Xchange C18 column (Acchrom Co., CO, USA) 250 mm × 4.6 mm, 5 μm, 0.5 mL/min). The organic portion was dissolved in MeOH at 10 mg/mL, filtered through a Gracepure C18 SPE cartridge and analyzed by LC–MS/MS. Ten μL aliquot of each sample was injected and eluted with a gradient program of MeOH-H_2O (0.1% formic acid) (0–20 min 10–100%, 21–25 min 100%; 1.0 mL/min). Mass spectra were obtained in positive ESI mode and with an automated fully dependent MS/MS scan from 100 to 1000 Da. MS/MS data were converted digitally to mzXML files using Filezilla software. The molecular networking was performed using the GNPS data analysis workflow using the spectral clustering algorithm [28]. The spectral networks were imported into Cytoscape 3.9.1 and visualized using the force-directed layout.

3.5. Spectroscopic and Spectrometric Data

Sclerotioloid A (**1**): yellow powder; UV (MeOH) λ_{max} (log ε) 221 (0.13), 288 (0.16) nm; HRESIMS (*m*/*z* 263.0794 $[M+Na]^+$, calcd for 263.0797); IR (KBr) ν_{max} 3749, 2926, 2361, 1557, 1398, 1053 cm^{-1}; ^1H and ^{13}C NMR data, see Table 1.

Sclerotioloid B (**2**): yellow powder; UV (MeOH) λ_{max} (log ε) 219 (0.03), 288 (0.05) nm; HRESIMS (*m*/*z* 337.1160 $[M+Na]^+$, calcd for 337.1164); IR (KBr) ν_{max} 2361, 1796, 1701, 1522, 1418, 1264 cm^{-1}; ^1H and ^{13}C NMR data, see Table 2.

X-ray Crystallographic Analysis of Compounds **2**. Colorless crystals of **2** suitable for X-ray diffraction were obtained from 95% MeOH (H_2O) by slow evaporation. The crystal data were collected at 293 K using an Agilent Gemini Ultra (Agilent, PA, USA) diffractometer with Cu Kα radiation (λ = 1.54184 Å). The structure was solved using CrysAlis^{Pro} version 1.171.41.121 (Rigaku Oxford Diffraction, 2021). Empirical absorption correction using spherical harmonics implemented in SCALE3 ABSPACK scaling algorithm.

Crystal Data for **2**. $C_{17}H_{19}N_2O_4$, *Mr* = 314.33, triclinic, space group $P-1$ with a = 8.8982(11) Å, b = 9.6133(10) Å, c = 10.9564(14) Å, α = 67.791(11), β = 71.857(11), γ = 82.76(1), V = 824.51(19) Å3, Z = 2, Dx = 1.266 g/cm^3, μ (Cu Kα) = 0.752 mm^{-1}, and F (000) = 332. Crystal dimensions: 0.12 × 0.12 × 0.11 mm^3. Independent reflections: 1915 (R_{int} = 0.0576). The final R1 value was 0.0576.

Sclerotioloid C (**3**): white powder; $[\alpha]^{20}_D = -11.3°$ (*c* 0.01, MeOH); UV (MeOH) λ_{max} (log ε) 226 (0.01), 318 (0.01) nm; HRESIMS (*m*/*z* 323.1370 $[M+K]^+$, calcd for 323.1162); IR (KBr) ν_{max} 2923, 2824, 1735, 1512, 1455, 1253, 1179 cm^{-1}; ^1H and ^{13}C NMR data, see Table 3.

3.6. Biological Assay

Cytotoxic activity against human cancer cell lines was evaluated following the CCK-8 assay [29]. Twenty cell lines were used, including human lung cancer cells (CCL-185), human breast cancer cells (HTB-22), human gastric cancer cells (TCP-1008), human colon cancer cells (CCL-247EMT), human hepatocellular carcinoma cells (HB-8065), human cervical cancer cells (CCL-2), human chronic myeloid leukemia cells (CRL-3344), human brain tumor cells (CRL-2020), normal human liver cells (CRL-2254), human renal clear cell adenocarcinoma cells (CRL-1932), human esophageal carcinoma cells (PDM-246), human bladder cancer cells (HTB-9), human prostate cancer cells (HTB-81), human thyroid cancer cells (CRL-3354), human pancreatic cancer cells (CRL-2151), human osteosarcoma cells (CRL-1543), human malignant melanoma cells (CRL-1619), human rhabdomyosarcoma cells (CRL-1598), human embryonic kidney cells (CRL-3216) and human gallbladder carcinoma cells (PDM-273). Adriamycin hydrochloride was used as a positive control.

Antibacterial activity was evaluated following the standards recommended by Pierce [30], with ampicillin sodium as a positive control. The antifungal bioassay was conducted following the standards recommended by the Clinical and Laboratory Standards Institute [30], with amphotericin B used as a positive control.

Compounds **1–9** were tested against 20 pathogenic bacteria, including *Enterobacter cloacae, E. hormaechei, Aeromonas salmonicida, Escherichia coli, Pseudomonas fulva, Photobacterium halotolerans, P. aeruginosa, Staphylococcus aureus, Bacillus subtilis, Vibrio Parahemolyticus* HUB183, *V. alginolyticus* HUB184, *V. anguillarum* HUB185, *S. aureus* ATCC 43300, *Enterococcus faecalis* ATCC 51299, *E. faecium* ATCC 35667, *S. aureus* ATCC 33591, *S. aureus* ATCC 29213, *S. aureus* ATCC 25923, *V. vulnificus* MCCC E1758, *V. campbellii* MCCC E333, and two pathogenic fungi *Candida albicans* ATCC 24,433 and *C. parapsilosis* ATCC 22019.

The DPPH scavenging assay was performed using the method described by Aquino et al. [31]. The reaction mixture consisted of freshly prepared DPPH dissolved in ethanol (100 μmol/L) mixed with different concentrations of the tested compound. The reaction mixture was incubated for 20 min at room temperature in the dark, and the optical density was recorded at 517 nm.

The bioassay for NO production inhibitory activity was conducted as described by Xia et al. [32]. The mouse macrophages were seeded in 96-well plates. In each well, LPS (1 μg/mL) was added after treating with or without the tested compound for 24 h. The NO production in the supernatant was detected by the Griess reaction. The absorbance at 540 nm was measured with a microplate reader. The NO concentration and the inhibitory rate were calculated as a calibration curve. Dexamethasone was used as the positive control. Experiments were performed in triplicate, and the data are described as mean ± SD of three independent experiments.

4. Conclusions

In conclusion, nine alkaloids, including three new ones, were obtained by using MS/MS-based molecular networking for continuous investigation of the marine-derived fungus, *A. sclerotiorum* ST0501. Sclerotioloid A (**1**) represents the first example of 2,5-diketopiperazine alkaloid with a rare terminal alkyne. Additionally, the absolute configuration of compound **2** was unambiguously determined by single crystal X-ray analysis. Compound **2** showed the inhibition of NO production induced by lipopolysaccharide (LPS). Compound **3** displayed weak proliferation inhibitory activity against human chronic myeloid leukemia cells K-562 and human renal clear cell adenocarcinoma cell 786-O. Therefore, the potential of the marine-derived fungus, *A. sclerotiorum* ST0501, to produce novel bioactive secondary metabolites is worthy of further exploration.

Supplementary Materials: The following supporting information can be downloaded at: https://www.mdpi.com/article/10.3390/md21040219/s1, Figures S1–S19: ^{1}H NMR, ^{13}C NMR, ^{1}H-^{1}H COSY, HSQC, HMBC, HRESIMS, UV and IR spectra of compounds **1** and **2**. Figure S20–S28: ^{1}H NMR, ^{13}C NMR, ^{1}H-^{1}H COSY, HSQC, HMBC, NOESY, HRESIMS, UV and IR spectra of compound **3**. Table S1: The activity results about anti-tumor of compound **3**. Table S2: The activity results about anti-inflammatory of compounds **1–3**.

Author Contributions: Conceptualization, Y.L. and G.-S.Y.; methodology, J.-Q.M.; data analysis, J.-Q.M., Y.-Y.Z.; bioassays, J.-Q.M.; writing—original draft preparation, J.-Q.M.; writing—review and editing, C.-Y.W., Y.L. and G.-S.Y. All authors have read and agreed to the published version of the manuscript.

Funding: This work was supported by grants from the National Natural Science Foundations of China (NOs. 81874300; 41830535), Shandong Provincial Natural Science Foundation (Major Basic Research Projects) (NO. ZR2019ZD18); Natural Science Foundation of Fujian Province (NO. 2021J011043) and Minjiang University (NOs. MYK19011; JAT190622); the Program of Open Studio for Druggability Research of Marine Natural Products, National Laboratory for Marine Science and Technology (Qingdao, China) Directed by Kai-Xian Chen and Yue-Wei Guo; the Taishan Scholars Program, China.

Institutional Review Board Statement: This article does not contain any studies with animals performed by any of the authors.

Acknowledgments: We appreciate Xiaomei Teng and Ping Zhang of Qingdao marine biopharmaceutical research institute for HRESIMS testing. We are grateful to Houting Liu of Liaocheng university for X-ray crystallographic analysis.

Conflicts of Interest: The authors declare no conflict of interest.

References

1. Carroll, A.R.; Carroll, A.R.; Davis, R.A.; Copp, B.R.; Davis, R.A.; Keyzers, R.A.; Prinsep, M.R. Marine natural products. *Nat. Prod. Rep.* **2022**, *39*, 1122–1171. [CrossRef]
2. Carroll, A.R.; Copp, B.R.; Davis, R.A.; Keyzers, R.A.; Prinsep, M.R. Marine natural products. *Nat. Prod. Rep.* **2021**, *38*, 362–413. [CrossRef] [PubMed]
3. Carroll, A.R.; Copp, B.R.; Davis, R.A.; Keyzers, R.A.; Prinsep, M.R. Marine natural products. *Nat. Prod. Rep.* **2020**, *37*, 175–223. [CrossRef] [PubMed]
4. Rateb, M.E.; Ebel, R. Secondary metabolites of fungi from marine habitats. *Nat. Prod. Rep.* **2011**, *28*, 290–344. [CrossRef] [PubMed]
5. Jin, L.; Quan, C.; Hou, X.; Fan, S. Potential pharmacological resources: Natural bioactive compounds from marine-derived fungi. *Mar. Drugs* **2016**, *14*, 76. [CrossRef] [PubMed]
6. Youssef, F.S.; Ashour, M.L.; Singab, A.N.B.; Wink, M. A comprehensive review of bioactive peptides from marine fungi and their biological significance. *Mar. Drugs* **2019**, *17*, 559. [CrossRef] [PubMed]
7. Natoli, M.; Herzig, P.; Pishali, B.E.; Buchi, M.; Ritschard, R.; Heinzelmann, V.; Trub, M.; Zippelius, A.; Kashyap, A.S.; Natoli, M. Plinabulin, a Distinct Microtubule-Targeting Chemotherapy, Promotes M1-Like Macrophage Polarization and Anti-tumor Immunity. *Front. Oncol.* **2021**, *11*, 644608. [CrossRef] [PubMed]
8. Hendriks, H.R.; Fiebig, H.H.; Giavazzi, R.; Langdon, S.P.; Jimeno, J.M.; Faircloth, G.T. High antitumour activity of ET743 against human tumour xenografts from melanoma, non-small-cell lung and ovarian cancer. *Ann. Oncol.* **1999**, *10*, 1233–1240. [CrossRef] [PubMed]
9. Ibrahim, S.R.M.; Abdallah, H.M.; Mohamed, G.A.; Deshmukh, S.K. Exploring Potential of *Aspergillus sclerotiorum*: Secondary Metabolites and Biotechnological Relevance. *Mycol. Prog.* **2022**, *22*, 8. [CrossRef]
10. Guo, X.; Meng, Q.; Liu, J.; Wu, J.; Jia, H.; Liu, D.; Gu, Y.; Liu, J.; Huang, J.; Fan, A. Sclerotiamides C-H, Notoamides from a Marine Gorgonian-Derived Fungus with Cytotoxic Activities. *J. Nat. Prod.* **2022**, *85*, 1067–1078. [CrossRef]
11. Wang, H.; Zheng, J.-K.; Qu, H.-J.; Liu, P.-P.; Wang, Y.; Zhu, W.-M. A new cytotoxic indole-3-ethenamide from the halotolerant fungus *Aspergillus sclerotiorum* PT06-1. *J. Antibiot.* **2011**, *64*, 679–681. [CrossRef] [PubMed]
12. Chao, R.; Hou, X.M.; Xu, W.F.; Hai, Y.; Wei, M.Y.; Wang, C.Y.; Gu, Y.C.; Shao, C.L. Targeted Isolation of Asperheptatides from a Coral-Derived Fungus Using LC-MS/MS-Based Molecular Networking and Antitubercular Activities of Modified Cinnamate Derivatives. *J. Nat. Prod.* **2021**, *84*, 11–19. [CrossRef] [PubMed]
13. Chen, M.; Shao, C.-L.; Fu, X.-M.; Xu, R.-F.; Zheng, J.-J.; Zhao, D.-L.; She, Z.-G.; Wang, C.-Y. Bioactive indole alkaloids and phenyl ether derivatives from a marine-derived *Aspergillus* sp. Fungus. *J. Nat. Prod.* **2013**, *76*, 547–553. [CrossRef]
14. Liu, L.; Zheng, Y.-Y.; Shao, C.-L.; Wang, C.-Y. Metabolites from marine invertebrates and their symbiotic microorganisms: Molecular diversity discovery, mining, and application. *Mar. Life Sci. Technol.* **2019**, *1*, 60–94. [CrossRef]
15. Wu, J.-S.; Yao, G.-S.; Shi, X.-H.; Rehman, S.U.; Fu, X.-M.; Zhang, X.-L.; Wang, C.-Y. Epigenetic Agents Trigger the Production of Bioactive Nucleoside Derivatives and Bisabolane Sesquiterpenes from the Marine-Derived Fungus *Aspergillus versicolor*. *Front. Microbiol.* **2020**, *11*, 85. [CrossRef] [PubMed]
16. Wu, J.-S.; Shi, X.-H.; Yao, G.-S.; Shao, C.-L.; Fu, X.-M.; Zhang, X.-L.; Guan, H.-S.; Wang, C.-Y. New thiodiketopiperazine and 3,4-dihydroisocoumarin derivatives from the marine-derived fungus *Aspergillus Terreus*. *Mar. Drugs* **2020**, *18*, 132. [CrossRef] [PubMed]
17. Zhang, Y.-H.; Geng, C.; Zhang, X.-W.; Zhu, H.-J.; Shao, C.-L.; Cao, F.; Wang, C.-Y. Discovery of bioactive indole-diketopiperazines from the marine-derived fungus *Penicillium brasilianum* aided by genomic information. *Mar. Drugs* **2019**, *17*, 514. [CrossRef]
18. Wang, C.-Y.; Liu, X.-H.; Zheng, Y.-Y.; Ning, X.-Y.; Zhang, Y.-H.; Fu, X.-M.; Li, X.; Shao, C.-L.; Wang, C.-Y. 2,5-Diketopiperazines from a Sponge-Derived Fungus *Aspergillus sclerotiorum*. *Front. Microbiol.* **2022**, *13*, 808532. [CrossRef]
19. Wang, B.; Park, E.M.; King, J.B.; Mattes, A.O.; Nimmo, S.L.; Clendinen, C.; Edison, A.S.; Anklin, C.; Cichewicz, R.H. Transferring Fungi to a Deuterium-Enriched Medium Results in Assorted, Conditional Changes in Secondary Metabolite Production. *J. Nat. Prod.* **2015**, *78*, 1415–1421. [CrossRef]
20. Whyte, A.C.; Gloer, J.B.; Wicklow, D.T.; Dowd, P.F. Sclerotiamide: A new member of the paraherquamide class with potent antiinsectan activity from the sclerotia of *Aspergillus sclerotiorum*. *J. Nat. Prod.* **1996**, *59*, 1093–1095. [CrossRef]
21. He, W.; Xu, Y.; Fu, P.; Zuo, M.; Liu, W.; Jiang, Y.; Wang, L.; Zhu, W. Cytotoxic Indolyl Diketopiperazines from the *Aspergillus* sp. GZWMJZ-258, Endophytic with the Medicinal and Edible Plant Garcinia multiflora. *J. Agric. Food Chem.* **2019**, *67*, 10660–10666. [CrossRef]
22. Qian-Cutrone, J.; Huang, S.; Shu, Y.-Z.; Vyas, D.; Fairchild, C.; Menendez, A.; Krampitz, K.; Dalterio, R.; Klohr, S.E.; Gao, Q. Stephacidin A and B: Two structurally novel, selective inhibitors of the testosterone-dependent prostate LNCaP cells. *J. Am. Chem. Soc.* **2002**, *124*, 14556–14557. [CrossRef]

23. Kato, H.; Yoshida, T.; Tokue, T.; Nojiri, Y.; Hirota, H.; Ohta, T.; Williams, R.M.; Tsukamoto, S. Notoamides A-D: Prenylated indole alkaloids isolated from a marine-derived fungus, *Aspergillus* sp. *Angew. Chem. Int. Ed.* **2007**, *46*, 2254–2256. [CrossRef]
24. Fu, P.; Liu, P.; Qu, H.; Wang, Y.; Chen, D.; Wang, H.; Li, J.; Zhu, W. α-Pyrones and diketopiperazine derivatives from the marine-derived actinomycete *Nocardiopsis dassonvillei* HR10-5. *J. Nat. Prod.* **2011**, *74*, 2219–2223. [CrossRef] [PubMed]
25. Shin, C.-G.; Kato, H.; Yonezawa, Y.; Hayakawa, M.; Yoshimura, J. Synthesis and structural assignment of naturally occurring 3-benzyl-6-benzylidene-2,5-piperazinedione. *Heterocycles* **1980**, *14*, 1767–1770. [CrossRef]
26. Marcuccio, S.M.; Elix, J.A. Pyrazine chemistry. V. Synthesis of methylanhydropicroroccellin and dimethylpicroroccellin. *Aust. J. Chem.* **1985**, *38*, 1785–1796. [CrossRef]
27. Sterns, M.; Patrick, J.M.; Patrick, V.A.; White, A.H. Conformational studies of some piperazine-2,5-diones: Crystal structures of three isomeric forms of methylanhydropicroroccellin (derivative of a lichen diketopiperazine) and a derivative bromohydrin. *Aust. J. Chem.* **1989**, *42*, 349–364. [CrossRef]
28. Wang, M.; Carver, J.J.; Phelan, V.V.; Sanchez, L.M.; Garg, N.; Peng, Y.; Nguyen, D.D.; Watrous, J.; Kapono, C.A.; Luzzatto-Knaan, T. Sharing and community curation of mass spectrometry data with Global Natural Products Social Molecular Networking. *Nat. Biotechnol.* **2016**, *34*, 828–837. [CrossRef]
29. Kunwar, A.; Barik, A.; Mishra, B.; Rathinasamy, K.; Pandey, R.; Priyadarsini, K.I. Quantitative cellular uptake, localization and cytotoxicity of curcumin in normal and tumor cells. *Biochim. Biophys. Acta Gen. Subj.* **2008**, *1780*, 673–679. [CrossRef]
30. Pierce, C.G.; Uppuluri, P.; Tristan, A.R.; Wormley, F.L., Jr.; Mowat, E.; Ramage, G.; Lopez-Ribot, J.L. A simple and reproducible 96-well plate-based method for the formation of fungal biofilms and its application to antifungal susceptibility testing. *Nat. Protoc.* **2008**, *3*, 1494–1500. [CrossRef]
31. Aquino, R.; Morelli, S.; Lauro, M.R.; Abdo, S.; Saija, A.; Tomaino, A. Phenolic constituents and antioxidant activity of an extract of Anthurium versicolor leaves. *J. Nat. Prod.* **2001**, *64*, 1019–1023. [CrossRef] [PubMed]
32. Xia, W.; Luo, P.; Hua, P.; Ding, P.; Li, C.; Xu, J.; Zhou, H.; Gu, Q. Discovery of a New Pterocarpan-Type Antineuroinflammatory Compound from Sophora tonkinensis through Suppression of the TLR4/NFκB/MAPK Signaling Pathway with PU.1 as a Potential Target. *ACS Chem. Neurosci.* **2019**, *10*, 295–303. [CrossRef] [PubMed]

marine drugs

Article

Antibacterial Indole Diketopiperazine Alkaloids from the Deep-Sea Cold Seep-Derived Fungus *Aspergillus chevalieri*

Li-Hong Yan [1,2], Feng-Yu Du [3], Xiao-Ming Li [1,4], Sui-Qun Yang [1,2,4], Bin-Gui Wang [1,2,4,5,*] and Xin Li [1,2,4,*]

1 CAS and Shandong Province Key Laboratory of Experimental Marine Biology, Institute of Oceanology, Chinese Academy of Sciences, Nanhai Road 7, Qingdao 266071, China
2 University of Chinese Academy of Sciences, Yuquan Road 19A, Beijing 100049, China
3 College of Chemistry and Pharmacy, Qingdao Agricultural University, Changcheng Road 700, Qingdao 266109, China
4 Laboratory of Marine Biology and Biotechnology, Qingdao National Laboratory for Marine Science and Technology, Wenhai Road 1, Qingdao 266237, China
5 Center for Ocean Mega-Science, Chinese Academy of Sciences, Nanhai Road 7, Qingdao 266071, China
* Correspondence: wangbg@ms.qdio.ac.cn (B.-G.W.); lixin@qdio.ac.cn (X.L.); Tel.: +86-532-8289-8553 (B.-G.W.); +86-532-8289-8890 (X.L.)

Abstract: A large body of fungal secondary metabolites has been discovered to exhibit potent antibacterial activities with distinctive mechanisms and has the potential to be an untapped resource for drug discovery. Here, we describe the isolation and characterization of five new antibacterial indole diketopiperazine alkaloids, namely 24,25-dihydroxyvariecolorin G (**1**), 25-hydroxyrubrumazine B (**2**), 22-chloro-25-hydroxyrubrumazine B (**3**), 25-hydroxyvariecolorin F (**4**), and 27-epi-aspechinulin D (**5**), along with the known analogue neoechinulin B (**6**) from a fungal strain of deep-sea cold seep-derived *Aspergillus chevalieri*. Among these compounds, **3** and **4** represented a class of infrequently occurring fungal chlorinated natural products. Compounds **1–6** showed inhibitory activities against several pathogenic bacteria with MIC values ranging from 4 to 32 μg/mL. It was revealed that compound **6** could induce structural damage to the *Aeromonas hydrophila* cells based on the observation by scanning electron microscopy (SEM), which led to the bacteriolysis and death of *A. hydrophila*, suggesting that neoechinulin B (**6**) might be a potential alternative to novel antibiotics development.

Keywords: *Aspergillus chevalieri*; deep-sea cold seep; indole diketopiperazine; antibacterial activity

Citation: Yan, L.-H.; Du, F.-Y.; Li, X.-M.; Yang, S.-Q.; Wang, B.-G.; Li, X. Antibacterial Indole Diketopiperazine Alkaloids from the Deep-Sea Cold Seep-Derived Fungus *Aspergillus chevalieri*. *Mar. Drugs* **2023**, *21*, 195. https://doi.org/10.3390/md21030195

Academic Editors: Jaime Rodríguez and Rosa Maria Vitale

Received: 23 January 2023
Revised: 21 March 2023
Accepted: 21 March 2023
Published: 22 March 2023

1. Introduction

A wide range of fungi-derived natural products have been proved to possess significant pharmaceutical applications, such as penicillin derived from *Penicillium* sp. and cephalosporin C derived from *Acremonium chrysogenum*. Therefore, much attention has been paid to fungal secondary metabolites with potent antibacterial activities and novel mechanisms [1–3]. Deep-sea cold seeps are special extreme environments characterized by fluid emissions rich in methane and hydrogen sulfide, developing a unique chemosynthetically-driven ecosystem that creates a unique habitat for microorganisms in sediments [4]. Recently, a batch of intriguing findings on secondary metabolites from deep-sea cold seep-derived fungi has highlighted the potential of these microbes as a source of new antibiotics [5–9].

In our continuous research for bioactive metabolites from deep-sea cold seep-derived fungi [5–9], the fungal strain *Aspergillus chevalieri* CS-122, which was isolated from the sulfate-reducing sediments collected at a cold seep in the northeast of the South China Sea, drew our attention. As a result, a couple of indole diketopiperazine alkaloids containing an unprecedented spiro-bicyclo skeleton were isolated and identified [9]. Further work on this fungus has now resulted in the isolation of five other new indole diketopiperazine alkaloids. Among these compounds, compounds **3** and **4** represented a class of infrequently occurring fungal chlorinated natural products. Their structures were determined by detailed analyses of spectroscopic data, NMR calculation with DP4+ probability analysis, and 2,2-dimethoxypropane derivatization. This paper describes the isolation, structure elucidation, and stereochemical assignment of compounds **1–5** as well as the antibacterial activities of **1–6**.

2. Results and Discussion

2.1. Structure Elucidation

The culture broth of the fungus *Aspergillus chevalieri* CS-122 was extracted with EtOAc, and the crude extract was subjected to multiple chromatographic methods (a combination of column chromatography on Si gel, Sephadex LH-20, and Lobar LiChroprep RP-18) to yield subfractions, which were further purified by preparative TLC and semipreparative HPLC to provide the compounds **1–6** (Figure 1).

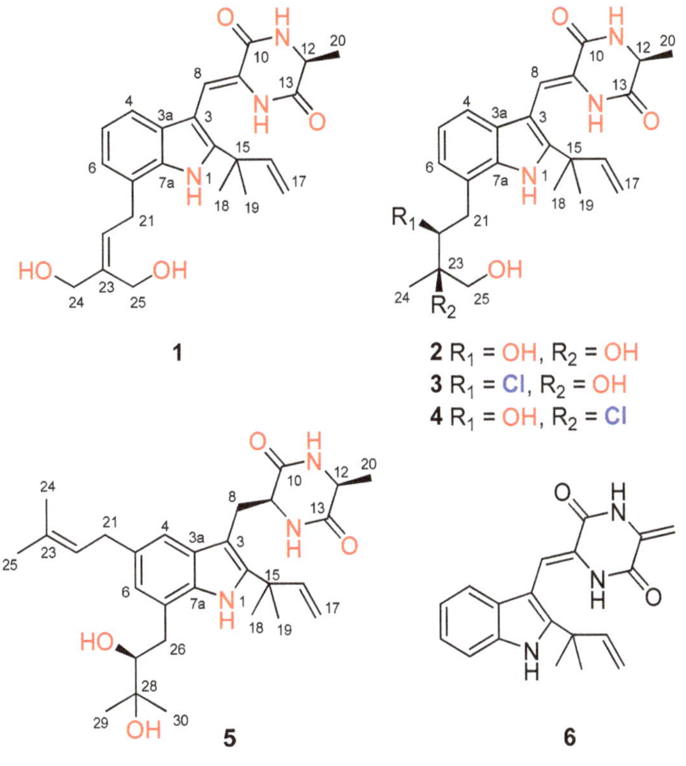

Figure 1. Structures of the isolated compounds **1–6**.

Compound **1** was obtained as a colorless amorphous powder. Its molecular formula was deduced as $C_{24}H_{29}N_3O_4$ on the basis of positive HRESIMS data, indicating 12 degrees of unsaturation. In the ^{1}H NMR spectrum (Table 1), the signals for the *ortho*-coupled aromatic protons (δ_H 7.04, dd, J = 7.4, 1.7 Hz, H-4; δ_H 6.96, t, J = 7.4 Hz, H-5; and δ_H 6.93, dd, J = 7.4, 1.7 Hz, H-6) were indicative of a 1,2,3-trisubstituted benzene ring system. Meanwhile, two trisubstituted olefins (δ_H 6.88, s, H-8 and δ_H 5.59, t, J = 7.8 Hz, H-22) and a vinyl group (δ_H 6.10, dd, J = 17.6, 10.3 Hz, H-16; δ_H 5.03, dd, J = 17.6, 1.3 Hz, H-17a; δ_H 5.03, dd, J = 10.3, 1.3 Hz, H-17b) were also observed. The ^{13}C NMR and DEPT data (Table 2) displayed 24 carbon signals, accounting for three methyls, four methylenes (two oxygenated and one olefinic), seven methines (six aromatic/olefinic), and ten quaternary carbons (two carbonyls and seven aromatic/olefinic) (Table 1). Detailed analysis of the 1D and 2D NMR data revealed that the structure of compound **1** was similar to variecolorin G, a previously reported indole diketopiperazine isolated from the fungal strain *Aspergillus variecolor* B-17 [10]. However, the signals for two methyls (δ_H 1.75/δ_C 25.6, CH_3-24 and δ_H 1.75/δ_C 17.8, CH_3-25) in the NMR spectra of variecolorin G were absent in those of **1**, while resonances for two oxygenated methylenes (δ_H 3.96/δ_C 63.5, CH_2-24 and δ_H 4.19/δ_C 57.1, CH_2-25) were observed in the NMR spectra of **1** (Tables 1 and 2). The key HMBC correlations from H-24 and H-25 to C-22 and C-23 (Figure 2) further determined the planar structure of **1**.

Table 1. ^{1}H NMR Spectroscopic Data for Compounds **1–5** (δ in ppm, J in Hz) [a].

No.	1	2	3	4	5
1-NH	10.58, s	10.46, s	10.13, s	10.10, s	9.89, s
4	7.04, dd, (7.4, 1.7)	7.05, m	7.09, m	7.08, dd, (7.3, 1.7)	7.07, s
5	6.96, t, (7.4)	6.94, overlap	6.99, overlap	6.98, overlap	
6	6.93, dd, (7.4, 1.7)	6.94, overlap	7.00, overlap	7.00, overlap	6.71, s
8	6.88, s	6.90, s	6.88, s	6.89, s	Ha, 3.01, dd, (14.5, 9.6) Hb, 3.34, m
9					3.96, m
11-NH	8.31, s	8.31, s	8.33, s	8.34, s	8.18, s
12	4.15, qd, (6.9, 1.6)	4.16, q, (6.9)	4.14, qd, (6.9, 1.4)	4.17, qd, (6.9, 1.9)	3.82, qd, (6.9, 2.3)
14-NH	8.62, s	8.56, s	8.74, s	8.57, s	7.41, s
16	6.10, dd, (17.6, 10.3)	6.09, dd, (17.3, 10.6)	6.11, dd, (17.2, 10.6)	6.10, dd, (17.3, 10.6)	6.16, dd, (17.4, 10.6)
17	Ha, 5.03, dd, (17.6, 1.3) Hb, 5.03, dd, (10.3, 1.3)	Ha, 5.05, dd, (17.3, 1.2) Hb, 5.06, dd, (10.6, 1.2)	Ha, 5.04, d, (17.2) Hb, 5.05, d, (10.6)	Ha, 5.06, dd, (17.3, 1.2) Hb, 5.07, dd, (10.6, 1.2)	Ha, 5.08, d, (17.4) Hb, 5.04, d, (10.6)
18	1.48, s	1.49, s	1.50, s	1.50, s	1.48, s
19	1.48, s	1.49, s	1.50, s	1.50, s	1.49, s
20	1.37, d, (6.9)	1.37, d, (6.9)	1.39, d, (6.9)	1.38, d, (6.9)	1.33, d, (6.9)
21	3.76, d, (7.8)	Ha, 2.76, dd, (14.6, 8.8) Hb, 3.25, d, (14.6)	Ha, 3.15, dd, (15.5, 10.6) Hb, 3.67, d, (15.5)	Ha, 2.93, dd, (14.6, 8.8) Hb, 3.31, d, (14.6)	3.31, d, (7.2)
22	5.59, t, (7.8)	3.70, d, (8.8)	4.37, d, (10.6)	3.97, t, (8.8)	5.32, t, (7.2)
22-OH				5.31, d, (7.2)	
23-OH			5.14, s		
24	3.96, d, (4.2)	1.12, s	1.28, s	1.56, s	1.71, s
24-OH	4.70, t, (4.2)				
25	4.19, d, (3.8)	Ha, 3.46, d, (10.9) Hb, 3.40, d, (10.9)	3.57, m	3.75, m	1.70, s
25-OH	5.32, t, (3.8)		5.26, t, (5.5)	5.54, t, (6.1)	
26					Ha, 2.65, dd, (14.5, 8.7) Hb, 3,13, d, (14.5)
27					3.42, m
27-OH					4.99, d, (5.7)
28-OH					4.69, s
29					1.16, s
30					1.17, s

[a] Measured at 500 MHz in DMSO-d_6.

Table 2. ^{13}C NMR Spectroscopic Data for Compounds **1–5** (δ in ppm, type) [a].

No.	1	2	3	4	5
2	143.8, C	143.5, C	143.6, C	143.5, C	141.0, C
3	103.9, C	103.8, C	104.2, C	104.0, C	104.9, C
3a	126.0, C	125.9, C	126.1, C	126.0, C	129.1, C
4	117.0, CH	116.8, CH	117.3, CH	117.0, CH	114.8, CH
5	121.1, CH	119.6, CH	119.5, CH	119.7, CH	131.4, C
6	119.7, CH	121.7, CH	121.3, CH	121.9, CH	122.2, CH
7	123.7, C	125.2, C	122.6, C	123.8, C	124.1, C
7a	133.9, C	134.3, C	133.9, C	134.2, C	132.7, C
8	110.3, CH	110.2, CH	110.1, CH	110.0, CH	31.3, CH_2
9	125.1, C	124.8, C	125.2, C	125.1, C	55.6, CH
10	159.9, C	159.9, C	159.8, C	159.8, C	167.4, C
12	50.5, CH	50.5, CH	50.6, CH	50.5, CH	50.3, CH
13	166.4, C	166.3, C	166.4, C	166.4, C	167.9, C
15	39.1, C	38.9, C	39.0, C	38.9, C	38.7, C
16	145.4, CH	145.1, CH	145.2, CH	145.1, CH	146.5, CH
17	111.4, CH_2	111.7, CH_2	111.7, CH_2	111.8, CH_2	111.1, CH2
18	27.6, CH_3	27.4, CH_3	27.6, CH_3	27.5, CH_3	27.9, CH_3

Table 2. *Cont.*

No.	1	2	3	4	5
19	27.6, CH_3	27.4, CH_3	27.5, CH_3	27.4, CH_3	27.9, CH_3
20	19.6, CH_3	19.6, CH_3	19.8, CH_3	19.6, CH_3	20.7, CH_3
21	29.6, CH_2	34.0, CH_2	33.8, CH_2	34.2, CH_2	34.2, CH_2
22	124.7, CH	74.8, CH	66.9, CH	77.7, CH	124.9, CH
23	139.2, C	74.2, C	74.2, C	74.0, C	130.2, C
24	63.5, CH_2	19.1, CH_3	21.0, CH_3	22.8, CH_3	17.7, CH_3
25	57.1, CH_2	67.6, CH_2	67.5, CH_2	68.1, CH_2	25.6, CH_3
26					34.5, CH_2
27					79.0, CH
28					72.2, C
29					23.6, CH_3
30					27.0, CH_3

[a] Measured at 125 MHz in DMSO-d_6.

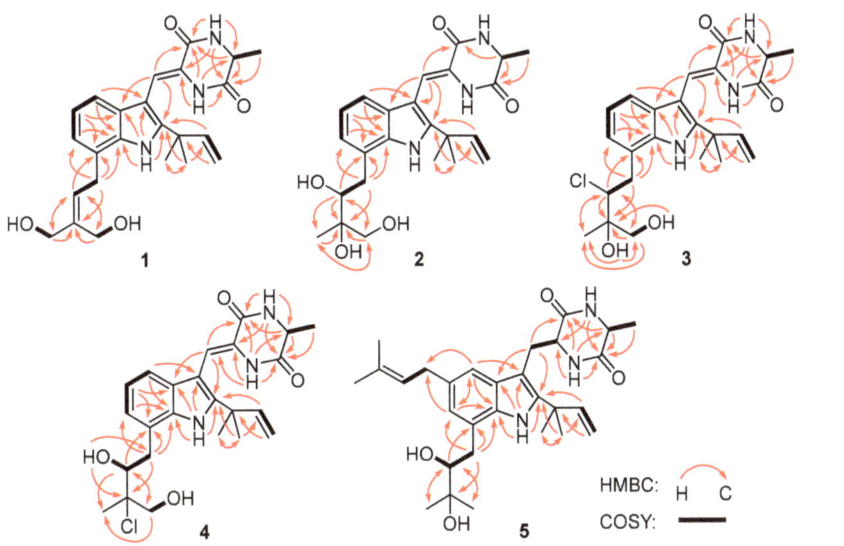

Figure 2. Key ^{1}H-^{1}H COSY (bold lines) and HMBC (red arrows) correlations of compounds **1–5**.

The geometry of the double bond at C-8 was determined to be *Z*, as the chemical shift of H-8 (δ_H 6.88) was observed downfield under the influence of the deshielding effect of the C=O group [10–13]. The absolute configuration of C-12 was assigned by chiral HPLC analysis of the acidic hydrolysate. The retention time of the acidic hydrolysate of **1** matched well with that of L-Ala (Figure S1), indicating the *S* configuration of C-12. Thus, the structure of **1** was determined, and it was named 24,25-dihydroxyvariecolorin G.

Compound **2** possessed the molecular formula $C_{24}H_{31}N_3O_5$ on the basis of HRESIMS data, indicating 11 degrees of unsaturation. Detailed analysis of the ^{1}H and ^{13}C NMR data of **2** (Tables 1 and 2) suggested that **2** was an indole diketopiperazine derivative, showing similarity to rubrumazine B, which was characterized from *Eurotium rubrum* MA-150, a fungus isolated from marine mangrove rhizospheric soil [12]. However, the signals for a methyl group (CH_3-25) resonating at δ_H 1.18 and δ_C 26.8 in the NMR spectra of rubrumazine B disappeared in those of **2**. Instead, signals for an oxygenated methylene (CH_2-25) resonating at δ_H 3.46/3.40 and δ_C 67.6 were observed in the NMR spectra of **2** (Tables 1 and 2). Meanwhile, compared to rubrumazine B, a smaller chemical shift for CH_3-24 (δ_H 1.12/δ_C 19.1) was also observed. The key HMBC correlations from H-25 to C-22, C-23, and C-24 (Figure 2) further constructed the planar structure of **2**, which was designed as the 25-hydroxylated derivative of rubrumazine B.

Compound **3,** isolated as a colorless amorphous powder, was found to have the molecular formula $C_{24}H_{30}ClN_3O_4$ on the basis of HRESIMS data, suggesting 11 degrees of unsaturation. Specifically, the existence of a chlorine group was deduced by the isotopic peaks at *m/z* 460 and 462 with a ratio of 3:1. Its NMR spectroscopic data were very similar to those of **2**, except for the signal of the oxygenated methine group (CH-22) resonating at δ_H 3.70/δ_C 74.8 in **2**, which shifted to δ_H 4.37/δ_C 66.9 in those of **3**. The above observation revealed that the 22-OH in compound **2** was replaced by a substituent with less electronegativity in compound **3**. Considering the molecular formula, all the atoms have been assigned except for a remaining chlorine atom, suggesting the substitution of a chlorine group at C-22. Compared to **2**, an obvious larger chemical shift of H-22 was observed, which might be attributed to the steric effect of the chlorine atom. Therefore, the planar structure of **3** was determined as 22-chloro-25-hydroxyrubrumazine B.

The HRESIMS data of compound **4** demonstrated its molecular formula to be $C_{24}H_{30}ClN_3O_4$, the same as that of compound **3**. Similarly, the isotopic peaks at *m/z* 460 and 462 with a ratio of 3:1 suggested the presence of a chlorine group. The NMR spectra of **4** showed close similarity to those of the chlorine-containing indole diketopiperazine variecolorin F [10]. However, the signals for a methyl (CH_3-25) at δ_H 1.63 and δ_C 29.5 in the NMR spectrum of variecolorin F were replaced by an oxygenated methylene (CH_2-25) resonating at δ_H 3.75 and δ_C 68.1 in those of **4**. This deduction was further supported by the COSY correlations from H-25 to 25-OH and the HMBC correlations from H-25 to C-22 and C-24. The structure of **4** was finally defined as 25-hydroxyvariecolorin F.

The configurations of Δ^8 and C-12 in compounds **2**–**4** were identical to those in **1** based on the same approaches used for **1** (Figures S9, S23, and S37). However, it was extremely difficult to assign the absolute configurations of C-22 and C-23 on the isopentyl moiety of **2**–**4** because of the flexible nature of the saturated side chain and the separation of C-22 and C-23 away from the chiral center C-12, which continued to puzzle the natural product chemists for a long time [14,15]. To determine the absolute configurations of C-22 and C-23 in compounds **2**–**4**, the relative configurations of C-22 and C-23 should be assigned first and then correlated with the absolute configuration of C-12. Quantum chemical calculations of NMR with DP4+ probability analysis is a recently developed and widely used strategy for structural elucidation and configuration assignment [16]. According to the previously reported examples, approximately 20% of compounds with stereoclusters separated through flexible spacers were assigned the configurations with the aid of DP4+ [16–18]. Hence, GIAO (gauge-including atomic orbitals) NMR calculations at mPW1PW91/6-31+G(d,p) level with DP4+ probability analyses were performed on compounds **2**–**4**. As a result, the experimental ^1H and ^{13}C NMR data of **2**–**4** matched well with those calculated for the isomers **2c**, **3c**, and **4c** (12*S**,22*S**,23*R**), as indicated by DP4+ probabilities of 99.6% (Table S1), 95.9% (Table S2), and 99.1% (Table S3), respectively. Among them, the relative configuration of compound **4** was further supported by the 2,2-dimethoxypropane derivatization, which produced acetonide **7** (Figure 3A), fixing the rotation of the single bond (C22-C23) through the formation of a six-membered ketal ring. The chemical shifts of two acetal methyl groups at δ_C 29.0 (C-26) and δ_C 18.7 (C-27) indicated the chair conformation of the six-membered ketal ring [19]. Subsequently, NOE correlations from H_3-27 to H-25a and H-22 in **7** suggested the co-face of these groups (Figure 3B), whereas correlations from H_3-24 to H-25b placed them on the opposite face, and this assignment is consistent with the results obtained by DP4+ probability analysis. Thus, the relative configurations of compounds **2**–**4** were speculated as 12*S**, 22*S**, and 23*R**, and the absolute configurations of compounds **2**–**4** were tentatively speculated as 12*S*, 22*S*, and 23*R* by correlating the stereo configurations of C-22 and C-23 with the absolute configuration of C-12.

To verify the above deduction, many efforts, including a modification of Mosher's method and the cultivation of single crystals, were made to further assign the configurations of C-22 and C-23 in **2**–**4**. Unfortunately, the presence of multiple -OH groups (three in **2** and two in **3** and **4**) and three -NH groups caused many side reactions during Mosher's esterification, which resulted in failure to obtain the aim products. Meanwhile, the cultivation of single crystals under various conditions has also not succeeded. Hence, continuous investigations on the absolute configurations of **2–4** are still required in our future work, which might involve the continuous cultivation of high-quality crystals and applying alternative protocols for NMR calculations (including different methods for conformational sampling and other functionals and bases for calculations) or performing RSCA and RDC experiments. Once the configurations of **2–4** are unambiguously assigned, we will report the findings in due course.

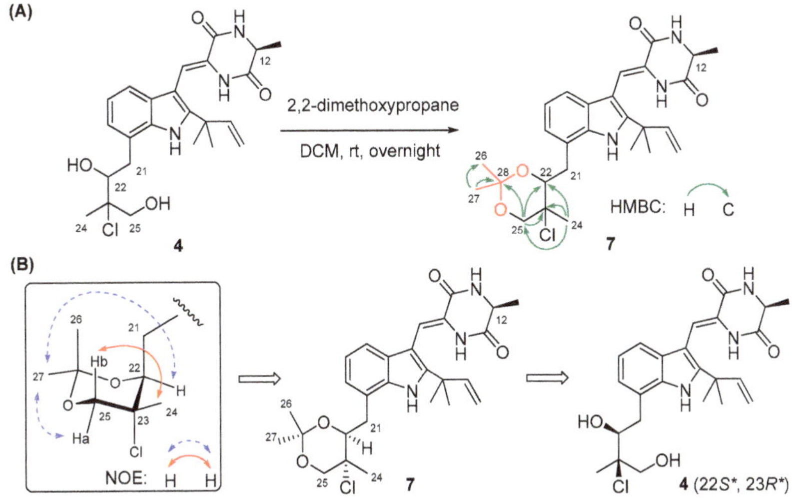

Figure 3. (**A**) The preparation of acetonide **7** and its HMBC correlations. (**B**) NOESY correlations of acetonide **7**.

Compound **5** had a molecular formula of $C_{29}H_{41}N_3O_4$ as determined by HRESIMS analysis. The ^1H and ^{13}C NMR data revealed that **5** is an indole diketopiperazine alkaloid, similar to previously reported aspechinulin D from the deep-sea-derived fungus *Aspergillus* sp. FS445 [20]. Compared to the NMR data of aspechinulin D (δ_C 75.8, C-28; δ_H/δ_C 1.62/27.6, CH_3-29; δ_H/δ_C 1.62/30.0, CH_3-30, measured in DMSO-d_6), obvious upfielded shifts for the oxygenated quaternary carbon (C-28) and two methyl groups (CH_3-29 and CH_3-30) in the NMR spectra of **5** were observed (δ_C 72.2, C-28; δ_H 1.16/δ_C 23.6, CH_3-29; δ_H 1.17/δ_C 27.0, CH_3-30), suggesting that **5** is a diastereomer of aspechinulin D, epimeric at C-27. Thus, compound **5** was named 27-epi-aspechinulin D.

2.2. Antimicrobial Activity

All the isolated compounds were examined to assess their antimicrobial activity against several human pathogenic and aqua-pathogenic bacteria (Table 3). Compounds **1–5** showed moderate activity against the human pathogen *Escherichia coli* and the aquatic bacterium *Vibrio harveyi* (MIC $\leq$ 32 μg/mL). Among them, compound **1** displayed significant inhibitory effects against *E. coli*, with a MIC value of 4 μg/mL, while compound **3** displayed noticeable inhibitory effects against *V. harveyi*, with a MIC value of 8 μg/mL. Moreover, compounds **2** (with hydroxyl groups at C-22 and C-23) and **5** (with hydroxyl groups at C-27 and C-28) exhibited broad-spectrum antibacterial activity against five tested bacterial strains (MIC $\leq$ 32 μg/mL), implying that the bishyroxylation at these positions made them broad-spectrum.

Additionally, compound **6** showed significant activity against aquatic *Aeromonas hydrophila*, with a MIC value of 4 μg/mL. The possible antibacterial mechanism of **6** was studied based on the observation by scanning electron microscopy (SEM) for the cells before and after treatment with compound **6**. For the *A. hydrophila* cells untreated with **6**

(Figure 4A) or treated only with DMSO (Figure 4B), it was observed that the cell surfaces were smooth and the cell structures were plump and intact. However, for the *A. hydrophila* cells treated with compound **6**, conglutination of the cells with a viscous substance attached was observed, while deep grooves (red arrow) and obvious pores (green arrow) on the surfaces of the cells were also observed (Figure 4C,D). Meanwhile, serious cell deformation and severe damage to cell membranes also appeared on a number of cells, which were accompanied by bacteriolysis and even complete disappearance, which resulted in large bubbles (Figure 4E,F, yellow arrow). The above observations revealed that compound **6** could induce structural damage to the *A. hydrophila* cells, which led to the bacteriolysis and death of *A. hydrophila*.

Table 3. Antimicrobial Activities of Compounds **1–6** (MIC, μg/mL).

Strain	Compound						
	1	**2**	**3**	**4**	**5**	**6**	**Chloramphenicol** [b]
Vibrio harveyi	16	32	8	32	16	-	2
Edwardsiella tarda	-[a]	16	-	-	32	-	8
Aeromonas hydrophila	-	32	-	-	32	4	2
Escherichia coli	4	16	32	32	32	8	2
Micrococcus luteus	-	32	-	-	16	-	2

[a] (–) = MIC > 32 μg/mL. [b] Chloramphenicol as positive control.

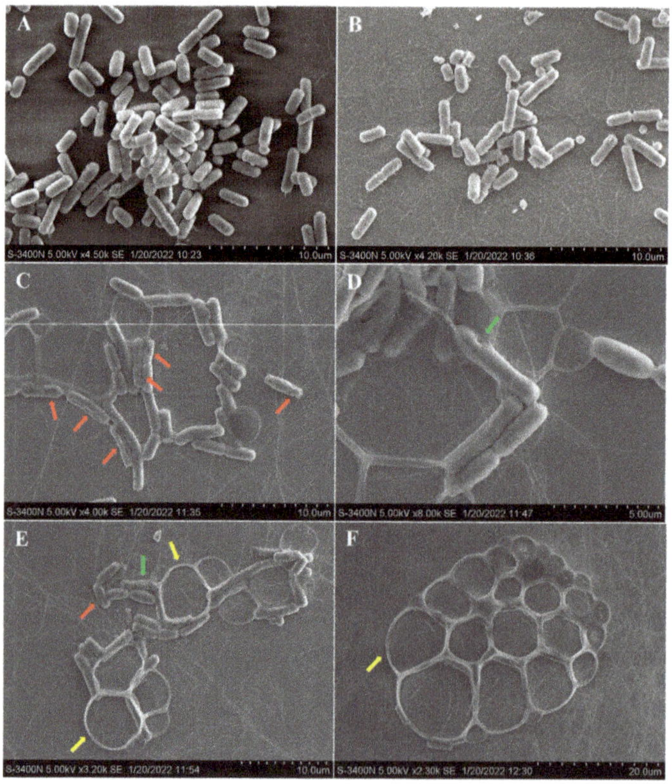

Figure 4. SEM observations of morphological changes of *Aeromonas hydrophila* following treatment with neoechunulin B (**6**). *A. hydrophila* cell morphology without any treatment (**A**), with DMSO treatment (**B**), and with compound **6** treatment (**C–F**). The red, green, and yellow arrows indicate the grooves, pores, and bubbles, respectively.

3. Experimental Section

3.1. General Experimental Procedures

Detailed information for the apparatus, reagents, solvents, and materials used in the present work is the same as that described in our previous publication [9].

3.2. Fungal Material

The fungus *Aspergillus chevalieri* CS-122 was isolated from the deep-sea cold seep sediment, which was collected in the northeast of the South China Sea (119°17′ E, 22°06′ N) in August 2018. The fungal strain was identified as *A. chevalieri* according to the β-tubulin gene sequence [21], which is the same (100%) as that of *A. chevalieri* (accession No. KU872171.1). The sequence data of CS-122 were deposited in GenBank with accession No. OM304365.1 (https://www.ncbi.nlm.nih.gov/nuccore/OM304356.1, accessed on 22 March 2023). This strain is stored at the Key Laboratory of Experimental Marine Biology, Institute of Oceanology, Chinese Academy of Sciences (IOCAS).

3.3. Fermentation, Extraction, and Isolation

The fungal strain *Aspergillus chevalieri* CS-122 was cultivated on potato dextrose agar medium at 28 °C for 7 days. Next, it was transferred into 1 L Erlenmeyer flasks containing rice solid medium (each flask contained 70 g rice, 0.3 g peptone from animal tissue, 0.5 g yeast extract, 0.2 g corn steep liquor, 0.1 g monosodium glutamate, and naturally sourced seawater) and incubated at room temperature for 30 days. Then, the solid fermented substrate was extracted three times with EtOAc. The combined extracts were concentrated under reduced pressure to provide a dark brown crude extract (256 g).

The total extract (256 g) was subjected to VLC (vacuum liquid chromatography) eluting with a gradient of petroleum ether (PE)/EtOAc (from 20:1 to 1:1) and CH_2Cl_2/MeOH (from 50:1 to 1:1) to yield ten fractions (Frs. 1–10). Fr. 8 (10.3 g), eluted with CH_2Cl_2/MeOH (10:1), was further purified by reversed-phase column chromatography over RP-18 with a MeOH/H_2O gradient (from 10:90 to 90:10) to afford nine subfractions (Frs. 8.1–8.9). Fr. 8.4 (2.1 g) was further purified by CC on silica gel eluting with a CH_2Cl_2/MeOH gradient (from 150:1 to 50:1) and then by preparative TLC as well as Sephadex LH-20 (MeOH) to yield **1** (8.6 mg) and **2** (4.9 mg). Fr. 8.5 (1.6 g) was split by CC on silica gel eluting with a CH_2Cl_2/MeOH gradient (from 150:1 to 30:1) to afford five subfractions (Frs. 8.5.1–8.5.5). Fr. 8.5.2 (57.7 mg) was further purified by semipreparative HPLC (42% MeCN–H_2O, 12 mL/min, 254 nm) to provide **4** (5.8 mg), while Fr. 8.5.5 (62.8 mg) was also purified by semipreparative HPLC (77% MeOH–H_2O, 10 mL/min, 254 nm) to provide **5** (6.2 mg). Fr. 8.6 (2.4 g) was fractionated by CC on silica gel eluting with a CH_2Cl_2/MeOH gradient (from 150:1 to 50:1) and then purified by semipreparative HPLC (80% MeOH–H_2O, 10 mL/min, 254 nm) to obtain **3** (5.6 mg). The isolation of compound **6** was described in our previous publication [9].

24,25-Dihydroxyvariecolorin G (**1**): colorless amorphous powder; [α]25 D –17.4 (*c* =0.23, MeOH); UV (MeOH) λ_{max} (log ε) 226 (3.26) nm, 252 (3.11) nm, 279 (2.90) nm, 335 (2.96) nm; ECD (0.35 mM, MeOH) λ_{max} (Δε) 214 (–5.52), 240 (+2.22), 342 (–1.46) nm; ^{1}H and ^{13}C NMR data, Tables 1 and 2; HRESIMS *m/z* 424.2225 $[M+H]^+$ (calcd for $C_{24}H_{30}N_3O_4$, 424.2231).

25-Hydroxyrubrumazine B (**2**): colorless amorphous powder; [α]25 D –23.8 (*c* =0.21, MeOH); UV (MeOH) λ_{max} (log ε) 225 (3.66) nm, 255 (3.28) nm, 282 (3.13) nm, 339 (3.22) nm; ECD (0.57 mM, MeOH) λ_{max} (Δε) 221 (–11.88), 248 (+0.52), 333 (–1.97) nm; ^{1}H and ^{13}C NMR data, Tables 1 and 2; HRESIMS *m/z* 442.2330 $[M+H]^+$ (calcd for $C_{24}H_{32}N_3O_5$, 442.2336).

22-Chloro-25-hydroxyrubrumazine B (**3**): colorless amorphous powder; [α]25 D –25.0 (*c* =0.16, MeOH); UV (MeOH) λ_{max} (log ε) 224 (3.56) nm, 257 (3.15) nm, 278 (2.98) nm, 335 (3.06) nm; ECD (0.35 mM, MeOH) λ_{max} (Δε) 206 (–9.54), 234 (+7.56), 328 (–3.46) nm; ^{1}H and ^{13}C NMR data, Tables 1 and 2; HRESIMS *m/z* 460.1987 $[M+H]^+$ (calcd for $C_{24}H_{30}ClN_3O_4$, 460.1998).

25-Hydroxyvariecolorin F (**4**): colorless amorphous powder; [α]25 D −53.8 (*c* =0.26, MeOH); UV (MeOH) λ_{max} (log ε) 226 (3.70) nm, 255 (3.30) nm, 279 (3.13) nm, 334 (3.22) nm; ECD (0.28 mM, MeOH) λ_{max} (Δε) 204 (−12.69), 240 (+2.38), 334 (−2.71) nm; ^{1}H and ^{13}C NMR data, Tables 1 and 2; HRESIMS *m/z* 460.1992 [M+H]$^+$ (calcd for $C_{24}H_{30}ClN_3O_4$, 460.1998).

27-epi-Aspechinulin D (**5**): colorless amorphous powder; [α]25 D −23.1 (*c* =0.13, MeOH); UV (MeOH) λ_{max} (log ε) 231 (3.61) nm, 280 (2.99) nm; ECD (0.26 mM, MeOH) λ_{max} (Δε) 226 (−9.63), 271 (+1.03) nm; ^{1}H and ^{13}C NMR data, Tables 1 and 2; HRESIMS m/z 496.3157 [M+H]$^+$ (calcd for $C_{29}H_{42}N_3O_4$, 496.3170).

3.4. Computational NMR Chemical Shift Calculation and DP4+ Analysis

All the theoretical calculations were conducted in the Gaussian09 program package. Conformational searches for possible isomers based on molecular mechanics with the MMFF method were performed using HyperChem 8.0 software. NMR shielding tensors were calculated using the GIAO method. The corresponding stable conformers whose Boltzmann distributions were higher than 2%, were further optimized at B3LYP/6-31G(d) in vacuo. Then, all the optimized conformers were subjected to the DFT method at mPW1PW91\6-31+G(d) with PCM level in DMSO to acquire the calculated shielding tensors. The calculated shielding tensors were later obtained according to the Boltzmann weighting of each conformer. Finally, the DP4+ analysis of the calculated shielding tensors and experimental chemical shifts were applied with the Excel formulas provided by the original authors [22].

3.5. Acidic Hydrolysis of Compounds 1–5

Compounds **1–5** (1 mg each) were dissolved in 10 mL of 6 N HCl and heated in a sealed tube at 110 °C for 24 h [12]. The solutions were then evaporated to dryness under reduced pressure. Each sample, including the standard amino acids L-Ala and D-Ala, was dissolved in 1 mL of eluting solvent (2 mM $CuSO_4 \cdot 5H_2O$ in 100 mL of H_2O). Chiral HPLC analysis, both alone and by co-injection with standards, was carried out using a Phenomenex-Chirex-3126 column (250 mm × 4.60 mm, 5 μm; flow rate 1.0 mL/min at 40 °C; detection at 254 nm).

3.6. 2,2-Dimethoxypropane Derivatization of Compound 4

Compound **4** (2 mg) was dissolved in 2 mL CH_2Cl_2. Next, 50 μL 2,2-dimethoxypropane and 10 mg Amberlyst-15H were added to the solution. The reaction mixture was stirred at room temperature overnight and then filtered, and the organic layer was evaporated at reduced pressure [23]. Further purification by preparative TLC afforded acetonide **7** (0.5 mg).

Acetonide **7**: colorless amorphous powder; Diagnostic ^{1}H NMR data (DMSO-d_6, 500 MHz) δ_H 4.41 (1H, br d, *J* = 9.6 Hz, H-22), 4.06 (1H, d, *J* = 11.3 Hz, H-25a), 3.76 (1H, d, *J* = 11.3 Hz, H-25b), 3.08 (1H, dd, *J* = 16.1, 9.6 Hz, H-21b), 1.78 (3H, s, H-24), 1.52 (6H, s, H-18/H-19), 1.43 (3H, s, H-27), 1.37 (3H, d, *J* = 6.9 Hz, H-20), 1.31 (3H, s, H-26); Diagnostic ^{13}C NMR data (DMSO-d_6, 125 MHz) δ_C 99.4 (C, C-28), 76.3 (CH, C-22), 70.5 (CH_2, C-25), 65.7 (C, C-23), 29.0 (CH_3, C-26), 27.5 (CH_3, C-18/C-19), 20.6 (CH_3, C-24), 19.6 (CH_3, C-20), 18.7 (CH_3, C-27); Key HMBC correlations from H_2-25b to C-28, from H_3-26 to C-27 and C-28, from H_3-27 to C-26 and C-28.

3.7. Antibacterial Assay

The antibacterial activities against human pathogenic bacteria (*Escherichia coli* and *Micrococcus luteus*) and aquatic pathogens (*Vibrio harveyi*, *Edwardsiella tarda*, *V. anguillarum*, and *Aeromonas hydrophilia*) were determined by a serial dilution technique using 96-well microtiter plates with minor modifications, as described in our previous report [5–9]. The bacteria were cultivated overnight at 37 °C in liquid LB medium and diluted with the LB broth to a concentration of 1.5 × 10 ^{8} CFU/mL. The tested compounds and positive control (chloramphenicol) were dissolved in DMSO to provide a stock solution. Then, 5 μL of the sample solutions with different concentrations, together with 95 μL of the bacterial suspension, were added to the 96-well plates and incubated at 37 °C for 12 h. The growth situation of the bacteria was measured by a multi-detection microplate reader (Infinite M1000 Pro, Tecan) at 600 nm. The human or aquatic pathogenic strains were offered by the Institute of Oceanology, Chinese Academy of Sciences.

3.8. Scanning Electron Microscopy (SEM)

The effect of neoechinulin B (**6**) on the morphological changes of *Aeromonas hydrophilia* was examined by scanning electron microscopy (SEM) [24]. Briefly, *Aeromonas hydrophilia* was incubated at 28 °C in LB broth at 140 rpm, with shaking for 12 h, to an OD_{600} of 0.2. Then, the bacteria were cultivated for an additional 12 h with 64 μg/mL neoechinulin B or DMSO. After cultivation, the culture broth was centrifuged and then washed with PBS (pH 7.2–7.4) three times. The collected bacteria were further fixed with 5% glutaraldehyde solution, washed twice with PBS, and then dehydrated for 15 min through a graded ethanol series (30%, 50%, 70%, 80%, 90% and 100%). The dried bacterial cells were transferred to isopentyl acetate for 20 min. The samples were gold coated and visualized under the scanning electron microscope (Hitachi, S-3400N).

4. Conclusions

In this study, we isolated and characterized five new compounds (**1–5**) from the cold seep sediment-derived fungus *Aspergillus chevalieri* CS-122, which are new members of the indole diketopiperazine alkaloids. Among them, compounds **3** and **4** were a kind of infrequently occurring fungal chlorinated natural products. Compounds **1–6** showed potent antibacterial activities against several aquatic pathogens. Among them, compound **6** significantly destroyed the cell morphology of *A. hydrophila* to exert a growth-inhibitory effect, the specific mechanism and target of which must be further investigated. These compounds possess the potential to be developed as antibiotic lead compounds for aquaculture.

Supplementary Materials: The following are available online at https://www.mdpi.com/article/10.3390/md21030195/s1. Figures S1–S69: The chiral HPLC analysis of the acidic hydrolysate, the 1D and 2D NMR spectra, HRESIMS, the ECDs of compounds **1–5**, the 1D and 2D NMR spectra of acetonide 7, and the optimized conformers for the candidate isomers and the results of DP4+ probability analysis of compounds **2–4**; Tables S1–S27: The DP4+ probability analysis, calculated shielding tensors of each conformer for the candidate isomers, and cartesian coordinates of the lowest energy conformers for the candidate isomers of compounds **2–4.**

Author Contributions: L.-H.Y. performed the experiments for the isolation, structure elucidation, and bioactivity evaluation and prepared the manuscript; F.-Y.D. finished the isolation and structure elucidation of compounds **3** and **4**; X.-M.L. performed the 1D and 2D NMR experiments; S.-Q.Y. contributed to the isolation of the fungus *Aspergillus chevalieri* CS-122; X.L. and B.-G.W. supervised the research work and revised the manuscript. All authors have read and agreed to the published version of the manuscript.

Funding: This work was financially supported by the Shandong Special Fund for Qingdao National Laboratory for Marine Science and Technology (2022QNLM030004-4), the National Natural Science Foundation of China (42076090 and U2006203), the Senior User Project of RV KEXUE (KEXUE2020GZ02), and the Shandong Provincial Natural Science Foundation (ZR2021ZD28 and ZR2019ZD18). X.L. acknowledges the Youth Innovation Promotion Association CAS (2023216).

Institutional Review Board Statement: Not applicable.

Acknowledgments: B.-G.W. acknowledges the support of the RV *KEXUE* of the National Major Science and Technology Infrastructure from the Chinese Academy of Sciences (for sampling) and the Oceanographic Data Center at IOCAS (for CPU time). The authors thank Yuan-Yuan Sun. at the Analytical Center of IOCAS for assistance with the SEM operations.

Conflicts of Interest: The authors declare no conflict of interest.

References

1. Yan, L.-H.; Li, X.-M.; Chi, L.-P.; Li, X.; Wang, B.-G. Six new antimicrobial metabolites from the deep-sea sediment-derived fungus *Aspergillus fumigatus* SD-406. *Mar. Drugs* **2022**, *20*, 4. [CrossRef] [PubMed]
2. Li, Y.-H.; Li, X.-M.; Li, X.; Yang, S.-Q.; Shi, X.-S.; Li, H.-L.; Wang, B.-G. Antibacterial alkaloids and polyketide derivatives from the deep sea-derived fungus *Penicillium cyclopium* SD-413. *Mar. Drugs* **2020**, *18*, 553. [CrossRef] [PubMed]
3. Shang, Z.; Salim, A.A.; Khalil, Z.; Quezada, M.; Bernhardt, P.V.; Capon, R.J. Viridicatumtoxins: Expanding on a rare tetracycline antibiotic scaffold. *J. Org. Chem.* **2015**, *80*, 12501–12508. [CrossRef]
4. Cui, H.; Su, X.; Chen, F.; Holland, M.; Yang, S.; Liang, J.; Su, P.; Dong, H.; Hou, W. Microbial diversity of two cold seep systems in gas hydrate-bearing sediments in the South China Sea. *Mar. Environ. Res.* **2019**, *44*, 230–239. [CrossRef]
5. Hu, X.-Y.; Li, X.-M.; Yang, S.-Q.; Li, X.; Wang, B.-G.; Meng, L.-H. New cytochalasin derivatives from deep-sea cold seep-derived endozoic fungus *Curvularia verruculosa* CS-129. *Chem. Biodivers.* **2022**, *19*, e202200550. [CrossRef] [PubMed]
6. Hu, X.-Y.; Wang, C.-Y.; Li, X.-M.; Yang, S.-Q.; Li, X.; Wang, B.-G.; Si, S.-Y.; Meng, L.-H. Cytochalasin derivatives from the endozoic *Curvularia verruculosa* CS-129, a fungus isolated from the deep-sea squat lobster *Shinkaia crosnieri* living in the cold seep environment. *J. Nat. Prod.* **2021**, *84*, 3122–3130. [CrossRef]
7. Song, Q.; Yang, S.-Q.; Li, X.-M.; Hu, X.-Y.; Li, X.; Wang, B.-G. Aromatic polyketides from the deep-Sea cold-seep mussel associated endozoic fungus *Talaromyces minioluteus* CS-138. *Mar. Drugs* **2022**, *20*, 529. [CrossRef] [PubMed]
8. Hu, X.-Y.; Li, X.; Yang, S.-Q.; Li, X.-M.; Wang, B.-G.; Meng, L.-H. Vercytochalasins A and B: Two unprecedented biosynthetically related cytochalasins from the deep-sea-sourced endozoic fungus *Curvularia verruculosa*. *Chin. Chem. Lett.* **2023**, *34*, 107516. [CrossRef]
9. Yan, L.-H.; Li, P.-H.; Li, X.-M.; Yang, S.-Q.; Liu, K.-C.; Wang, B.-G.; Li, X. Chevalinulins A and B, proangiogenic alkaloids with a spiro[bicyclo[2.2.2]octane-diketopiperazine] skeleton from deep-sea cold-seep-derived fungus *Aspergillus chevalieri* CS-122. *Org. Lett.* **2022**, *24*, 2684–2688. [CrossRef]
10. Wang, W.-L.; Lu, Z.-Y.; Tao, H.-W.; Zhu, T.-J.; Fang, Y.-C.; Gu, Q.-Q.; Zhu, W.-M. Isoechinulin-type alkaloids, variecolorins A–L, from halotolerant *Aspergillus variecolor*. *J. Nat. Prod.* **2007**, *70*, 1558–1564. [CrossRef]
11. Marchelli, R.; Dossena, A.; Pochini, A.; Dradi, E. The structures of five new didehydropeptides related to neoechinulin, isolated from *Aspergillus amstelodami*. *J. Chem. Soc. Perkin Trans.* **1977**, *1*, 713–717. [CrossRef]
12. Meng, L.-H.; Du, F.-Y.; Li, X.-M.; Pedpradab, P.; Xu, G.-M.; Wang, B.-G. Rubrumazines A–C, indolediketopiperazines of the isoechinulin class from *Eurotium rubrum* MA-150, a fungus obtained from marine mangrove-derived rhizospheric soil. *J. Nat. Prod.* **2015**, *78*, 909–913. [CrossRef] [PubMed]
13. Du, F.-Y.; Li, X.; Li, X.-M.; Zhu, L.-W.; Wang, B.G. Indolediketopiperazine alkaloids from *Eurotium cristatum* EN-220, an endophytic fungus isolated from the marine alga *Sargassum thunbergia*. *Mar. Drugs* **2017**, *15*, 24. [CrossRef]
14. Wei, X.; Feng, C.; Wang, S.-Y.; Zhang, D.-M.; Li, X.-H.; Zhang, C.-X. New indole diketopiperazine alkaloids from soft coral-associated epiphytic fungus *Aspergillus* sp. EGF 15-0-3. *Chem. Biodivers.* **2020**, *17*, e2000106. [CrossRef]
15. Li, Y.-F.; Wu, X.-B.; Niaz, S.-I.; Zhang, L.-H.; Huang, Z.-J.; Lin, Y.-C.; Li, J.; Liu, L. Effect of culture conditions on metabolites produced by the crinoid-derived fungus *Aspergillus ruber* 1017. *Nat. Prod. Res.* **2017**, *31*, 1299–1304. [CrossRef] [PubMed]
16. Marcarino, M.O.; Cicetti, S.; Zanardi, M.M.; Sarotti, A.M. A critical review on the use of DP4+ in the structural elucidation of natural products: The good, the bad and the ugly. A practical guide. *Nat. Prod. Rep.* **2022**, *39*, 58–76. [CrossRef]
17. Ryu, M.-J.; Hillman, P.-F.; Lee, J.; Hwang, S.; Lee, E.-Y.; Cha, S.-S.; Yang, I.; Oh, D.-C.; Nam, S.-J.; Fenical, W. Antibacterial meroterpenoids, merochlorins G-J from the marine bacterium *Streptomyces* sp. *Mar. Drugs* **2021**, *19*, 618. [CrossRef] [PubMed]
18. Zu, W.-Y.; Tang, J.-W.; Hu, K.; Zhou, Y.-F.; Gou, L.-L.; Su, X.-Z.; Lei, X.; Sun, H.-D.; Puno, P.-T. Chaetolactam A, an Azaphilone Derivative from the Endophytic Fungus *Chaetomium* sp. g1. *J. Org. Chem.* **2021**, *86*, 475–483. [CrossRef]
19. Rychnovsky, C.D.; Rogers, B.N.; Richardson, T.I. Configurational assignment of polyene macrolide antibiotics using the [^{13}C] acetonide analysis. *Acc. Chem. Res.* **1998**, *31*, 9–17. [CrossRef]
20. Liu, Z.; Chen, Y.; Li, S.; Hu, C.; Liu, H.; Zhang, W. Indole diketopiperazine alkaloids from the deep-sea-derived fungus *Aspergillus* sp. FS445. *Nat. Prod. Res.* **2021**, *36*, 5213–5221. [CrossRef]
21. Glass, N.L.; Donaldson, G.C. Development of primer sets designed for use with the PCR to amplify conserved genes from filamentous ascomycetes. *Appl. Environ. Microbiol.* **1995**, *61*, 1323–1330. [CrossRef] [PubMed]
22. Grimblat, N.; Zanardi, M.M.; Sarotti, A.M. Beyond DP4: An improved probability for the stereochemical assignment of isomeric compounds using quantum chemical calculations of NMR shifts. *J. Org. Chem.* **2015**, *80*, 12526–12534. [CrossRef] [PubMed]
23. Lombardo, M.; Morganti, S.; Trombini, C. 3-Bromopropenyl esters in organic synthesis: Indium- and zinc-mediated entries to alk-1-ene-3,4-diols. *J. Org. Chem.* **2003**, *68*, 997–1006. [CrossRef]
24. Yang, H.; Lu, B.; Zhou, D.; Zhao, L.; Song, W.; Wang, L. Identification of the first cathelicidin gene from skin of Chinese giant salamanders *Andrias davidianus* with its potent antimicrobial activity. *Dev. Comp. Immunol.* **2017**, *77*, 141–149. [CrossRef] [PubMed]

marine drugs

Communication

Potential α-Glucosidase Inhibitors from the Deep-Sea Sediment-Derived Fungus *Aspergillus insulicola*

Weibo Zhao [1,2,†], Yanbo Zeng [2,3,†], Wenjun Chang [2,3], Huiqin Chen [2], Hao Wang [2], Haofu Dai [2] and Fang Lv [1,*]

1 Beijing Key Laboratory for Separation and Analysis in Biomedicine and Pharmaceuticals, School of Life Science, Beijing Institute of Technology, Beijing 100081, China
2 Hainan Provincial Key Laboratory for Functional Components Research and Utilization of Marine Bio-Resources, Institute of Tropical Bioscience and Biotechnology, Chinese Academy of Tropical Agricultural Sciences & Key Laboratory for Biology and Genetic Resources of Tropical Crops of Hainan Province, Hainan Institute for Tropical Agricultural Resources, Haikou 571101, China
3 Zhanjiang Experimental Station of Chinese Academy of Tropical Agricultural Sciences, Zhanjiang 524013, China
* Correspondence: lvfangbeijing@bit.edu.cn
† These authors contributed equally to this work.

Abstract: Three new phenolic compounds, epicocconigrones C–D (**1–2**) and flavimycin C (**3**), together with six known phenolic compounds: epicocconigrone A (**4**); 2-(10-formyl-11,13-dihydroxy-12-methoxy-14-methyl)-6,7-dihydroxy-5-methyl-4-benzofurancarboxaldehyde (**5**); epicoccolide B (**6**); eleganketal A (**7**); 1,3-dihydro-5-methoxy-7-methylisobenzofuran (**8**); and 2,3,4-trihydroxy-6-(hydroxymethyl)-5-methylbenzyl-alcohol (**9**), were isolated from fermentation cultures of a deep-sea sediment-derived fungus, *Aspergillus insulicola*. Their planar structures were elucidated based on the 1D and 2D NMR spectra and HRESIMS data. The absolute configurations of compounds **1–3** were determined by ECD calculations. Compound **3** represented a rare fully symmetrical isobenzofuran dimer. All compounds were evaluated for their α-glucosidase inhibitory activity, and compounds **1**, **4–7**, and **9** exhibited more potent α-glucosidase inhibitory effect with IC_{50} values ranging from 17.04 to 292.47 μM than positive control acarbose with IC_{50} value of 822.97 μM, indicating that these phenolic compounds could be promising lead compounds of new hypoglycemic drugs.

Keywords: marine fungus; *Aspergillus insulicola*; phenolic compounds; ECD calculations; α-glucosidase inhibition

Citation: Zhao, W.; Zeng, Y.; Chang, W.; Chen, H.; Wang, H.; Dai, H.; Lv, F. Potential α-Glucosidase Inhibitors from the Deep-Sea Sediment-Derived Fungus *Aspergillus insulicola*. *Mar. Drugs* **2023**, *21*, 157. https://doi.org/10.3390/md21030157

Academic Editor: Emiliano Manzo

Received: 4 February 2023
Revised: 23 February 2023
Accepted: 24 February 2023
Published: 26 February 2023

1. Introduction

According to the International Diabetes Federation, 537 million people worldwide were diagnosed with diabetes mellitus in 2021, and about 90 percent of them were type 2 diabetes mellitus (T2DM) [1,2]. T2DM is a chronic metabolic disease that is characterized by postprandial hyperglycemia in the case of insulin resistance and relative lack of insulin [3]. The inhibition of α-glucosidase can reduce the cleavage of glucose from disaccharides or oligosaccharides to inhibit postprandial hyperglycemia [4]. Therefore, α-glucosidase is a common therapeutic target for the treatment of T2DM [5]. Currently available α-glucosidase inhibitors, such as acarbose, voglibose and miglitol, have been used to treat T2DM patients. Nevertheless, the use of these drugs has been associated with serious side effects, such as abdominal distension and diarrhea [6,7]. For this reason, the search for natural, efficient and non-toxic α-glucosidase inhibitors provides an attractive strategy for the development of new hypoglycemic drugs.

Phenolic compounds have been proved to be effective α-glucosidase inhibitors [8–11]. Marine phenolic compounds are far less researched than those from terrestrial sources, which could suggest great potential in the ocean to develop novel diabetes drugs [12]. Some marine phenolic compounds isolated from seaweed [13,14] and seagrass [15] have been confirmed to have wonderful α-glucosidase inhibitory activity. In order to find more marine phenolic compounds with α-glucosidase inhibitory activity, our team studied marine fungi from the South China Sea. *Aspergillus insulicola*, a fungi previously not extensively studied, had great development and utilization value. Previous chemical studies of *A. insulicola* have discovered many peptides [16–18] and nitrobenzoyl sesquiterpenoids [19,20], which showed significant biological activities, including anti-bacteria [16] and cytotoxic [19,20]. During our ongoing research in finding new compounds with potential bioactivities [21–23], a chemical investigation of the deep-sea sediment-derived fungus *A. insulicola* led to the isolation and identification of three new phenolic compounds, epicocconigrones C–D (**1–2**) and flavimycin C (**3**), together with six known phenolic compounds (**4–9**) (Figure 1). All compounds were investigated for their α-glucosidase inhibitory activity. Herein, we describe the structure elucidation of the new metabolites as well as the α-glucosidase inhibitory activity of the isolated compounds.

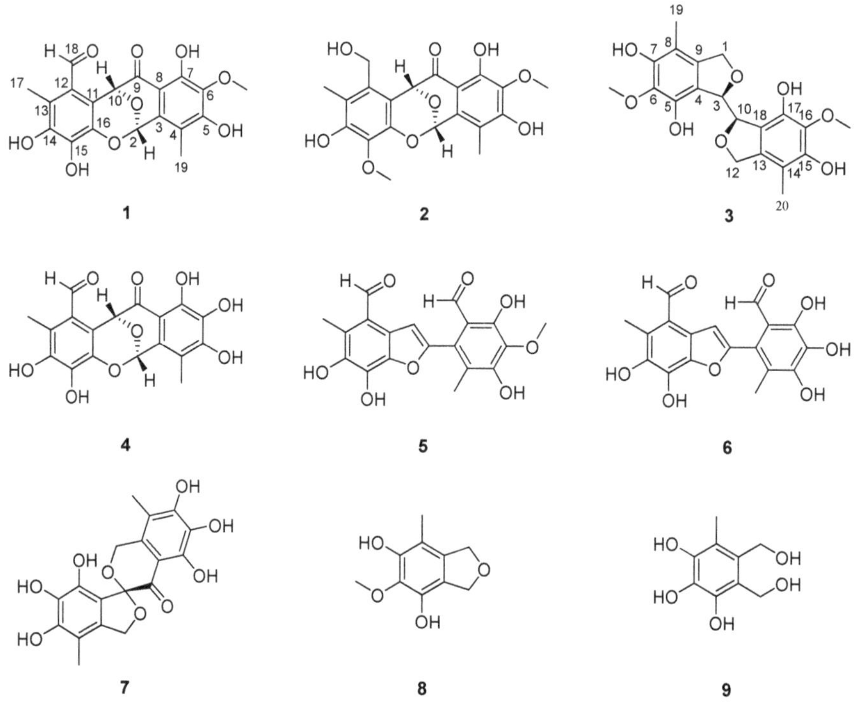

Figure 1. Structures of compounds **1–9** from *Aspergillus insulicola*: epicocconigrones C–D (**1–2**); flavimycin C (**3**); epicocconigrone A (**4**); 2-(10-formyl-11,13-dihydroxy-12-methoxy-14-methyl)-6,7-dihydroxy-5-methyl-4-benzofurancarboxaldehyde (**5**); epicoccolide B (**6**); eleganketal A (**7**); 1,3-dihydro-5-methoxy-7-methylisobenzofuran (**8**); and 2,3,4-trihydroxy-6-(hydroxymethyl)-5-methylbenzyl-alcohol (**9**).

2. Results and Discussion

2.1. Structure Elucidation of New Compounds 1–3

Epicocconigrone C (**1**) was isolated as a yellow solid, and its molecular formula was determined to be $C_{19}H_{16}O_9$ with 12 degrees of unsaturation by HRESIMS data at *m/z* 411.0698 (calcd. 411.0687 for $C_{19}H_{16}O_9Na$, $[M + Na]^+$), which was supported by the ^{13}C NMR and DEPT spectral data. The IR spectrum of **1** featured typical absorption bands for hydroxyl (3413 cm^{-1}) and conjugated ketone (1670 cm^{-1}). The ^1H NMR spectrum (Table 1) of **1** revealed two methyls (δ_H 2.26 and δ_H 2.31), one methoxy (δ_H 3.70), two oxymethines (δ_H 6.38 and δ_H 6.83), one aldehyde proton (δ_H 10.34), and one hydroxyl proton (δ_H 11.33). The ^{13}C NMR (Table 2) and DEPT spectra showed 19 well-resolved carbon atom signals, including one ketone carbonyl (δ_C 196.9), one aldehydic carbonyl (δ_C 191.2), two oxygenated tertiary carbons (δ_C 89.8 and δ_C 68.6), one methoxy carbon (δ_C 60.2), two methyls (δ_C 11.8 and δ_C 10.2), and twelve olefinic quaternary carbons at δ_C 156.9—104.5, accounting for 8 degrees of unsaturation. Thus, compound **1** was thought to possess a tetracyclic skeleton. The strong heteronuclear multiple-bond correlation (HMBC) correlations from H-17 (δ_H 2.31) to C-12 (δ_C 121.7), C-13 (δ_C 121.9), and C-14 (δ_C 144.3), from H-18 (δ_H 10.34) to C-11 (δ_C 112.6), C-12, C-13, and C-14, as well as the weak signals from H-17 to C-11, C-15 (δ_C 138.4), and C-16 (δ_C 135.8) confirmed the existence of ring A (Figure 2). The HMBC correlations from H-19 (δ_H 2.26) to C-3 (δ_C 130.8), C-4 (δ_C 115.8), and C-5 (δ_C 156.9), as well as the HMBC correlations from 7-OH (δ_H 11.33) to C-6, C-7 (δ_C 153.6), and C-8 (δ_C 104.5) established the substitution of the aromatic ring D. Furthermore, the HMBC correlations from H-2 (δ_H 6.83) to C-10 (δ_C 68.6) and C-16, from H-10 (δ_H 6.38) to C-2 (δ_C 89.8), C-11, C-12 and C-16 suggested the presence of two oxygen bridges between C-16/C-2 and C-2/C-10 in ring B, which could be confirmed by the low field chemical shift signal of CH-2 (δ_C 89.8, δ_H 6.83). Ring C was established by the HMBC correlations from H-2 to C-4 and C-8, and from H-10 to C-8 and C-9 (δ_C 196.9). The comprehensive NMR analysis indicated that **1** shared the same oxygen-bridged skeleton with epicocconigrone A (**4**) [24], with the exception that the appearance of 6-OCH_3 in **1** replaced 6-OH in **4**, which was supported by the HMBC correlation from 6-OCH_3 (δ_H 3.70) to C-6 (δ_C 134.7). Thus, the planar structure of **1** was elucidated as shown (Figure 1), named epicocconigrone C. In the nuclear Overhauser effect spectroscopy (NOESY) spectrum of **1**, the correlation between H-2 and H-10 was indicative of their *cis* relationship (Figure 2). The absolute configuration of **1** was confirmed by the ECD calculation. Its experimental ECD curve for the absolute configurations of 2*S* and 10*R* was consistent with the calculated ECD curve of (2*S*, 10*R*) (Figure 3).

Table 1. ^1H NMR data of epicocconigrones C–D (**1–2**) and flavimycin C (**3**) (δ in ppm, J in Hz) in DMSO-d_6.

Position	**1** [a]	**2** [a]	**3** [b]
1			4.54 (d, 15.0) 4.65 (d, 15.0)
2	6.83, s	6.76, s	
3			4.30, s
10	6.38, s	5.65, s	4.30, s
12			4.54 (d, 15.0) 4.65 (d, 15.0)
17	2.31, s	2.09, s	
18	10.34, s	4.81 (d, 12.1) 4.32 (d, 12.1)	
19	2.26, s	2.18, s	1.88, s
20			1.88, s
6-OCH_3	3.70, s	3.57, s	3.64, s
14-OCH_3			3.64, s
15-OCH_3		3.69, s	
5-OH			8.55, s
7-OH	11.33, s	11.46, s	8.68, s
14-OH		8.92, s	
15-OH			8.68, s
17-OH			8.55, s

[a] Recorded at 500 MHz; [b] Recorded at 600 MHz.

Table 2. ^{13}C NMR (125 MHz) data of epicocconigrones C-D (**1–2**) and flavimycin C (**3**) in DMSO-d_6.

Position	1	2	3
1			65.8, CH_2
2	89.8, CH	89.9, CH	
3	130.8, C	131.0, C	66.1, CH
4	115.8, C	115.8, C	112.2, C
5	156.9, C	158.9, C	147.0, C
6	134.7, C	134.6, C	134.3, C
7	153.6, C	153.6, C	147.8, C
8	104.5, C	103.4, C	109.7, C
9	196.9, C	196.8, C	129.7, C
10	68.6, CH	70.3, CH	66.1, CH
11	112.6, C	108.3, C	
12	121.7, C	132.1, C	65.8, CH_2
13	121.9, C	118.1, C	129.7, C
14	144.3, C	148.5, C	109.7, C
15	138.4, C	134.6, C	147.8, C
16	135.8, C	140.3, C	134.3, C
17	11.8, CH_3	11.0, CH_3	147.0, C
18	191.2, CH	55.8, CH_2	112.2, C
19	10.2, CH_3	10.3, CH_3	9.5, CH_3
20			9.5, CH_3
6-OCH_3	60.2, CH_3	60.3, CH_3	60.2, CH_3
14-OCH_3			60.2, CH_3
15-OCH_3		60.0, CH_3	

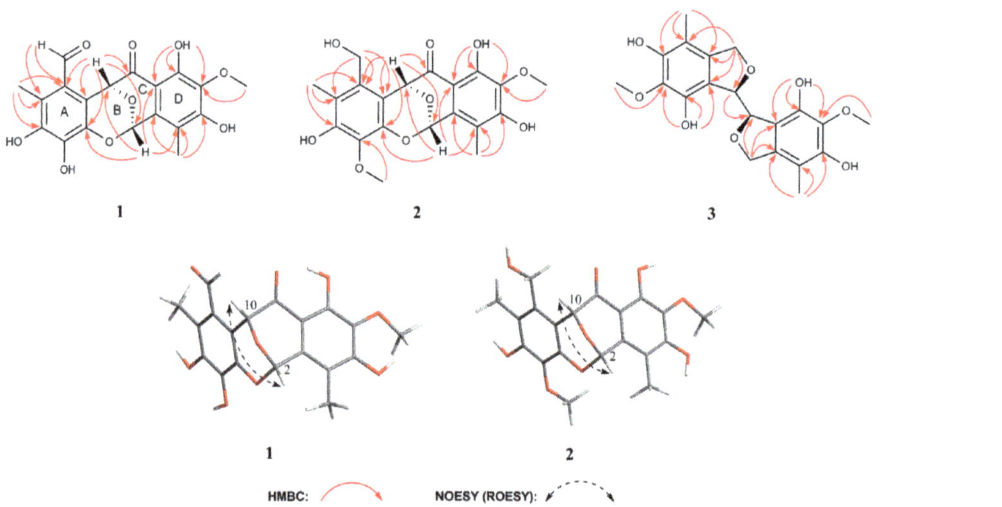

Figure 2. Key HMBC correlations of compounds **1–3** and key NOESY/ROESY correlations of compounds **1–2**.

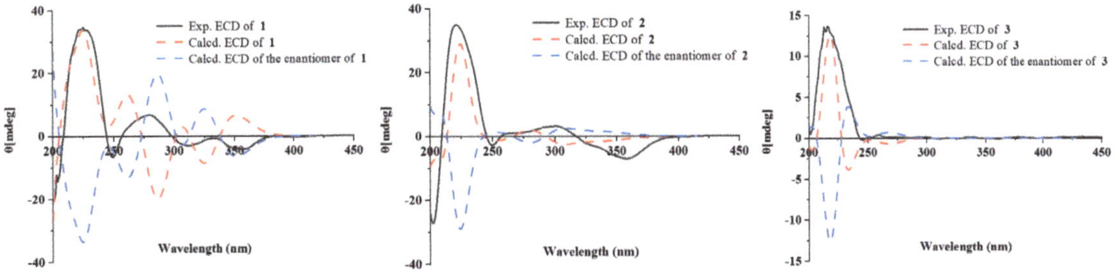

Figure 3. Experimental and calculated ECD spectra of compounds **1**–**3**.

Epicocconigrone D (**2**) was obtained as a yellow solid. The molecular formula of **2** was determined as $C_{20}H_{20}O_9$ with 11 unsaturated degrees by HRESIMS data at *m/z* 427.1004 (calcd. 427.1000 for $C_{20}H_{20}O_9Na$, $[M + Na]^+$), which was supported by the ^{13}C NMR and DEPT spectral data. The IR spectrum of **2** featured typical absorption bands for hydroxyl (3446 cm^{-1}) and conjugated ketone (1626 cm^{-1}). The ^1H NMR spectrum (Table 1) of **2** indicated two methyl groups (δ_H 2.09 and δ_H 2.18), two methoxy groups (δ_H 3.57 and δ_H 3.69), one methylene (δ_H 4.32, d, J = 12.1 Hz; 4.81 d, J = 12.1 Hz), two oxymethines (δ_H 5.65 and δ_H 6.76), and two hydroxyl protons (δ_H 8.92 and δ_H 11.46). The ^{13}C NMR (Table 2) and DEPT spectra revealed 20 carbon atom signals, including one ketone carbonyl (δ_C 196.8), two oxygenated tertiary carbons (δ_C 89.9 and δ_C 70.3), two methoxy carbons (δ_C 60.3 and δ_C 60.0), one methylene (δ_C 55.8), two methyls (δ_C 11.0 and δ_C 10.3), and twelve olefinic quaternary carbons. Detailed analysis of 2D NMR spectra of **2** revealed that it had a similar structure to **1**. The major differences in **2** were a hydroxymethylene group and a methoxy group substituted at C-12 and C-15, instead of the aldehyde group and the hydroxyl group, respectively, when compared to **1** (Figure 2), which were further confirmed by the HMBC correlations from H_2-18 (δ_H 4.32, 4.81) to C-11 (δ_C 108.3), C-12 (δ_C 132.1), and C-13 (δ_C 118.1), and from 15-OCH_3 (δ_H 3.69) to C-15 (δ_C 134.6). Thus, the planar structure of **2** was elucidated as shown (Figure 1), named epicocconigrone D. The ROESY correlation between H-2 and H-10 indicated their *cis* orientation (Figure 2). The absolute configuration of **2** was understood to be 2*S*, 10*R* by comparing the experimental and simulated ECD curves (Figure 3).

Flavimycin C (**3**) was isolated as a white solid. It had a molecular formula of $C_{20}H_{22}O_8$ with 10 degrees of unsaturation, as determined by HRESIMS data at *m/z* 391.1385 (calcd. 391.1387 for $C_{20}H_{23}O_8$, $[M + H]^+$), which was supported by the ^{13}C NMR and DEPT spectral data. The IR spectrum of **3** featured typical absorption bands for hydroxyl (3449 cm^{-1}). The ^1H NMR spectrum (Table 1) of **3** exhibited one methyl (δ_H 1.88), one methoxy (δ_H 3.64), one methine (δ_H 4.30), one methylene (δ_H 4.54 d, J = 15.0 Hz; 4.65 d, J = 15.0 Hz), and two hydroxyl protons (δ_H 8.55, δ_H 8.68). The ^{13}C NMR (Table 2) and DEPT spectra displayed 10 well-resolved carbon atom signals, dividing into six quaternary carbons that were assigned to one benzene ring, one methylene (δ_C 65.8), one methine (δ_C 66.1), one methoxy carbon (δ_C 60.2), and one methyl (δ_C 9.5). The NMR data of **3** were very similar to those of **8** except for the absence of the methylene signal, and instead, the presence of the methine signal of C-3 (δ_H 4.30/δ_C 66.1) in **3**. Combined with molecular formula, **3** was deduced to be a symmetrical dimeric derivative. The above data suggested **3** was a symmetrical dimer of **8**, connecting at C-3/C-10 between the two units (Figure 2), which was further confirmed by the HMBC correlation from H-3 to C-10. Thus, the planar structure of **3** was confirmed as shown (Figure 1), and named flavimycin C. The ^1H and ^{13}C NMR spectra (Tables 1 and 2) of this aromatic polyketide dimer only exhibited a set of signals of aromatic polyketide monomer. There were three possible absolute configurations of two chiral carbons C-3 and C-10 in **3**. The obvious negative optical activity ($[\alpha]^{20}_D$ = −70.0) and the Cotton effect indicated that compound **3** was not a mesomer, which implied the possibility of 3*R*, 10*S* was excluded. Consequently, the absolute configurations of C-3 and C-10 were the same

(3*S*, 10*S* or 3*R*, 10*R*). The absolute configuration of **3** was understood to be 3*R*, 10*R* by comparing the experimental and simulated ECD curves (Figure 3).

The known compounds: epicocconigrone A (**4**) [24]; 2-(10-formyl-11,13-dihydroxy-12-methoxy-14-methyl)-6,7-dihydroxy-5-methyl-4-benzofurancarboxaldehyde (**5**) [25]; epicoccolide B (**6**) [26]; eleganketal A (**7**) [27]; 1,3-dihydro-5-methoxy-7-methylisobenzofuran (**8**) [28]; and 2,3,4-trihydroxy-6-(hydroxymethyl)-5-methylbenzyl-alcohol (**9**) [29] were identified by comparing their NMR data with those reported in the literature.

The new compounds **1–3** are all aromatic polyketide dimers, particularly compounds **1** and **2** simultaneously featuring consistent 6/6/6/6 heterotetracyclic ring cores and compounds **1–3** co-occurrence in the same marine-derived fungus suggest that they should originate from the same biogenetic pathway. A plausible biosynthetic pathway toward the formation of compounds **1–3** can be proposed by detailed analysis of their structures (Scheme 1).

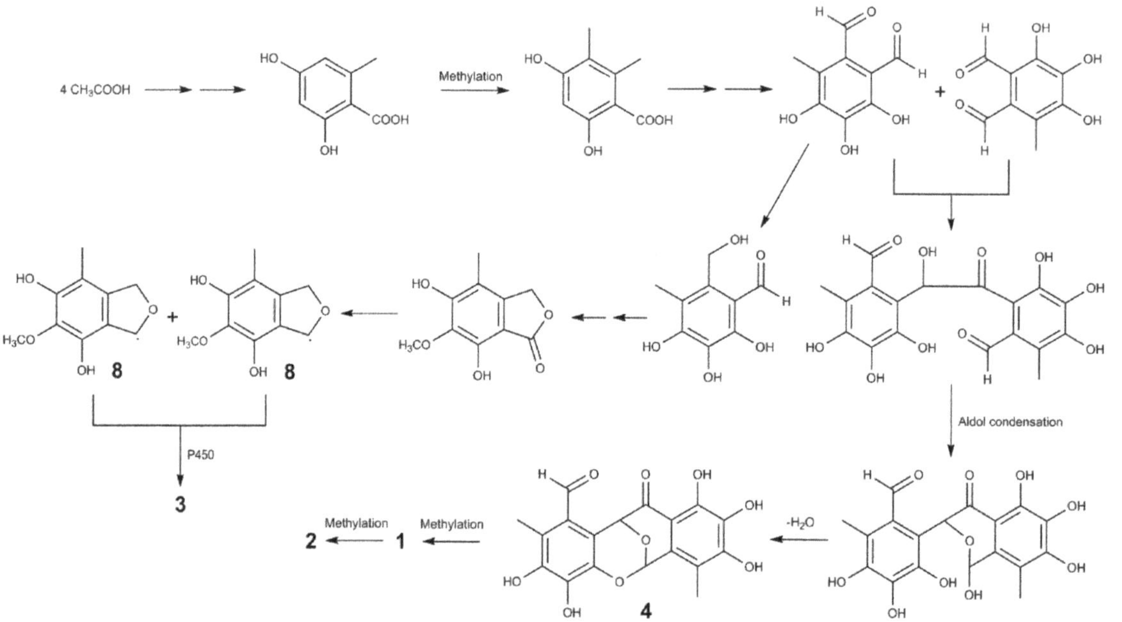

Scheme 1. Putative biosynthetic pathways toward the formation of compounds **1–3**.

2.2. *In Vitro Evaluation of α-Glucosidase Inhibitory Activity*

All compounds were tested for their α-glucosidase inhibitory activities using a reported method [30], with acarbose as the positive control. The results revealed that compounds **1**, **4–7**, and **9** showed more potent inhibitory activity (IC_{50} values ranging from 17.04 ± 0.28 to 292.47 ± 5.87 μM) than acarbose (IC_{50}, 822.97 ± 7.10 μM) (Table 3). The potent α-glucosidase inhibitory activity of epicocconigrone A (**4**) and epicoccolide B (**6**) has been already reported [31]. It could be noted herein that the number of hydroxyl groups of polyhydroxy phenolic compounds was important for α-glucosidase inhibitory activity, as reflected by the low IC_{50} values of compounds **4** and **6**, while structures with fewer hydroxyl groups (compounds **1** and **5**) exhibited little activity.

Table 3. α-Glucosidase inhibitory activities of compounds **1–9**.

Compounds	$IC_{50} \pm SD$ (μM) [a]
1	292.47 ± 5.87
2	–
3	–
4	25.69 ± 0.30
5	40.07 ± 4.64
6	17.04 ± 0.28
7	49.53 ± 2.45
8	–
9	130.63 ± 2.87
Acarbose [b]	822.97 ± 7.10

[a] Values represent means ± SD based on three parallel experiments. [b] Positive control.–no activity at a concentration of 200 μM.

3. Materials and Methods

3.1. Fungal Material and Fermentation

The fungal strain *A. insulicola* was isolated from deep-sea sediments, which were collected from the South China Sea at the depth of 2500 m. After grinding, the sample (1.0 g) was diluted to 10^{-2} g/mL with sterile H_2O, 100 μL of which was spread on potato dextrose agar medium (200.0 g potato, 20.0 g glucose, and 20.0 g agar per liter of seawater) plates containing chloramphenicol as a bacterial inhibitor. It was identified by its morphological characteristics and ITS gene sequences (GenBank accessing No. ON413861), the used primers of which were ITS1 (TCCGTAGGTGAACCTGCGG) and ITS4 (TCCTCCGCTTATTGATATGC). A reference culture of *A. insulicola* was deposited at the Hainan Provincial Key Laboratory for Functional Components Research and Utilization of Marine Bio-resources, Haikou, China.

3.2. Culture Conditions

The fungal strain *A. insulicola* was cultured in potato dextrose broth medium (consisting of 200.0 g/L potato, 20.0 g/L glucose, and 1000.0 mL deionized water), and incubated on a rotary shaker (150 rpm) for 72 h at 28 °C. Thereafter, 3 mL of seed broth was transferred to fifty 1000 mL Erlenmeyer flasks containing solid rice medium (each flask contained 80 g rice and 120 mL seawater), used for fermentation. The flasks were incubated under static conditions at room temperature for 30 days.

3.3. General Experimental Procedures

Optical rotation was measured using a Modular Circular Polarimeter 5100 polarimeter (Anton Paar, Austria). The NMR spectra were measured on Bruker Avance 500 NMR spectrometer (Bruker, Bremen, Germany) and Bruker DRX-600 spectrometer (Bruker Biospin AG, Fällanden, Germany) using TMS as an internal standard. HRESIMS were determined with an API QSTAR Pulsar mass spectrometer (Bruker, Bremen, Germany). ECD and UV spectra were recorded on a MOS-500 spectrometer (Biologic, France). IR data were measured on a Nicolet 380 infrared spectrometer (Thermo Electron Corporation, Madison, WI, USA). Analytic HPLC was performed with an Agilent Technologies 1260 Infinity II equipped with an Agilent DAD G1315D detector (Agilent, Palo Alto, CA, USA), the separation columns were (COSMOSIL-packed C_{18}, 5 mm, 4.6 mm × 250 mm). Semi-preparative HPLC was performed on reversed-phased columns (COSMOSIL-packed C_{18}, 5 mm, 10 mm × 250 mm). Silica gel (60–80, 200–300 and 300–400 mesh, Qingdao Marine Chemical Co. Ltd., Qingdao, China) and Sephadex LH-20 (Merck, Germany) were used for column chromatography. TLC was conducted on precoated silica gel GF254 plates (Qingdao Marine Chemical Co. Ltd., Qingdao, China), and spots were detected by spraying with 10% H_2SO_4 in EtOH followed by heating.

3.4. Extraction and Isolation

After the fermentation of the strain, the cultures were extracted with EtOAc, then filtered with filter paper. After repeating the procedure three times, the EtOAc extract was evaporated under a reduced pressure to obtain a crude extract (124.0 g). The crude extract was dispersed in water and extracted with petroleum ether, ethyl acetate and *n*-butanol three times, respectively. After vacuum concentration, the petroleum ether extract (11.3 g), ethyl acetate extract (34.0 g) and n-butanol extract (20.0 g) were obtained, respectively. Then, the EtOAc extract (34.0 g) was subjected to silica gel vacuum liquid chromatography using step gradient elution with $CHCl_3$/MeOH (1:0, 200:1, 150:1, 100:1, 80:1, 50:1, 20:1, 10:1, 0:1, *v*/*v*) to obtain 13 fractions (Fr.1–Fr.13). Fr.4 (425.0 mg) was applied to Sephadex LH-20 gel chromatography eluted with $CHCl_3$/MeOH (1:1, *v*/*v*) to give six subfractions (Fr.4.1–Fr.4.6). Fr.4.3 (150.5 mg) was subjected to silica gel column chromatography (petroleum ether/EtOAc, 10:1, *v*/*v*) to afford nine subfractions (Fr.4.3.1–Fr.4.3.9). Fr.4.3.9 (40.5 mg) was separated by semi-preparative HPLC, eluting with 45% MeOH/H_2O to yield compound **2** (t_R 11.5 min, 4.5 mg), and Fr.4.3.7 (29.8 mg) was separated by semi-preparative HPLC, eluting with 35% MeOH/H_2O to give compound **8** (t_R 15.3 min, 4.2 mg). Fr.6 (1.15 g) was applied to ODS chromatography eluting with MeOH/H_2O (10%–100%) to give thirteen subfractions (Fr.6.1–Fr.6.13). Fr.6.11 (91.5 mg) was subjected to Sephadex LH-20 (eluted with 100% MeOH) and then purified by semi-preparative HPLC (eluted with 48% MeOH/H_2O) to give compound **1** (t_R 21.9 min, 11.1 mg). Fr.6.12 (58.9 mg) was subjected to Sephadex LH-20 (eluted with 100% MeOH) and then purified by semi-preparative HPLC (eluted with 65% MeOH/H_2O) to give compound **5** (t_R 10.0 min, 4.2 mg). Fr.6.6 (72.3 mg) was purified on silica gel (petroleum ether/EtOAc, 3:2, *v*/*v*) to yield compound **3** (7.5 mg). Fr.9 (10.0 g) was subjected to Sephadex LH-20 gel chromatography eluted with MeOH to give ten subfractions (Fr.9.1–Fr.9.10). Fr.9.7 (2.1 g) was subjected to silica gel column chromatography (CH_2Cl_2/MeOH, 100:1, *v*/*v*), and subsequently purified by semi-preparative HPLC, eluting with 50 % MeOH/H_2O to yield compounds **4** (t_R 12.0 min, 5.1 mg) and **6** (t_R 18.5 min, 2.7 mg). Fr.9.6 (1.67 g) was separated by Sephadex LH-20 column chromatography eluted with MeOH and then purified by silica gel column chromatography eluting with petroleum ether/EtOAc (3:1; *v*/*v*) to obtain compound **9** (5.1 mg). Fr.9.8 (0.8 g) was subjected to silica gel column chromatography (CH_2Cl_2/MeOH, 35:1, *v*/*v*), and subsequently purified by semi-preparative HPLC, eluting with 55 % MeOH/H_2O to yield compound **7** (t_R 6.8 min, 8.1 mg).

Epicocconigrone C (**1**): Yellow film. $[\alpha]_D^{20}$ = +98.0 (*c* 0.10, MeOH); UV (MeOH) λ_{max} (logε): 237 (4.31) nm; 261 (3.91) nm; 309 (4.27) nm; 359 (4.06) nm; IR (KBr) υ_{max} (cm^{-1}): 3413, 1669, 1466, 1395, 1355, 1296, 1117. ^1H and ^{13}C NMR data see Tables 1 and 2; HRESIMS $[M + Na]^+$ *m/z* 411.0698 (calcd. for $C_{19}H_{16}O_9Na$, 411.0687).

Epicocconigrone D (**2**): Yellow film. $[\alpha]_D^{20}$ = +57.0 (*c* 0.10, MeOH); UV (MeOH) λ_{max} (logε): 234 (4.26) nm; 260 (4.04) nm; 309 (4.21) nm; 365 (4.19) nm; IR (KBr) υ_{max} (cm^{-1}): 3446, 2931, 1626, 1469, 1359, 1226, 1154, 1115. ^1H and ^{13}C NMR data see Tables 1 and 2; HRESIMS $[M + Na]^+$ *m/z* 427.1004 (calcd. for $C_{20}H_{20}O_9Na$, 427.1000).

Flavimycin C (**3**): White film. $[\alpha]_D^{20}$ = −70.0 (*c* 0.10, MeOH); UV (MeOH) λ_{max} (logε): 232 (4.15) nm; 284 (3.56) nm; IR (KBr) υ_{max} (cm^{-1}): 3449, 2928, 1606, 1478, 1376, 1264, 1110, 1027. ^1H and ^{13}C NMR data see Tables 1 and 2; HRESIMS $[M + H]^+$ *m/z* 391.1385 (calcd. for $C_{20}H_{23}O_8$, 391.1387).

3.5. ECD Calculation

The conformers of compounds were generated using the Confab [32] program ebbed in the Openbabel 3.1.1 software, and further optimized with xtb at GFN2 level [33]. The conformers with population over 1% were subjected to geometry optimization using the Gaussian 16 package [34] at B3LYP/6-31G(d) level and proceeded to calculation of excitation energies, oscillator strength, and rotatory strength at B3LYP/TZVP level in the polarizable continuum model (PCM, methanol). The ECD spectra were Boltzmann-weighted and generated using SpecDis 1.71 software [35].

3.6. α-Glucosidase Inhibitory Activity

All the assays were carried out under 0.1 M sodium phosphate buffer (PH = 6.8). The samples were dissolved with DMSO and diluted into a series of gradient concentrations (final concentrations of 6.25, 12.5, 25, 50, 100, 200, 400, and 800 μM). The 10 μL sample was mixed with 100 μL α-glucosidase solution (0.2 U/mL, Sigma) and shaken well, then added to a 96-well plate and placed at 37 °C for 15 min. Subsequently, 40 μL of 2.5 mM 4-nitrophenyl-α-D-glucopyranoside was added and further incubated at 37 °C for 15 min. Finally, the OD value of each well was detected at 405 nm wavelength of microplate reader. Acarbose was used as a positive control. The control was prepared by adding DMSO instead of the sample in the same way as the test. The blank was prepared by adding sodium phosphate buffer instead of 4-nitrophenyl-α-D-glucopyranoside using the same method. The percentage inhibition was calculated using the following equation:

$$\%\ \text{inhibition} = [(OD_{control} - OD_{sample})/(OD_{control} - OD_{blank})] \times 100$$

4. Conclusions

In summary, two new tetracyclic cores of integrastatins, named epicocconigrones C–D (**1–2**), one new dimeric isobenzofuran, named flavimycin C (**3**), and six known compounds (**4–9**) were isolated from fermentation cultures of the deep-sea sediment-derived fungus *A. insulicola*. The biological evaluation revealed compounds **1**, **4–7**, **9** exhibited significant α-glucosidase inhibitory with IC_{50} values ranging from 17.04 ± 0.28 to 292.47 ± 5.87 μM, among which compound **6** was the most potent α-glucosidase inhibitor, with an IC_{50} value 48-fold stronger than positive control acarbose. Comparing the structure of compounds **1**, **4**, **5** and **6** revealed the α-glucosidase inhibitory activity was greatly enhanced after the hydroxyl group replaced the methoxy group, which further confirmed that polyhydroxy phenolic compounds were efficient α-glucosidase inhibitors, and provided a reference value for the synthesis of novel α-glucosidase inhibitors. In conclusion, the study has enriched the structural diversity of phenolic compounds and provided a promising lead toward the development of novel α-glucosidase inhibitors.

Supplementary Materials: The following supporting information can be downloaded at: https://www.mdpi.com/article/10.3390/md21030157/s1, Figures S1–S30: 1D, 2D NMR, MS, UV, and IR spectra of compounds **1–3**.

Author Contributions: Conceptualization, Y.Z. and F.L.; methodology, Y.Z. and F.L.; software, Y.Z., F.L. and H.D.; validation, W.Z., Y.Z., W.C., F.L. and H.D.; formal analysis, Y.Z. and H.W.; investigation, W.Z. and W.C.; resources, Y.Z., F.L. and H.D.; data curation, W.Z. and W.C.; writing—original draft preparation, W.Z., Y.Z., H.W. and F.L.; writing—review and editing, Y.Z., F.L., H.W. and H.C.; visualization, W.Z., Y.Z. and F.L.; supervision, Y.Z., F.L. and H.D.; project administration, Y.Z., F.L. and H.D.; funding acquisition, Y.Z. and F.L. All authors have read and agreed to the published version of the manuscript.

Funding: This work was financially supported by Natural Science Foundation of Hainan (322MS131, 220RC702), National Natural Science Foundation of China (41776093), and Financial Fund of the Ministry of Agriculture and Rural Affairs, P. R. China (NFZX2023).

Institutional Review Board Statement: Not applicable.

Informed Consent Statement: Not applicable.

Data Availability Statement: The authors declare that all relevant data supporting the results of this study are available within the article and its Supplementary Materials file, or from the corresponding authors upon request.

Conflicts of Interest: The authors declare no conflict of interest.

References

1. Kashtoh, H.; Baek, K.H. Recent updates on phytoconstituent alpha-glucosidase inhibitors: An approach towards the treatment of type two diabetes. *Plants* **2022**, *11*, 2722. [CrossRef] [PubMed]
2. Malik, A.; Ardalani, H.; Anam, S.; McNair, L.M.; Kromphardt, K.J.K.; Frandsen, R.J.N.; Franzyk, H.; Staerk, D.; Kongstad, K.T. Antidiabetic xanthones with α-glucosidase inhibitory activities from an endophytic *Penicillium canescens*. *Fitoterapia* **2020**, *142*, 104522. [CrossRef]
3. Du, X.P.; Wang, X.; Yan, X.; Yang, Y.F.; Li, Z.P.; Jiang, Z.D.; Ni, H. Hypoglycaemic effect of all-trans astaxanthin through inhibiting α-glucosidase. *J. Funct. Foods* **2020**, *74*, 104168. [CrossRef]
4. Jiang, L.L.; Wang, Z.; Wang, X.Y.; Wang, S.J.; Cao, J.; Liu, Y. Exploring the inhibitory mechanism of piceatannol on α-glucosidase relevant to diabetes mellitus. *RSC Adv.* **2020**, *10*, 4529–4537. [CrossRef] [PubMed]
5. Attjioui, M.; Ryan, S.; Ristic, A.K.; Higgins, T.; Goni, O.; Gibney, E.; Tierney, J.; O'Connell, S. Kinetics and mechanism of α-glucosidase inhibition by edible brown algae in the management of type 2 diabetes. *P. Nutr. Soc.* **2020**, *79*, E633. [CrossRef]
6. Fallah, Z.; Tajbakhsh, M.; Alikhani, M.; Larijani, B.; Faramarzi, M.A.; Hamedifar, H.; Mohammadi-Khanaposhtani, M.; Mahdavi, M. A review on synthesis, mechanism of action, and structure-activity relationships of 1,2,3-triazole-based α-glucosidase inhibitors as promising anti-diabetic agents. *J. Mol. Struct.* **2022**, *1255*, 132469. [CrossRef]
7. Shah, M.; Bashir, S.; Jaan, S.; Nawaz, H.; Nishan, U.; Abbasi, S.W.; Jamal, S.B.; Khan, A.; Afridi, S.G.; Iqbal, A. Computational analysis of plant-derived terpenes as α-glucosidase inhibitors for the discovery of therapeutic agents against type 2 diabetes mellitus. *S. Afr. J. Bot.* **2021**, *143*, 462–473. [CrossRef]
8. Zahid, H.F.; Ali, A.; Ranadheera, C.S.; Fang, Z.X.; Dunshea, F.R.; Ajlouni, S. In vitro bioaccessibility of phenolic compounds and alpha-glucosidase inhibition activity in yoghurts enriched with mango peel powder. *Food Biosci.* **2022**, *50*, 102011. [CrossRef]
9. Abdelli, I.; Benariba, N.; Adjdir, S.; Fekhikher, Z.; Daoud, I.; Terki, M.; Benramdane, H.; Ghalem, S. In silico evaluation of phenolic compounds as inhibitors of A-amylase and A-glucosidase. *J. Biomol. Struct. Dyn.* **2020**, *39*, 816–822. [CrossRef]
10. Machida, S.; Mukai, S.; Kono, R.; Funato, M.; Saito, H.; Uchiyama, T. Synthesis and comparative structure–activity study of carbohydrate-based phenolic compounds as α-glucosidase inhibitors and antioxidants. *Molecules* **2019**, *24*, 4340. [CrossRef]
11. Catarino, M.D.; Silva, A.M.S.; Mateus, N.; Cardoso, S.M. Optimization of phlorotannins extraction from *fucus vesiculosus* and evaluation of their potential to prevent metabolic disorders. *Mar. Drugs* **2019**, *17*, 162. [CrossRef] [PubMed]
12. Mateos, R.; Pérez-Correa, J.R.; Domínguez, H. Bioactive properties of marine phenolics. *Mar. Drugs* **2020**, *18*, 501. [CrossRef] [PubMed]
13. Zhu, Y.X.; Chen, W.Q.; Kong, L.; Zhou, B.X.; Hua, Y.; Han, Y.; Li, J.J.; Ji, J.; Fu, M.; Liu, W.W.; et al. Optimum conditions of ultrasound-assisted extraction and pharmacological activity study for phenolic compounds of the alga *Chondrus ocellatus*. *J. Food Process. Preserv.* **2022**, *46*, E16400. [CrossRef]
14. Naveen, J.; Baskaran, R.; Baskarana, V. Profiling of bioactives and in vitro evaluation of antioxidant and antidiabetic property of polyphenols of marine algae Padina tetrastromatica. *Algal Res.* **2021**, *55*, 102250. [CrossRef]
15. Rengasamy, K.R.R.; Sadeer, N.B.; Zengin, G.; Mahomoodally, M.F.; Cziáky, Z.; Jekő̋d, J.; Diuzheva, A.; Abdallah, H.H.; Kim, D.H. Biopharmaceutical potential, chemical profile and in silico study of the seagrass–*Syringodium isoetifolium* (Asch.) Dandy. *S. Afr. J. Bot.* **2019**, *127*, 167–175. [CrossRef]
16. Sun, C.X.; Zhang, Z.P.; Ren, Z.L.; Yu, L.; Zhou, H.; Han, Y.X.; Shah, M.; Che, Q.; Zhang, G.J.; Li, D.H.; et al. Antibacterial cyclic tripeptides from Antarctica-sponge-derived fungus *Aspergillus insulicola* HDN151418. *Mar. Drugs* **2020**, *18*, 532. [CrossRef]
17. Wu, Q.X.; Jin, X.J.; Draskovic, M.; Crews, M.S.; Tenney, K.; Valeriote, F.A.; Yao, X.J.; Crews, P. Unraveling the numerous biosynthetic products of the marine sediment-derived fungus, *Aspergillus insulicola*. *Phytochem. Lett.* **2012**, *5*, 114–117. [CrossRef]
18. Wu, Q.X.; Crews, M.S.; Draskovic, M.; Sohn, J.; Johnson, T.A.; Tenney, K.; Valerick, F.A.; Yao, X.J.; Bjeldanes, L.F.; Crerws, P. Azonazine, a novel dipeptide from a Hawaiian marine sediment-derived fungus, *Aspergillus insulicola*. *Org. Lett.* **2010**, *12*, 4458–4461. [CrossRef]
19. Sun, C.X.; Liu, X.Y.; Sun, N.; Zhang, X.M.; Shah, M.; Zhang, G.J.; Che, Q.; Zhu, T.J.; Li, J.; Li, D.H. Cytotoxic nitrobenzoyl sesquiterpenoids from an Antarctica sponge-derived *Aspergillus insulicola*. *J. Nat. Prod.* **2022**, *85*, 987–996. [CrossRef]
20. Zhao, H.Y.; Anbuchezhian, R.; Sun, W.; Shao, C.L.; Zhang, F.L.; Yin, Y.; Yu, Z.S.; Li, Z.Y.; Wang, C.Y. Cytotoxic nitrobenzoyloxy-substituted sesquiterpenes from sponge-derived endozoic fungus *Aspergillus insulicola* MD10-2. *Curr. Pharm. Biotechno.* **2016**, *17*, 271–274. [CrossRef]
21. Wang, S.; Zeng, Y.B.; Yin, J.J.; Chang, W.J.; Zhao, X.L.; Mao, Y. Two new azaphilones from the marine-derived fungus *Penicillium sclerotiorum* E23Y-1A. *Phytochem. Lett.* **2022**, *47*, 76–80. [CrossRef]
22. Wang, Z.; Zeng, Y.B.; Zhao, W.B.; Dai, F.H.; Chang, W.J.; Lv, F. Structures and biological activities of brominated azaphilones produced by *Penicillium sclerotiorum* E23Y-1A. *Phytochem. Lett.* **2022**, *52*, 138–142. [CrossRef]
23. Zeng, Y.B.; Wang, Z.; Chang, W.J.; Zhao, W.B.; Wang, H.; Chen, H.Q.; Dai, F.H.; Lv, F. New azaphilones from the marine-derived fungus *Penicillium sclerotiorum* E23Y-1A with their anti-inflammatory and antitumor activities. *Mar. Drugs* **2023**, *21*, 75. [CrossRef]
24. El Amrani, M.; Lai, D.; Debbab, A.; Aly, A.H.; Siems, K.; Seidel, C.; Schnekenburger, M.; Gaigneaux, A.; Diederich, M.; Feger, D.; et al. Protein Kinase and HDAC Inhibitors from the Endophytic Fungus *Epicoccum nigrum*. *J. Nat. Prod.* **2014**, *77*, 49–56. [CrossRef]
25. Laszlo, V.; Michael, K.; Astrid, M.E.; Luigi, T. 2-Phenyl-benzofuran derivatives, method for the production thereof and their use. D.E. Patent 10,351,315, 2005.

26. Talontsi, F.M.; Dittrich, B.; Schüffler, A.; Sun, H.; Laatsch, H. Epicoccolides: Antimicrobial and antifungal polyketides from an endophytic fungus *Epicoccum* sp. associated with *Theobroma cacao*. *Eur. J. Org. Chem.* **2013**, *2013*, 3174–3180. [CrossRef]
27. Luan, Y.P.; Wei, H.J.; Zhang, Z.P.; Che, Q.; Liu, Y.K.; Zhu, T.J.; Mándi, A.; Kurtán, T.; Gu, Q.Q.; Li, D.H. Eleganketal A, a highly oxygenated dibenzospiroketal from the marine-derived fungus *Spicaria elegans* KLA03. *J. Nat. Prod.* **2014**, *77*, 1718–1723. [CrossRef] [PubMed]
28. Lee, N.H.; Gloer, J.B.; Wicklow, D.T. Isolation of chromanone and isobenzofuran derivatives from a fungicolous isolate of *Epicoccum purpurascens*. *Bull. Korean Chem. Soc.* **2007**, *28*, 877–879. [CrossRef]
29. Nishihara, Y.; Takase, S.; Tsujii, E.; Hatanaka, H.; Hashimot, S. New anti-influenza agents, FR198248 and its derivatives. II. Characterization of FR198248, its related compounds and some derivatives. *J. Antibiot.* **2001**, *24*, 297–303. [CrossRef] [PubMed]
30. Yang, L.; Yang, Y.L.; Dong, W.H.; Li, W.; Wang, P.; Cao, X.; Yuan, J.Z.; Chen, H.Q.; Mei, W.L.; Dai, H.F. Sesquiterpenoids and 2-(2-phenylethyl) chromones respectively acting as α-glucosidase and tyrosinase inhibitors from agarwood of an *Aquilaria* plant. *J. Enzym. Inhib. Med. CH.* **2019**, *34*, 853–862. [CrossRef] [PubMed]
31. Yan, Z.Y.; Huang, C.Y.; Guo, H.X.; Zheng, S.Y.; He, J.R.; Lin, J.; Long, Y.H. Isobenzofuranone monomer and dimer derivatives from the mangrove endophytic fungus *Epicoccum nigrum* SCNU-F0002 possess α-glucosidase inhibitory and antioxidant activity. *Bioorg. Chem.* **2020**, *94*, 103407. [CrossRef] [PubMed]
32. O'Boyle, N.M.; Vandermeersch, T.; Flynn, C.J.; Maguire, A.R.; Hutchison, G.R. Confab-systematic generation of diverse low-energy conformers. *J. Cheminform.* **2011**, *3*, 8. [CrossRef] [PubMed]
33. Bannwarth, C.; Ehlert, S.; Grimme, S.J. GFN2-xTB an accurate and broadly parametrized self-consistent tight-binding quantum chemical method with multipole electrostatics and density-dependent dispersion contributions. *Chem. Theory Comput.* **2019**, *15*, 1652–1671. [CrossRef] [PubMed]
34. Frisch, M.J.; Trucks, G.W.; Schlegel, H.B.; Scuseria, G.E.; Robb, M.A.; Cheeseman, J.R.; Scalmani, G.; Barone, V.; Petersson, G.A.; Nakatsuji, H.; et al. *Gaussian 16, Revision, C.01*; Gaussian, Inc.: Wallingford, CT, USA, 2019.
35. Bruhn, T.; Schaumloffel, A.; Hemberger, Y.; Bringmann, G. SpecDis: Quantifying the comparison of calculated and experimental electronic circular dichroism spectra. *Chirality* **2013**, *25*, 243–249. [CrossRef] [PubMed]

marine drugs

Article

New Azaphilones from the Marine-Derived Fungus *Penicillium sclerotiorum* E23Y-1A with Their Anti-Inflammatory and Antitumor Activities

Yanbo Zeng [1,2,3,*,†], Zhi Wang [1,2,†], Wenjun Chang [1,3], Weibo Zhao [1,2], Hao Wang [1], Huiqin Chen [1], Haofu Dai [1,*] and Fang Lv [2,*]

1 Hainan Provincial Key Laboratory for Functional Components Research and Utilization of Marine Bio-resources, Institute of Tropical Bioscience and Biotechnology, Chinese Academy of Tropical Agricultural Sciences & Key Laboratory for Biology and Genetic Resources of Tropical Crops of Hainan Province, Hainan Institute for Tropical Agricultural Resources, Haikou 571101, China
2 Beijing Key Laboratory for Separation and Analysis in Biomedicine and Pharmaceuticals, School of Life Science, Beijing Institute of Technology, Beijing 100081, China
3 Zhanjiang Experimental Station of Chinese Academy of Tropical Agricultural Sciences, Zhanjiang 524013, China
* Correspondence: zengyanbo@itbb.org.cn (Y.Z.); daihaofu@itbb.org.cn (H.D.); lvfangbeijing@bit.edu.cn (F.L.)
† These authors contributed equally to this work.

Abstract: Nine new azaphilones, including penicilazaphilones I–N (**1**, **2** and **6**–**9**), *epi*-geumsanol D (**3**) and penidioxolanes C (**4**) and D (**5**) were isolated from the culture of the marine-derived fungus *Penicillium sclerotiorum* E23Y-1A. The structures of the isolates were deduced from extensive spectroscopic data (1D and 2D NMR), high-resolution electrospray ionization mass spectrometry (HRESIMS), and electronic circular dichroism (ECD) calculations. All the azaphilones from *P. sclerotiorum* E23Y-1A were tested for their anti-inflammatory and antitumor activities. Penicilazaphilone N (**9**) showed moderate anti-inflammatory activity with an IC_{50} value of 22.63 ± 2.95 μM, whereas penidioxolane C (**4**) exhibited moderate inhibition against human myeloid leukemia cells (K562), human liver cancer cells (BEL-7402), human gastric cancer cells (SGC-7901), human non-small cell lung cancer cells (A549), and human hela cervical cancer cells, with IC_{50} values of 23.94 ± 0.11, 60.66 ± 0.13, 46.17 ± 0.17, 60.16 ± 0.26, and 59.30 ± 0.60 μM, respectively.

Keywords: marine-derived fungus; *Penicillium sclerotiorum*; azaphilones; structure elucidation; anti-inflammatory activity; antitumor activity

Citation: Zeng, Y.; Wang, Z.; Chang, W.; Zhao, W.; Wang, H.; Chen, H.; Dai, H.; Lv, F. New Azaphilones from the Marine-Derived Fungus *Penicillium sclerotiorum* E23Y-1A with Their Anti-Inflammatory and Antitumor Activities. *Mar. Drugs* **2023**, *21*, 75. https://doi.org/10.3390/md21020075

Academic Editor: Concetta Imperatore

Received: 25 November 2022
Revised: 16 January 2023
Accepted: 19 January 2023
Published: 22 January 2023

1. Introduction

Marine-derived fungi represent an important source of marine natural products (MNPs). More than 33% of the new MNPs (approximately 2500) were reported from marine-derived fungi between 2016 and 2020 [1]. The interest in highly diverse structural classes and the number of various bioactivities of their metabolites keeps growing considerably [2–5]. Especially in some cases, marine fungi form symbiotic relationships with other organisms through the horizontal gene transfer of a mitochondrial intron from a fungus to marine organisms, consequentially fungi-derived natural products represent the characteristics of metabolites in symbiotic organisms as well [6–8].

Penicillium species, an important part of marine-derived fungi mainly derived from sediments, sponges, mangrove and algae, have a complex genetic background and abundant secondary metabolites. To date, more than 580 new natural products have been identified from marine-derived *Penicillium* fungi, including azaphilones, polyketides, alkaloids, terpenoids, and macrolides, which have displayed anti-inflammatory, antibacterial, anticancer, and other activities with potential application in the pharmaceutical and medical fields [9–12]. Azaphilones, a class of structurally diverse fungal metabolites mainly from *Penicillium* fungi, and defined as polyketides possessing a highly oxygenated pyranoquinone bicyclic core and a quaternary carbon center, have been found to exhibit various biological properties, such as antibacterial, antifungal, cytotoxic, anti-inflammatory, and enzyme inhibitory activities [13]. In our continuing interest in finding new compounds with potential bioactivities [14,15], a chemical investigation of the fungus *Penicillium sclerotiorum* E23Y-1A derived from the marine sponge *Holoxea* sp. was performed to yield a further nine new azaphilones, including penicilazaphilones I–N (**1**, **2** and **6**–**9**), *epi*-geumsanol D (**3**), and penidioxolanes C (**4**) and D (**5**) (Figure 1). The structures of the new compounds were determined by comprehensive analyses of NMR spectra, HRESIMS data, and ECD calculation. By screening of the inhibitory effects on NO production in the LPS-induced RAW 264.7 macrophages, the anti-inflammatory activity of these compounds was evaluated, and the results showed that compound **9** exhibited moderate inhibitory activity with an IC_{50} value of 22.63 ± 2.95 μM. Cytotoxic activities showed that penidioxolane C (**4**) exhibited moderate inhibition against K562, BEL-7402, SGC-7901, A549 and Hela cancer cells with IC_{50} values of 23.94 ± 0.11, 60.66 ± 0.13, 46.17 ± 0.17, 60.16 ± 0.26, and 59.30 ± 0.60 μM, respectively. This article describes the isolation, structure elucidation, as well as anti-inflammatory activity and cytotoxicities of these new azaphilones.

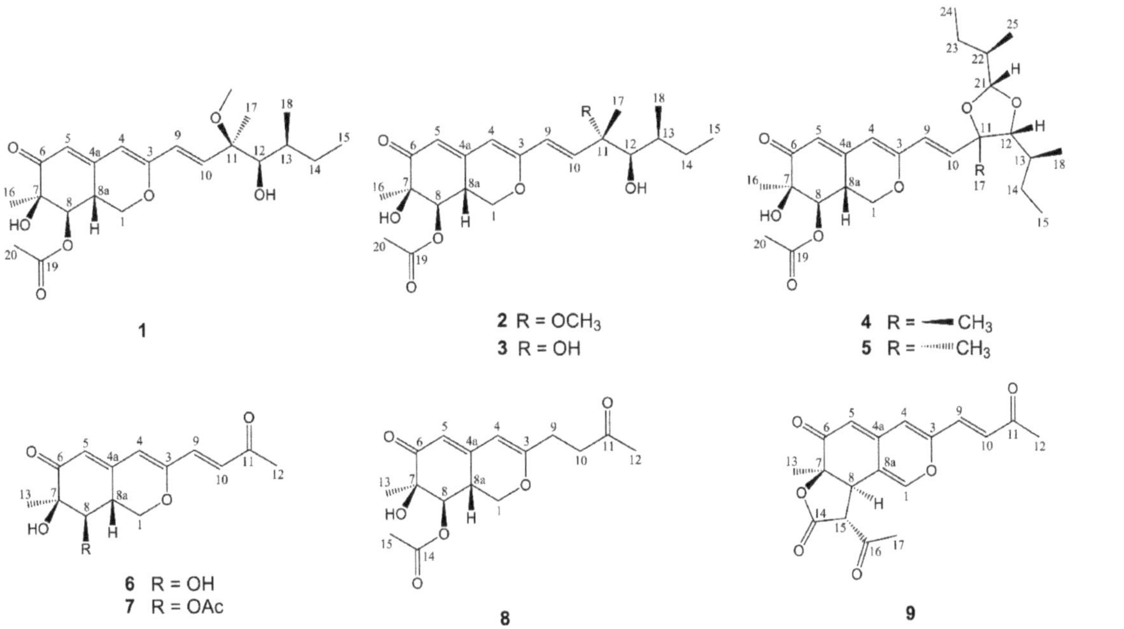

Figure 1. Structures of compounds **1**–**9** isolated from *P. sclerotiorum* E23Y-1A.

2. Results and Discussion

2.1. Structure Elucidation of New Compounds

Penicilazaphilone I (**1**) was obtained as a yellow oil. The molecular formula of **1** was established as $C_{22}H_{32}O_7$ with seven degrees of unsaturation according to HRESIMS data at *m/z* 431.2038 (calcd. 431.2040 for $C_{22}H_{32}O_7Na$, $[M + Na]^+$), which was confirmed by the ^{13}C NMR and DEPT spectral data. The ^{13}C NMR and DEPT data of compound **1** (Tables 1 and 2) suggested the presence of 22 carbons, comprising six methyl carbons (δ_C 50.4, 20.8, 20.0, 16.4, 14.0 and 11.9), two methylene carbons (δ_C 68.0 and 29.0), eight methine carbons (δ_C 139.5, 125.0, 116.9, 105.0, 79.5, 74.5, 35.4 and 34.7) and six quaternary carbons (δ_C 194.6, 170.6, 159.6, 151.1, 80.4 and 74.4). The HMBC correlations from H-1 (δ_H 4.35, 3.80) to C-3 (δ_C 159.6) and C-4a (δ_C 151.1), from H-4 (δ_H 5.67) to C-5 (δ_C 116.9) and C-8a (δ_C 34.7), from H-5 (δ_H 5.80) to C-7 (δ_C 74.4), from H-8 (δ_H 4.98) to C-4a, the ketocarbonyl carbon

C-6 (δ_C 194.6) and C-7, and from H-8a (δ_H 3.35) to C-4a, established a bicyclic core moiety [16]. Further analysis of HMBC correlations from H_3-16 (δ_H 1.36) to C-6 and the oxygenated carbon C-8 (δ_C 74.5), from H_3-20 (δ_C 2.21) to the ketocarbonyl carbon C-19 (δ_C 170.6), and from H-8 to C-19, indicated a methyl group and an acetoxy group attached at C-7 and C-8, respectively. These analysis of NMR spectra confirmed the characteristic signals of an azaphilone skeleton for compound **1**. The ^{1}H-^{1}H COSY spectrum of compound **1** showed two separated correlations of H-9/H-10 and H-12/H-13/H-14 (H_3-18)/H_3-15 in the side chain as well as one correlation of H-8/H-8a/H-1 in the bicyclic core (Figure 2). The ^{1}H-^{1}H COSY correlations of H-9/H-10 and H-12/H-13/H-14 (H_3-18)/H_3-15, associated with HMBC correlations from H-10 (δ_H 6.31) to C-9 (δ_C 125.0) and C-11 (δ_C 80.4), from H-12 to C-10 (δ_C 139.5) and C-14 (δ_C 29.0), from H_3-18 (δ_H 0.90) to C-12 (δ_C 79.5) and C-14, together with the correlations from H_3-OCH_3 (δ_H 3.16) to C-11, H_3-17 (δ_H 1.29) to C-10, C-11 and C-12, established a 3-methoxy-3,5-dimethylhept-1-ene-4-ol unit. In addition, HMBC correlations from H-9 to C-3 and C-4 showed that the whole side chain was attached at C-3 of the azaphilone unit. All arrangements of 1D and 2D NMR data of compound **1** led to the construction of the planar structure of compound **1**.

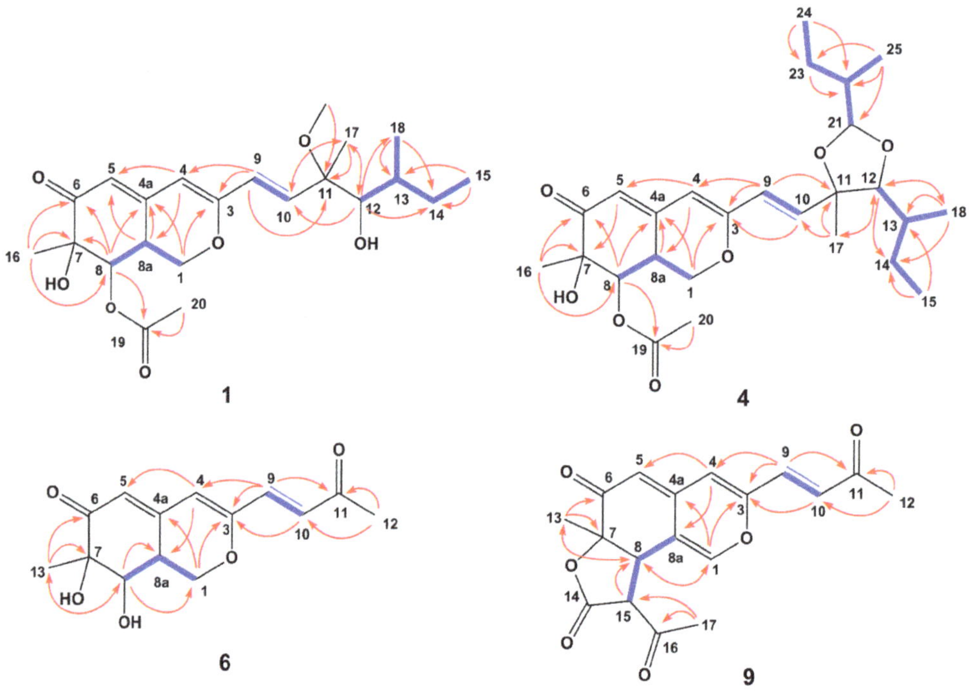

Figure 2. ^{1}H-^{1}H COSY (blue bold lines) and key HMBC (red arrows) correlations of compounds **1**, **4**, **6** and **9**.

The relative configuration of 7*R**, 8*R**, 8a*R**of compound **1** was suggested by NOESY correlations and ^{1}H-^{1}H coupling constants [17]. A large coupling constant ($^3J_{H\text{-}8/\ H\text{-}8a}$ = 10.0 Hz) suggested that H-8 and H-8a were positioned on the opposite face [17]. Meanwhile, the NOESY correlation between H_3-16 and H-8 indicated their cofacial orientation (Figure 3). As to azaphilones possessing a branched C_7 side chain fused at C-3, the absolute configuration of C-13 was suggested as *S* by the common biosynthetic pathway of the aliphatic side chain [13,18,19], which was unambiguously defined by X-ray single-crystal diffraction [20], hydrolysis [21], or ECD calculation [19]. To determine the relative configuration of C-11 and C-12 in the side chain of **1**, we performed theoretical NMR chemical shifts calculations of four diastereomers (Figure 4) of **1** at mPW1PW91-SCRF/6-311G(d,p)//B3LYP-D3BJ/6-31G(d) theoretical level in methanol with the GIAO method [22]. The calculated ^{13}C and ^{1}H NMR chemical shifts of (11*R**, 12*R**, 13*S**)-**1** showed the best agreement with the experimental values. Furthermore, DP4+ analysis [23] predicted that (11*R**, 12*R**, 13*S**)-**1** was the most likely candidate with 91.85% probability (Figure S1, Tables S1 and S2). Considering these observations, the relative stereochemistry of **1** was defined as depicted in Figure 1. The absolute configuration of (7*R*, 8*R*, 8a*R*, 11*R*, 12*R*, 13*S*) of compound **1** was established by analyzing the CD curve (Figure 5) and biosynthetic considerations. Cotton effects at 232 nm, 261 nm and 377 nm were completely consistent with those of penicilazaphilone F [14] because the substitution at C-5 in azaphilone had no effect on the configuration [24]. To further verify the assigned absolute configuration of compound **1**, its theoretical ECD curve was calculated. The calculated ECD curve was consistent with the experimental one (Figure 6), confirming the stereochemical assignment for compound **1**.

In order to conveniently clarify the configurations of co-metabolites (compounds **2–9**) of compound **1**, the molecular orbital (MO) analysis of the most populated conformer of compound **1** was carried out to reveal that the positive Cotton effect around 266 nm was assigned to $n \rightarrow \pi^*$ electron transition from acetate carbonyl to conjugated azaphilone unit, dominated by MO 105→MO 111 transition, and the negative Cotton effect around 375 nm was related to $\pi \rightarrow \pi^*$ electron transition of conjugated azaphilone unit from MO 110 (HOMO)→MO 111 (LUMO) transition (Figure S2). These two characteristic Cotton effects did not involve the electron transition of the functional groups on the side chain. Therefore, the ECD curves actually established the configurations of stereocenters on the azaphilone moiety, while the configurations of stereocenters on the side chain of co-metabolites were established by their NOESY correlation, ^{1}H-^{1}H coupling constants, chemical shifts and Cotton effects with those of the known analogues, in combination with biosynthetic considerations.

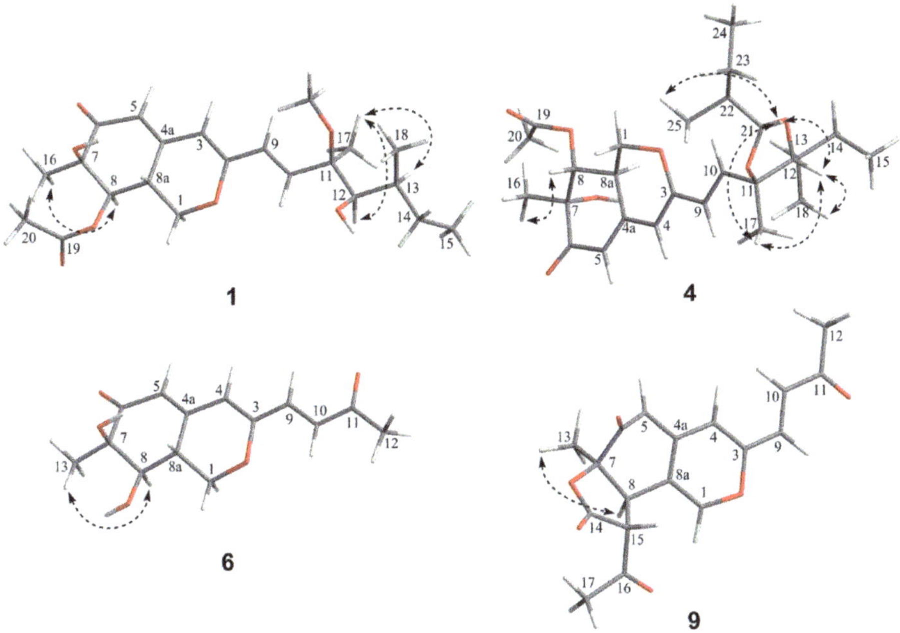

Figure 3. Key NOE correlations of compounds **1**, **4**, **6** and **9**.

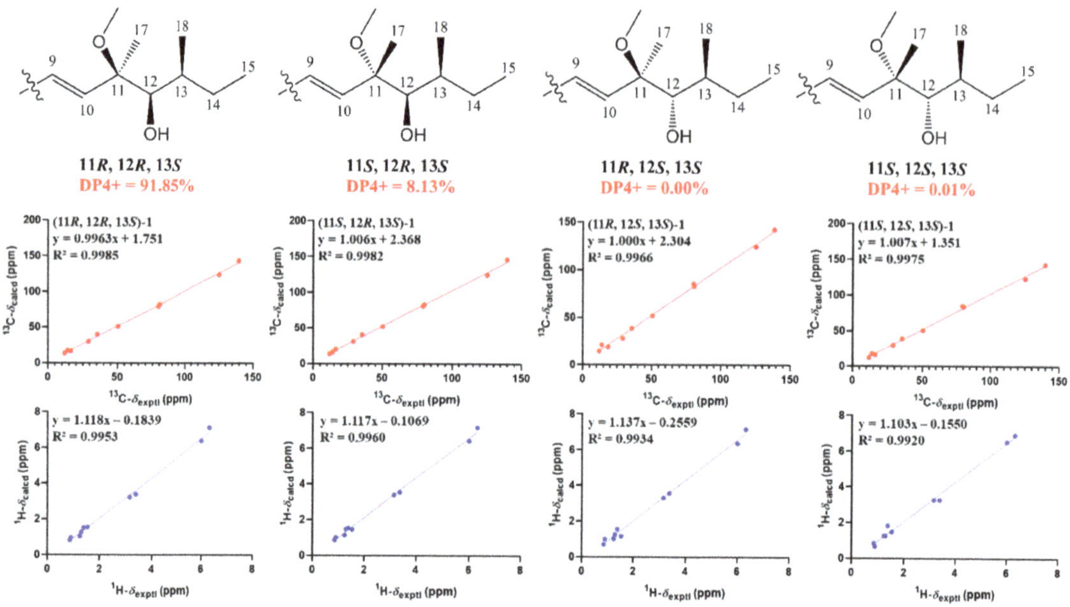

Figure 4. Linear regression analysis between experimental and calculated ^{13}C and ^{1}H NMR chemical shifts of isomers of **1**.

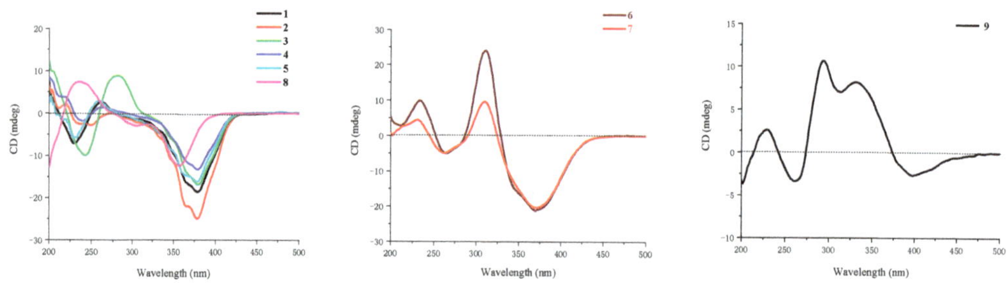

(**a**) (**b**) (**c**)

Figure 5. CD curves of compounds: (**a**) **1**–**5** and **8**; (**b**) **6** and **7**; (**c**) **9**.

Penicilazaphilone J (**2**) was obtained as a yellow oil and assigned the same molecular formula of $C_{22}H_{32}O_7$ as that of compound **1** by HRESIMS data at *m/z* 839.4156 (calcd. for 839.4188, $C_{44}H_{64}O_{14}Na$, $[2M + Na]^+$) and the ^{13}C NMR and DEPT spectral data. Careful comparation of their 1D and 2D NMR spectra revealed that compound **2** had the same planar structure as that of compound **1**. However, the obvious difference at C-17 ($\triangle\delta_C$ = 2.1 ppm) was observed (Table 2) and the NOESY spectrum of compound **2** exhibited a NOE correlation between H_3-OCH_3 and H-12 instead of the correlation of H_3-17 and H-12 of compound **1**, which suggested H_3-OCH_3 and H-12 of compound **2** were located on the same side. However, relative configuration assignment only based on NOESY correlations is usually not reliable enough in conformationally flexible molecules such as the acyclic part of organic compounds. To irrefutably determine the relative configuration of C-11 and C-12 on the side chain of **2**, the quantum chemistry–nuclear magnetic resonance calculations were employed. Following the DP4+ protocol, the calculated results showed that the experimentally observed NMR data for **2** gave a better match of the 11*S**, 12*R**, 13*S** isomer with 99.67% probability, while the isomer 11*R**, 12*R**, 13*S** resulted at 0.3% probability (Figures S3 and S4, Tables S3 and S4). Consequently, the relative configuration of the chain side of compound **2** was established as 11*S**, 12*R**, 13*S**. Among the hitherto reported azaphilones possessing a branched C_7 side chain fused at C-3, the absolute configuration of C-13 in the aliphatic side chain was *S*, which suggests the aliphatic branch of the azaphilones originates from a shared biosynthetic pathway [13]. Thus, the absolute configuration of C-13 of compound **2** was assigned the *S* configuration. Consequently, the absolute configuration of the chain side of compound **2** was established as 11*S*, 12*R*, 13*S*. As compounds **1** and **2** were co-metabolites showing similar Cotton effects (Figure 5), the absolute stereochemistry of the azaphilone core moiety in compound **2** was proposed to be 7*R*, 8*R*, and 8a*R*, the same as that of compound **1**. Therefore, compound **2** was identified as a chiral isomer of compound **1** at C-11.

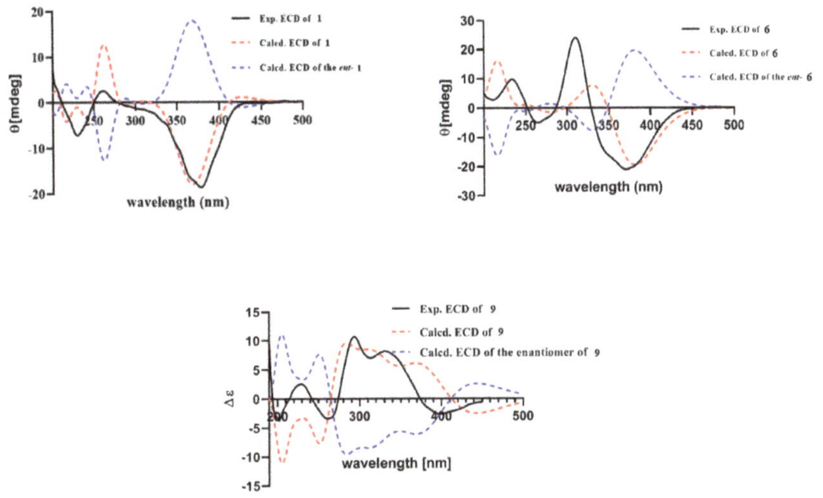

Figure 6. Experimental (black bold) and calculated (dash) ECD spectra of compounds **1**, **6** and **9**.

Epi-geumsanol D (**3**) was obtained as a yellow oil, its molecular formula $C_{21}H_{30}O_7$ was assigned by analysis of HRESIMS at *m/z* 417.1863 (calcd. for 417.1884, $C_{21}H_{30}O_7Na$, $[M + Na]^+$) in combination with NMR data. Careful comparison of the ^1H and ^{13}C NMR data of compounds **3** and **2**, found that compound **3** was highly similar to compound **2** except for the absence of a methoxyl group at C-11 (Tables 1 and 2). A large coupling constant ($^3J_{H\text{-}8/H\text{-}8a}$ = 10.0 Hz) suggested that H-8 and H-8a were positioned on the opposite face [17]. Meanwhile, the NOESY correlation between H_3-16 and H-8 indicated their cofacial orientation. The $^3J_{H\text{-}12/H\text{-}13}$ Value (2.1 Hz) was indicative of a *gauche* relationship of H-12 and H-13. The NOESY correlation between H-12 and H-13, in association with the absence of an NOESY correlation between H_3-17 and H-12 suggested the relative configuration at the side chain of compound **3** to be 11*S**, 12*R**, and 13*S** [25–27]. In order to irrefutably confirm the relative configuration at the side chain of compound **3**, the NMR calculation was performed and followed by a DP4+ analysis. The calculated chemical shifts of (11*S**, 12*R**, 13*S**)-**3** showed best agreement with the experimental values among the possible diastereomers and (11*S**, 12*R**, 13*S**)-**3** possessed 100% DP4+ probability (Figures S5 and S6, Tables S5 and S6), indicating that the relative configuration at the side chain of compound **3** to be 11*S**, 12*R**, and 13*S**. Furthermore, similar Cotton effects at 245 nm, 280 nm and 377 nm to compound **2** (Figure 5), in combination with biosynthetic considerations, allowed the absolute configuration of compound **3** to be 7*R*, 8*R*, 8a*R*, 11*S*, 12*R*, and 13*S*.

Table 1. ^{1}H NMR data of compounds **1–9** in $CDCl_3$.

No.	1 [a]	2 [a]	3 [b]	4 [a]	5 [a]	6 [a]	7 [a]	8 [a]	9 [a]
1	4.35, dd (10.8, 5.2)	4.35, dd (10.8, 5.2)	4.35, dd (10.7, 5.2)	4.36, dd (10.7, 5.3)	4.35, dd (10.8, 5.2)	4.86, dd, (10.9, 5.5)	4.39, dd, (10.8, 5.3)	4.28, dd, (10.7, 5.3)	7.41, s
	3.80, dd (13.7,10.8)	3.80, dd (13.6, 10.8)	3.81, dd (13.6,10.7)	3.79, dd (13.6, 10.7)	3.80, dd (13.6, 10.7)	3.81, dd, (13.4, 10.9)	3.82, dd, (13.6, 10.9)	3.73, dd, (13.8, 10.8)	
4	5.67, s	5.67, s	5.67, s	5.67, s	5.68, s	5.94, s	5.96, s	5.52, s	6.37, s
5	5.80, brs	5.79, s	5.79, d (2.1)	5.80, d (2.1)	5.79, d (2.1)	5.88, brs	5.91, brs	5.70, s	5.53, s
8	4.98, d (10.0)	4.97, d (10.1)	4.98, d (10.0)	4.98, d (10.1)	4.98, d (10.0)	3.47, d (9.3)	4.99, d (10.0)	4.95, d (9.9)	3.85, d (12.2)
8a	3.35, m	3.35, m	3.35, m	3.34, m	3.34, m	3.04, m	3.37, m	3.27, m	
9	6.00, d (15.9)	5.98, d (16.1)	6.19, d (15.6)	6.12, d (15.5)	6.18, d (15.4)	6.85, d (15.6)	6.84, d (15.7)	2.51, t (7.3)	6.93, d (15.8)
10	6.31, d (15.9)	6.42, d (16.1)	6.43, d (15.5)	6.37, d (15.5)	6.39, d (15.4)	6.70, d (15.7)	6.65, d (15.7)	2.68, m	6.76, d (15.7)
12	3.39, d (2.7)	3.45, s	3.49, d (2.1)	3.38, d (8.8)	3.30, d (9.6)	2.33, s	1.36, s	2.17, s	2.36, s
13	1.53, m	1.47, m	1.53, m	1.47, m	1.64, m	1.49, s	1.49, s	1.34, s	1.58, s
14	1.38, m 1.24, m	1.40, m 1.28, m	1.41, m 1.30, m	1.62, m 1.07, m	1.39, m 1.01, m				
15	0.85, t (7.4)	0.87, t (7.4)	0.90, t (7.4)	0.88, t (7.4)	0.90, t (7.3)		2.23, s	2.20, s	3.75, d (12.3)
16	1.35, s	1.34, s	1.35, s	1.36, s	1.36, s				
17	1.29, s	1.31, s	1.35, s	1.41, s	1.24, s				2.48, s
18	0.90, d (6.8)	0.83, d (6.5)	0.86, d (6.8)	0.97, d (6.6)	1.04, d (6.5)				
20	2.21, s	2.20, s	2.21, s	2.21, s	2.21, s				
OCH_3	3.16, s	3.16, s							
21				4.71, d (5.0)	4.74, d (4.0)				
22				1.62, m	1.60, m				
23				1.60, m 1.20, m	1.60, m 1.20, m				
24				0.93, t (7.3)	0.93, t (7.3)				
25				0.98, d (6.6)	0.94, d (6.6)				

[a] in 600 MHz; [b] in 400 MHz.

Penidioxolane C (**4**) was obtained as a yellow oil. The molecular formula, $C_{26}H_{38}O_7$ with eight degrees of unsaturation, was established by HRESIMS at *m/z* 463.2683 (calcd. for 463.2690, $C_{26}H_{39}O_7$, $[M + H]^+$) in combination with its NMR data. The ^{13}C NMR and DEPT data of compound **4** (Tables 1 and 2) suggested the presence of 26 carbons, comprising seven methyl carbons (δ_C 25.2, 20.8, 20.1, 16.0, 14.1, 11.6 and 11.2), three methylene carbons (δ_C 68.1, 25.6 and 24.8), ten methine carbons (δ_C 138.9, 123.2, 116.7, 105.3, 104.8, 90.4, 74.5, 38.8, 35.2 and 34.8) and six quaternary carbons (δ_C 194.4, 170.6, 160.0, 151.2, 80.8 and 74.5). Investigation of ^{1}H-^{1}H COSY correlations of H-8/H-8a/H-1 and HMBC correlations from H-1 to C-3/ C-4a, from H-4 to C-5/C-8a, from H-5 to C-7, from H-8 to C-6, from H-8a to C-4a, together with HMBC correlations from H_3-16 to C-6/C-8, from H_3-20 to C-19, and from H-8 to C-19 indicated that compound **4** possessed the same azaphilone skeleton as that of compound **1**. However, the side chain of compound **4** included the additional consecutive ^{1}H-^{1}H COSY correlations of H-21/H-22/H_2-23 (H_3-25)/H_3-24 excepting H-9/H-10 and H-12/H-13/H-14 (H_3-18)/H_3-15 were identical to compound **1**. This observation was further confirmed by the HMBC correlations from H-24 to C-23, from H-23 to C-22, from H-21 to C-22, and from H-25 to C-21, no long range HMBC correlations from H-21 to C-11 and C-12 were found, and the connection position with the azaphilone unit was deduced by downfield chemical shifts of C-11 and C-12 [28] (Figure 2). All arrangements of 1D and 2D NMR data allowed the construction of the planar structure of compound **4**. The relative configuration of compound **4** was suggested by ^{1}H-^{1}H coupling constants [16], together with NOESY spectral analyses. A large coupling constant ($^3J_{8/8a}$ = 10.1 Hz) and NOESY correlation between H_3-16 and H-8 indicated H-8 should be placed on the opposite orientation of H-8a with a cofacial orientation of H_3-16 [17]. The C-7, C-8 and C-8a chirality centers in the bicyclic core were assigned as 7*R**, 8*R**, 8a*R**. The relative configuration of 1,3-dioxolane part of compound **4** was determined as 11*S**, 12*R**, 13*S**, 21*R**, and 22*R** by NOESY correlations of H-12 (δ_H 3.38)/H_3-17 (δ_H 1.41), H-12 (δ_H 3.38)/H_3-18 (δ_H 0.97), H-12 (δ_H 3.38)/H-21 (δ_H 4.71), H-21 (δ_H 4.71)/H_3-17 (δ_H 1.41), and H-21 (δ_H 4.71)/ H_3-25 (δ_H 0.98) (Figure 3), which was irrefutably confirmed by NMR calculation followed by DP4+ analysis. The calculated chemical shifts of (11*S**, 12*R**, 13*S**, 21*R**, 22*R**)-**4** showed best agreement with the experimental values among the possible diastereomers and (11*S**, 12*R**, 13*S**, 21*R**, 22*R**)-**4** possessed 99.83% DP4+ probability (Figures S7 and S8, Tables S7 and S8). Thus, the relative stereochemistry of **4** was unambiguously defined as depicted in Figure 1. The absolute configuration of C-13 was suggested as *S* by the common biosynthetic pathway of the aliphatic side chain in azaphilones. Consequently, the absolute configuration of the 1,3-dioxolane part of compound **4** was assigned as 11*S*, 12*R*, 13*S*, 21*R*, and 22*R*. As compound **4** was a co-metabolite of compound **1** with similar Cotton effects (Figure 5), the stereochemistry of the azaphilone core moiety in compound **4** was proposed to be 7*R*, 8*R*, and 8a*R* as well. Therefore, the absolute configuration of compound **4** was 7*R*, 8*R*, 8a*R*, 11*S*, 12*R*, 13*S*, 21*R*, and 22*R*.

Table 2. ^{13}C NMR data (125 MHz) of compounds **1**–**9** in $CDCl_3$.

No.	1	2	3	4	5	6	7	8	9
1	68.0	68.0	68.1	68.1	68.0	69.1	68.2	68.3	147.1
3	159.6	159.6	160.0	160.0	160.0	158.1	157.7	166.4	153.5
4	105.0	104.8	104.7	104.8	105.0	110.9	110.7	101.3	114.3
4a	151.1	151.1	151.1	151.2	151.0	149.8	149.2	150.9	142.8
5	116.9	116.9	116.7	116.7	116.7	119.1	119.5	115.5	109.1
6	194.6	194.6	194.4	194.4	194.4	196.7	194.5	194.5	191.8
7	74.4	74.4	74.4	74.5	74.5	74.5	74.6	74.4	82.8
8	74.5	74.5	74.6	74.5	74.5	74.4	74.3	74.5	42.9
8a	34.7	34.7	36.4	34.8	34.8	36.7	34.7	34.4	114.4
9	125.0	125.0	122.3	123.2	122.5	135.0	134.8	28.5	131.9
10	139.5	138.4	140.0	138.9	140.8	130.3	130.4	40.1	131.0
11	80.4	80.6	76.1	80.8	80.7	197.8	197.7	206.6	196.8
12	79.5	80.2	79.4	90.4	88.8	28.7	28.5	30.1	29.0
13	35.4	35.7	34.8	35.2	34.5	20.5	19.9	20.1	23.3
14	29.0	29.2	28.9	25.6	25.4		170.5	170.6	168.3
15	11.9	12.0	12.0	11.2	11.1		20.8	20.8	57.2
16	20.0	20.0	20.1	20.1	20.1				200.0
17	16.4	18.5	26.7	25.2	21.3				30.3
18	14.0	13.8	13.4	16.0	16.4				
19	170.6	170.6	170.6	170.6	170.5				
20	20.8	20.8	20.8	20.8	20.8				
OCH_3	50.4	50.6							
21				105.3	105.8				
22				38.8	38.7				
23				24.8	24.5				
24				11.6	11.8				
25				14.1	13.6				

The molecular formula of penidioxolane D (**5**) was identical to that of compound **4** as determined by HRESIMS at *m/z*: 463.2688 (calcd. for 463.2690, $C_{26}H_{39}O_7$, $[M + H]^+$). Detailed analysis of NMR spectra of compound **5** revealed that it shared the same planar structure as that of compound **4**. The major changes in the ^{13}C NMR data of compound **5** were the shielded signal of C-12 (−1.6 ppm) and the de-shielded signals of C-10 (+1.9 ppm) as well as the shielded signal of H-17 (−0.17 ppm) in ^{1}H NMR data when compared with compound **4**. Moreover, in the NOESY spectrum of compound **5**, combined with the correlation of H-12 (δ_H 3.30) and H-21 (δ_H 4.74), the lack of the correlations of H_3-17 (δ_H 1.24) and H-12, and H_3-17 and H-21, implied H_3-17 was the opposite orientation of H-12 and H-21 in compound **5**. These observations suggested the configuration of C-11 of compound **5** was different from that of compound **4**. Consequently, the configuration of C-11 of compound **5** was confirmed as *R*.

Penicilazaphilone K (**6**) was obtained as a yellow oil. The molecular formula, $C_{14}H_{16}O_5$, was established by the protonated molecular ion peak at *m/z* 265.1073 (calcd. for $C_{14}H_{17}O_5$, 265.1071, $[M + H]^+$) in the HRESIMS, indicating seven degrees of unsaturation. Analysis of ^{1}H NMR spectrum (Table 1) of compound **6** revealed the presence of two methyls (δ_H 2.33, 1.49), one methylene (δ_H 4.86, 3.81), two methines (δ_H 3.47, 3.04), and four olefinic protons (δ_H 6.85, 6.70, 5.94, 5.88). The ^{13}C NMR and DEPT spectra of compound **6** revealed 14 carbon resonances, including two methyls (δ_C 28.7, 20.5), one methylene (δ_C 69.1), two methines (δ_C 74.4, 36.7), six olefinic carbons (δ_C 158.1, 149.8, 135.0, 130.3, 119.1, 110.9), and three quaternary carbons (δ_C 197.8, 196.7, 74.5) (Table 2). In addition to a 1-butene-3-one unit (four carbons), the remaining 10 carbons could be assigned to a typical azaphilone skeleton with a methyl group at C-7 confirmed by HMBC correlations from H-1 to C-3 and C-4a, from H-4 to C-5 and C-8a, from H-8 to C-8a and C-1, and from H_3-13 to C-6 and C-8. The key ^{1}H-^{1}H COSY and HMBC correlations of compound **6** are shown in Figure 2. The relative configuration of 7*R**, 8*R**, 8a*R** was established by NOESY correlation and ^{1}H-^{1}H coupling constant. A large coupling constant ($^3J_{8/8a}$ = 9.3 Hz) suggested that H-8 and H-8a are positioned on the opposite face [17]. The NOESY correlation between H_3-13 and H-8, meanwhile, indicated their cofacial orientation (Figure 3). Furthermore, the absolute configuration of compound **6** was established by its CD spectrum (Figure 5). The calculated ECD curve was consistent with the experimental one (Figure 6), confirming the stereochemical assignment of (7*R*, 8*R*, 8a*R*) for compound **6**.

Penicilazaphilone L (**7**) was obtained as a yellow oil. The molecular formula $C_{16}H_{18}O_6$ with eight degrees of unsaturation was established by analysis of HRESIMS at *m/z* 307.1180 (calcd. for 307.1176, $C_{16}H_{19}O_6$, $[M + H]^+$) in combination with its NMR data. The NMR data of compound **7** were quite similar to those of compound **6**, with notable difference in the ^{1}H, ^{13}C, and HSQC NMR data, being the presence of an additional acetoxy group (δ_C 170.5, δ_C 20.8, δ_H 2.23) (Tables 1 and 2). The assignment of an acetoxy group was performed by the HMBC correlation from H_3-15 (δ_H 2.23) to the ester carbonyl C-14 (δ_C 170.5). The HMBC correlation from the oxymethine proton H-8 (δ_H 4.99) to C-14 indicated the acetoxy group connected with the oxymethine carbon C-8. The analysis of NOESY correlations and CD curve (Figure 5) determined that the configuration of compound **7** was 7*R*, 8*R*, 8a*R*, the same as that of compound **6**.

Penicilazaphilone M (**8**) was obtained as a yellow oil. The molecular formula $C_{16}H_{20}O_6$ with seven degrees of unsaturation was established by the analysis of HRESIMS (*m/z* 331.1141 calcd. for 331.1152, $C_{16}H_{20}NaO_6$, $[M + Na]^+$) in combination with its NMR data. Analysis of the 1D and 2D NMR data of compound **8** allowed construction of a structure similar to that of compound **7** with the disappearance of one double bond ($\Delta^{9,10}$). A large coupling constant ($^3J_{8/8a}$ = 9.9 Hz) and a NOESY correlation between H_3-13 and H-8 allowed the determination of the relative configuration of compound **7** [17]. The absolute configuration of (7*R*, 8*R*, 8a*R*) of compound **8** was confirmed by its CD curve (Figure 5) showing the same characteristics as that of compound **1**.

Penicilazaphilone N (**9**) was obtained as a yellow amorphous solid. The molecular formula $C_{18}H_{16}O_6$ with 11 degrees of unsaturation was established by analysis of HRESIMS (*m/z* 351.0835 $[M + Na]^+$) and NMR data (Tables 1 and 2). The azaphilone skeleton of compound **9** was established by comparisons of NMR data with those of compound **6**. In addition to the appearance of a double bond ($\Delta^{1,8a}$), further analysis of ^{13}C NMR and

HSQC spectra with those of compound **6** determined the presence of four additional carbon resonances, including two carbonyls (δ_C 200.0 and 168.3), one methine (δ_C 57.2), and one methyl (δ_C 30.3). The ^{1}H-^{1}H COSY correlation between H-8 (δ_H 3.85) and H-15 (δ_H 3.75), along with HMBC correlations from H-8 to C-7 (δ_C 82.8) and from H-15 to C-14 (δ_C 168.3), confirmed the presence of the γ-lactone ring (Figure 2). Furthermore, HMBC correlations from H$_3$-17 (δ_H 2.48) to C-15 (δ_C 57.2) and C-16 (δ_C 200.0) determined an acetyl group attached to C-15. The relative configuration of 7*R**, 8*R**, 15*R** was established by a large coupling constant ($^3J_{8/15}$ = 12.2 Hz) and NOESY correlation between H-8 and H$_3$-13, which suggested that H-8 and H-15 are positioned on the opposite face while H-8 and H$_3$-13 have a cofacial orientation (Figure 3). The absolute configurations of compound **9** could not be established by direct comparison with the CD curve of compound **9,** as for other compounds, due to the presence of the γ-lactone moiety. Therefore, the theoretical ECD curve of compound **9** was calculated (Figure 5). Its calculated ECD curve of (7*R*, 8*R*, and 15*R*) configuration was consistent with the experimental one (Figure 6), which confirmed the absolute configurations of **9** as shown.

All of the isolated compounds (**1–9**) possessed a common azaphilone framework possessing a highly oxygenated pyranoquinone bicyclic core and a quaternary carbon center and their co-occurrence in the same fungus suggested that they should originate from the same biogenetic pathway. A plausible biosynthetic pathway toward the formation of compounds **1–9** can be proposed by detailed analysis of their structures (Scheme 1).

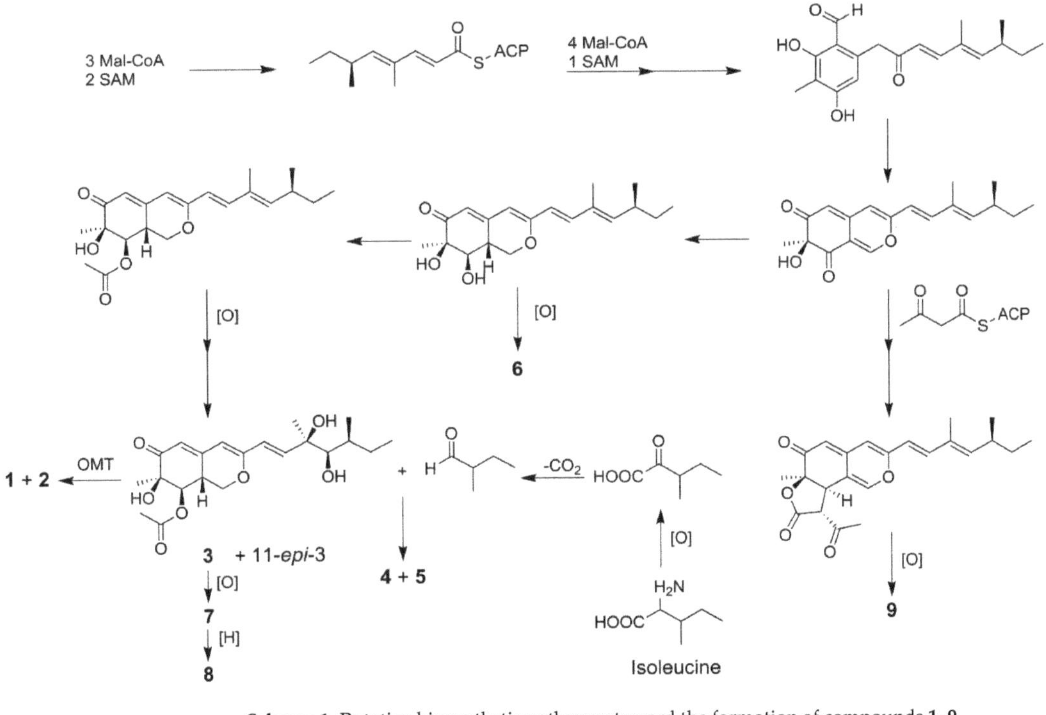

Scheme 1. Putative biosynthetic pathways toward the formation of compounds **1–9**.

2.2. Bioassay

All compounds were tested for anti-inflammatory activity and antitumor activities in vitro. The anti-inflammatory activity test showed that compound **9** exhibited moderate inhibition of nitric oxide production in LPS-stimulated RAW264.7 cells with an IC$_{50}$ value of 22.63 ± 2.95 µM, and compounds **1**, **6** and **7** exhibited weak inhibition with IC$_{50}$ values of 50.71 ± 8.41, 65.30 ± 7.21 and 31.84 ± 2.79 µM, respectively (Figure 7). Quercetin as a positive control showed an IC$_{50}$ value of 11.19 ± 0.38 µM. Further research on cell proliferation was carried out by MTT method, and the result indicated compounds **1**, **6**, **7** and **9** had no cytotoxicity against RAW264.7 at a concentration of 100 μM.

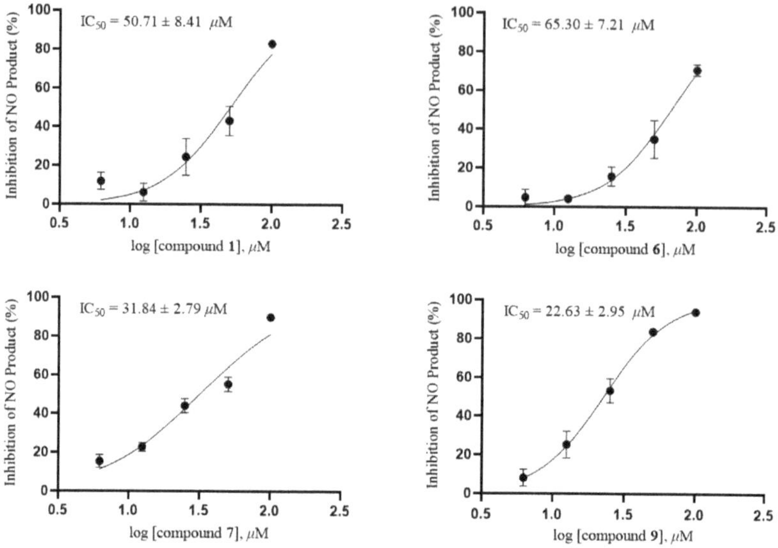

Figure 7. IC_{50} of compounds **1**, **6**, **7** and **9** on NO production in RAW264.7 cells.

The isolated azaphilones were evaluated for their cytotoxicities against human myeloid leukemia cells (K562), human liver cancer cells (BEL-7402), human gastric cancer cells (SGC-7901), human non-small cell lung cancer cells (A549), and human hela cervical cancer cells. The results were presented in Table 3. Penidioxolane C (**4**) exhibited moderate inhibition against K562, BEL-7402, SGC-7901, A549, and Hela cancer cells with IC_{50} values of 23.94 ± 0.11, 60.66 ± 0.13, 46.17 ± 0.17, 60.16 ± 0.26, and 59.30 ± 0.60 μM, respectively. Cisplatin as a positive control showed IC_{50} values of 3.08 ± 0.05, 4.02 ± 0.06, 4.11 ± 0.02, 1.93 ± 0.02, and 11.29 ± 0.15 μM to the abovementioned cancer cells, respectively.

Table 3. Cytotoxic activities of compounds **1–9**.

Compound	IC_{50} ± SD (μM) [a]				
	K562	BEL-7402	SGC-7901	A549	Hela
1	–	–	–	–	–
2	–	–	–	–	–
3	–	–	–	–	–
4	23.94 ± 0.11	60.66 ± 0.13	46.17 ± 0.17	60.16 ± 0.26	59.30 ± 0.60
5	–	–	–	–	–
6	–	–	–	–	–
7	–	–	–	–	–
8	–	–	–	–	–
9	–	–	–	–	–
cisplatin [b]	3.08 ± 0.05	4.02 ± 0.06	4.11 ± 0.02	1.93 ± 0.02	11.29 ± 0.15

[a] Values represent means ± SD based on three parallel experiments. [b] positive control. —no activity at a concentration of 20 μg/mL.

3. Materials and Methods

3.1. General Experimental Procedures

Optical rotation was performed on a Perkin–Elmer 241MC polarimeter (PerkinElmer, Fremont, CA, USA). NMR data were measured using Bruker Avance III 400 and Bruker DRX 600 instruments (Bruker Biospin AG, Fällanden, Germany). All chemical shifts (δ) were given in ppm referenced to TMS and coupling constants (J) given in Hz. HRESIMS were recorded on an Agilent G6520 Q-TOF mass spectrometer. The ECD curves and UV data were collected on a Jasco J-810 spectropolarimeter (JASCO, Tokyo, Japan). IR absorptions were obtained on a Nicolet 380 FT-IR instrument (Thermo, Waltham, MA, USA) using KBr pellets. HPLC was carried out using an Agilent 1260 series liquid chromatography (Agilent Technology Co., Ltd., Santa Clara, CA, USA) equipped with a DAD G1315D detector and an Agilent Eclipse XDB-C_{18} column (5 μm, 9.4 × 250 mm). Column chromatography was performed on a Sephadex LH-20 (Merck, Darmstadt, Germany) and silica gel (200–300 and 300–400 mesh; Qingdao Haiyang Chemical Group Co., Ltd., Qingdao, China).

3.2. Fungal Identification, Fermentation, and Extract

The fungus was isolated from the marine sponge Holoxea sp. collected in March 2019 from Quanfu Island, Hainan, China. It was authenticated as Penicillium sclerotiorum E23Y-1A (GenBank accession No. MW090660) via DNA amplification and sequencing of the internal transcribed spacer region of the rRNA gene, along with morphological characteristics. The fungus strain was deposited at the Hainan Provincial Key Laboratory for Functional Components Research and Utilization of Marine Bio-resources, Haikou, China.

The fungus *P. sclerotiorum* E23Y-1A was cultured in eighty-three 1000 mL Erlenmeyer flasks at room temperature for 28 days. Each flask contained 80 g rice and 120 mL water with 3.3% NaBr. After fermentation, the cultures were extracted four times with EtOAc, and 192.0 g dark brown gum was obtained.

3.3. Isolation and Purification

All extract was fractionated by vacuum liquid chromatography using silica gel as a stationary phase eluted with a step gradient of petroleum ether-EtOAc (1:0, 20:1, 10:1, 8:2, 7:3, 6:4 5:5, 3:7 and 0:1; *v*/*v*) and CH_2Cl_2-MeOH (9:1, 8:2 and 0:1; *v*/*v*) to obtain fractions 1–25. Fraction 14 (30.0 g) was separated by Sephadex LH-20 column chromatography eluted with MeOH and then purified by silica gel column chromatography eluted with CH_2Cl_2-MeOH (4:1; *v*/*v*) to give compound **9** (2.9 mg). Fraction 18 (5.3 g) was purified by Sephadex LH-20 column chromatography eluted with petroleum ether-CH_2Cl_2-MeOH (2:1:1, *v*/*v*/*v*) to afford subfractions N1-N14. Subfraction N5 (46.9 mg) was separated by silica gel column chromatography eluted with CH_2Cl_2-acetone (80:1; *v*/*v*) and further purified by semi-preparative HPLC using MeOH-H_2O (80:20, *v*/*v*) as mobile phase to afford compounds **4** (2.9 mg, t_R = 19.0 min) and **5** (2.7 mg, t_R = 20.0 min). Subfraction N7 (327.0 mg) was separated by silica gel column chromatography eluted with CH_2Cl_2-acetone (70:1; *v*/*v*) and further purified by semi-preparative HPLC using MeOH-H_2O (60:40, *v*/*v*) as mobile phase to afford compounds **1** (8.6 mg, t_R = 16.0 min) and **2** (12.6 mg, t_R = 17.0 min). Subfraction N10 (1.9 g) was separated by silica gel column chromatography eluted with CH_2Cl_2-acetone (70:1; *v*/*v*) and further purified by semi-preparative HPLC to afford compound **3** (5.5 mg, t_R = 13.8 min) by MeOH-H_2O (55:45, *v*/*v*), compounds **7** (3.5 mg, t_R = 16.0 min) and **8** (2.2 mg, t_R = 18.0 min) by MeOH-H_2O (42:58, *v*/*v*), and compound **6** (3.5 mg, t_R = 17.0 min) by MeOH-H_2O (35:65, *v*/*v*).

Penicilazaphilone I (**1**): yellow oil; $[\alpha]^{25}{}_D$ −193.1 (*c* 0.1, MeOH); UV (MeOH) λ_{max} (logε) 245 (2.96), 350 (3.63) nm; IR (KBr) ν_{max} 3419, 2931, 1745, 1661, 1586, 1379, 1229, 1060 cm^{-1}; HRESIMS *m/z*: 431.2038 $[M + Na]^+$ (calcd. for 431.2040, $C_{22}H_{32}O_7Na$); ^1H NMR (600 MHz) and ^{13}C NMR (125 MHz) data in $CDCl_3$; see Tables 1 and 2.

Penicilazaphilone J (**2**): yellow oil; $[\alpha]^{25}{}_D$ −223.3 (*c* 0.1, MeOH); UV (MeOH) λ_{max} (logε) 245 (3.03), 350 (3.75) nm; IR (KBr) ν_{max} 3445, 2930, 1744, 1659, 1586, 1378, 1230, 1061 cm^{-1}; HRESIMS *m/z*: 839.4156 $[2M + Na]^+$ (calcd. for 839.4188, $C_{44}H_{64}O_{14}Na$); ^1H NMR (600 MHz) and ^{13}C NMR (125 MHz) data in $CDCl_3$; see Tables 1 and 2.

Epi-geumsanol D (**3**): yellow oil; $[\alpha]^{25}{}_D$ −114.8 (*c* 0.1, MeOH); UV (MeOH) λ_{max} (logε) 246 (2.93), 350 (3.53) nm; IR (KBr) ν_{max} 3433, 2930, 1641, 1584, 1380, 1232, 1062 cm^{-1}; HRESIMS *m/z*: 417.1863 $[M + Na]^+$ (calcd. for 417.1884, $C_{21}H_{30}O_7Na$); ^1H NMR (400 MHz) and ^{13}C NMR (125 MHz) data in $CDCl_3$; see Tables 1 and 2.

Penidioxolane C (**4**): yellow oil; $[\alpha]^{25}{}_{D}$ −113.0 (*c* 0.1, MeOH); UV (MeOH) λ_{max} (logε) 244 (2.90), 351 (3.57) nm; IR (KBr) ν_{max} 3451, 2932, 1746, 1587, 1376, 1225, 1103 cm^{-1}; HRESIMS *m/z*: 463.2683 $[M + H]^+$ (calcd. for 463.2690, $C_{26}H_{39}O_7$); ^1H NMR (600 MHz) and ^{13}C NMR (125 MHz) data in $CDCl_3$; see Tables 1 and 2.

Penidioxolane D (**5**): yellow oil; $[\alpha]^{25}{}_{D}$ −170.3 (*c* 0.1, MeOH); UV (MeOH) λ_{max} (logε) 245 (2.95), 349 (3.63) nm; IR (KBr) ν_{max} 3417, 2928, 1648, 1587, 1376, 1229, 1101 cm^{-1}; HRESIMS *m/z*: 463.2688 $[M + H]^+$ (calcd. for 463.2690, $C_{26}H_{39}O_7$); ^1H NMR (600 MHz) and ^{13}C NMR (125 MHz) data in $CDCl_3$; see Tables 1 and 2.

Penicilazaphilone K (**6**): yellow oil; $[\alpha]^{25}{}_{D}$ −161.1 (*c* 0.1, MeOH); UV (MeOH) λ_{max} (logε) 245 (3.01), 349 (3.54) nm; IR (KBr) ν_{max} 3420, 2924, 1708, 1594, 1338, 1063 cm^{-1}; HRESIMS *m/z*: 265.1073 $[M + H]^+$ (calcd. for 265.1071, $C_{14}H_{17}O_5$); ^1H NMR (600 MHz) and ^{13}C NMR (125 MHz) data in $CDCl_3$; see Tables 1 and 2.

Penicilazaphilone L (**7**): yellow solid; $[\alpha]^{25}{}_{D}$ −81.5 (*c* 0.1, MeOH); UV (MeOH) λ_{max} (logε) 246 (2.82), 350 (3.43) nm; IR (KBr) ν_{max} 3431, 2925, 1661, 1602, 1376, 1231, 1056 cm^{-1}; HRESIMS *m/z*: 307.1180 $[M + H]^+$ (calcd. for 307.1176, $C_{16}H_{19}O_6$); ^1H NMR (600 MHz) and ^{13}C NMR (125 MHz) data in $CDCl_3$; see Tables 1 and 2.

Penicilazaphilone M (**8**): yellow oil; $[\alpha]^{25}{}_{D}$ −215.0 (*c* 0.1, MeOH); UV (MeOH) λ_{max} (logε) 213 (2.73), 323 (3.38) nm; IR (KBr) ν_{max} 3427, 2923, 1717, 1601, 1232, 1038 cm^{-1}; HRESIMS *m/z*: 331.1141 $[M + Na]^+$ (calcd. for 331.1152, $C_{16}H_{20}O_6Na$); ^1H NMR (600 MHz) and ^{13}C NMR (125 MHz) data in $CDCl_3$; see Tables 1 and 2.

Penicilazaphilone N (**9**): yellow solid; $[\alpha]^{25}{}_{D}$ +68.2 (*c* 0.1, MeOH); UV (MeOH) λ_{max} (logε) 229 (3.24), 290 (3.28), 372 (3.20) nm; IR (KBr) ν_{max} 3460, 2925, 1773, 1612, 1250, 1090 cm^{-1}; HRESIMS *m/z*: 351.0835 $[M + Na]^+$ (calcd. for 351.0839, $C_{18}H_{16}O_6Na$); ^1H NMR (600 MHz) and ^{13}C NMR (125 MHz) data in $CDCl_3$; see Tables 1 and 2.

3.4. NMR Calculation

Conformers of compounds were generated using the Confab [29] program ebbed in the Openbabel 3.1.1 software. All the conformers were further optimized with xtb at GFN2 level [30] and the conformers with population over 1% were subjected to geometry optimization using the Gaussian 16 package [31] at B3LYP/6-31G (d) level. The obtained conformers within an energy window of 3 kcal/mol were kept. Then, these conformers were refined by re-optimizations at B3LYP-D3BJ/6-311G (d,p) with IEFPCM solvent model in chloroform. Frequency analysis of all optimized conformations were also performed at the same level of theory to exclude the imaginary frequencies. NMR shielding tensors were calculated with the GIAO method at mPW1PW91/6-31 + G (d,p) level with IEFPCM solvent modeling in chloroform. The shielding constants were converted into chemical shifts by referencing to TMS at 0 ppm (δcal = σTMS − σcal), where the σTMS (the shielding constant of TMS) was calculated at the same level. Considering the almost same chemical shift of the azaphilone moiety, we only used the chemical shift of the side chain of **1–4** for the analysis to decrease the systematic error. For each candidate, the parameters a and b of the linear regression δcal = aδexp + b; the correlation coefficient, R2; the mean absolute error (MAE) defined as Σn | δcal − δexp | /n; the corrected mean absolute error, CMAE, defined as Σn | δcorr − δexp | /n, where δcorr = (δcal − b)/a, were calculated. DP4+ probability analysis was performed using the calculated NMR shielding tensors [23].

3.5. ECD Calculation

The conformers of compounds were generated using the Confab [29] program ebbed in the Openbabel 3.1.1 software, and further optimized with xtb at GFN2 level [30]. The conformers with population over 1% were subjected to geometry optimization using the Gaussian 16 package [31] at B3LYP/6-31G (d) level and proceeded to calculation of excitation energies, oscillator strength, and rotatory strength at B3LYP/TZVP level in the polarizable continuum model (PCM, methanol). The ECD spectra were Boltzmann weighted and generated using SpecDis 1.71 software [32].

3.6. Anti-Inflammatory Activity Test

All compounds were evaluated for their inhibitory effects on NO production in LPS-stimulated RAW264.7 macrophages (Stem Cell Bank of the Chinese Academy of Sciences, Shanghai, China) using the Griess assay [33]. The cells were cultured in DMEM medium (Thermo Fisher scientific, Waltham, MA, USA) in a humidified 5% CO_2/95% air atmosphere at 37 °C. Each compound was diluted in half by concentration gradients (200 μM, 100 μM, 50 μM, 25 μM, 12.5 μM). Quercetin (Sigma Company, St. Louis, MO, USA) was used as positive controls and media with DMSO as negative control. The effects on cell viability of compounds was measured using MTT method.

3.7. Cytotoxic Detection

All compounds were assayed for their cytotoxic activities against five human tumor cell lines: human myeloid leukemia cells (K562), human liver cancer cells (BEL-7402), human gastric cancer cells (SGC-7901), human non-small cell lung cancer cells (A549), and human hela cervical cancer cells which were bought from the Cell Bank of Type Culture Collection of Shanghai Institute of Cell Biology, Chinese Academy of Sciences, using modified MTT methods [34]. Briefly, the abovementioned human tumor cell lines were cultured in RPMI-1640 medium with 10% FBS under a humidified atmosphere of 5% CO_2 and 95% air at 37 °C, and 198 μL of cell suspension was plated in 96-well microtiter plates. An amount of 2 μL of the test solutions in DMSO was added to each well and further incubated for 36 h after being incubated for 24 h. An amount of 20 μL of MTT solution (5 mg/mL in RPMI-1640 medium) was added to each well and incubated for 4 h. Finally, 150 μL of medium containing MTT was gently replaced by DMSO and pipetted to dissolve any formazan crystals formed. Absorbance was tested on a Multiskan FC photometric microplate reader (Thermo Fisher Scientific) at 570 nm. Cisplatin was used as a positive control.

4. Conclusions

In summary, the investigation of *P. sclerotiorum* E23Y-1A derived from the marine sponge *Holoxea* sp. resulted in the isolation of nine new azaphilones (**1–9**), of which compounds **1–5** were identified as azaphilones possessing a branched C_7 side chain fused at C-3. Compounds **6–9** were characteristic of the branded C_4 side chain fused C-3, which are uncommonly found in nature. The bioassay of anti-inflammatory activity results revealed that compound **9** exhibited moderate inhibition of nitric oxide production in LPS-stimulated RAW264.7 cells, with an IC_{50} value of 22.63 ± 2.95 μM, and compounds **1**, **6** and **7** exhibited weak inhibition with IC_{50} values of 50.71 ± 8.41, 65.30 ± 7.21, and 31.84 ± 2.79 μM, respectively. The test of cytotoxic activities showed that penidioxolane C (**4**) exhibited moderate inhibition against K562, BEL-7402, SGC-7901, A549, and Hela cancer cells with IC_{50} values of 23.94 ± 0.11, 60.66 ± 0.13, 46.17 ± 0.17, 60.16 ± 0.26, and 59.30 ± 0.60 μM, respectively.

Supplementary Materials: The following supporting information can be downloaded at: https://www.mdpi.com/article/10.3390/md21020075/s1, Figure S1, S4, S6 and S8: The results of the DP4+ analysis of **1–4**; Figure S2: Key MOs in important transitions of most populated conformer of **1**; Tables S1–S8. Experimental and calculated ^{13}C NMR and ^{1}H NMR chemical shifts of **1–4**; Figures S3, S5 and S7: Linear regression analysis between experimental and calculated ^{13}C and ^{1}H NMR chemical shifts of isomers of **2–4**; Figures S9–S18: ^{1}H-^{1}H COSY, key HMBC, and NOESY correlations of **2**, **3**, **5**, **7** and **8**; Figures S19–S108: 1D, 2D NMR, HRESIMS, UV, and IR spectra of **1–9**.

Author Contributions: Conceptualization, Y.Z., F.L. and H.D.; methodology, Y.Z., F.L. and H.D.; software, Y.Z., F.L. and H.D.; validation, Y.Z., Z.W., W.C., W.Z., F.L. and H.D.; formal analysis, Y.Z., Z.W. and H.W.; investigation, Z.W., W.C. and W.Z.; resources, Y.Z., F.L. and H.D.; data curation, Z.W., W.C. and W.Z.; writing—original draft preparation, Y.Z., Z.W., H.W. and F.L.; writing—review and editing, Y.Z., F.L., H.W. and H.C.; visualization, Y.Z., Z.W. and F.L.; supervision, Y.Z., F.L. and H.D.;

project administration, Y.Z., F.L. and H.D.; funding acquisition, Y.Z. and H.D. All authors have read and agreed to the published version of the manuscript.

Funding: This work was financially supported by Natural Science Foundation of Hainan (322MS131), National Natural Science Foundation of China (41776093), and Financial Fund of the Ministry of Agriculture and Rural Affairs, P. R. China (NFZX2021).

Institutional Review Board Statement: Not applicable.

Informed Consent Statement: Not applicable.

Data Availability Statement: The authors confirm that the data supporting the findings of this study are available within the article and its supplementary materials.

Conflicts of Interest: The authors declare no conflict of interest.

References

1. Carroll, A.R.; Copp, B.R.; Davis, R.A.; Keyzers, R.A.; Prinsep, M.R. Marine natural products. *Nat. Prod. Rep.* **2022**, *39*, 1122–1171. [CrossRef] [PubMed]
2. Julianti, E.; Abrian, I.A.; Wibowo, M.S.; Azhari, M.; Tsurayya, N.; Izzati, F.; Juanssilfero, A.B.; Bayu, A.; Rahmawati, S.I.; Putra, M.Y. Secondary metabolites from marine-derived fungi and actinobacteria as potential sources of novel colorectal cancer drugs. *Mar. Drugs* **2022**, *20*, 67. [CrossRef] [PubMed]
3. Al-Saleem, M.S.M.; Hassan, W.H.B.; El Sayed, Z.I.; Abdel-Aal, M.M.; Abdel-Mageed, W.M.; Abdelsalam, E.; Abdelaziz, S. Metabolic profiling and in vitro assessment of the biological activities of the ethyl acetate extract of *Penicillium chrysogenum* "Endozoic of *Cliona* sp. Marine Sponge" from the Red Sea (Egypt). *Mar. Drugs* **2022**, *20*, 326. [CrossRef] [PubMed]
4. Xing, C.P.; Chen, D.; Xie, C.L.; Liu, Q.; Zhong, T.H.; Shao, Z.; Liu, G.; Luo, L.Z.; Yang, X.W. Anti-food allergic compounds from *Penicillium griseofulvum* MCCC 3A00225, a deep-sea-derived fungus. *Mar. Drugs* **2021**, *19*, 224. [CrossRef] [PubMed]
5. McCauley, E.P.; Piña, I.C.; Thompson, A.D.; Bashir, K.; Weinberg, M.; Kurz, S.L.; Crews, P. Highlights of marine natural products having parallel scaffolds found from marine-derived bacteria, sponges, and tunicates. *J. Antibiot.* **2020**, *73*, 504–525. [CrossRef]
6. Liu, L.; Zheng, Y.Y.; Shao, C.L.; Wang, C.Y. Metabolites from marine invertebrates and their symbiotic microorganisms: Molecular diversity discovery, mining, and application. *Mar. Life Sci. Tech.* **2019**, *1*, 60–94. [CrossRef]
7. Pang, K.L.; Overy, D.P.; Jones, E.B.G.; Calado, M.D.L.; Burgaud, G.; Walker, A.K.; Johnson, J.A.; Kerr, R.G.; Cha, H.J.; Bills, G.F. 'Marine fungi' and 'marine-derived fungi' in natural product chemistry research Toward a new consensual definition. *Fungal Biol. Rev.* **2016**, *30*, 163–175. [CrossRef]
8. Ding, B.; Yin, Y.; Zhang, F.L.; Li, Z.Y. Recovery and phylogenetic diversity of culturable fungi associated with marine sponges *Clathrina luteoculcitella* and *Holoxea* sp. in the South China Sea. *Mar. Biotechnol.* **2011**, *13*, 713–721. [CrossRef]
9. Ma, H.G.; Liu, Q.; Zhu, G.L.; Liu, H.S.; Zhu, W.M. Marine natural products sourced from marine-derived *Penicillium* fungi. *J. Asian Nat. Prod. Res.* **2016**, *18*, 92–115. [CrossRef]
10. Yang, X.L.; Liu, J.P.; Mei, J.H.; Jiang, R.; Tu, S.Z.; Deng, H.F.; Liu, J.; Yang, S.M.; Li, J. Origins, structures, and bioactivities of secondary metabolites from marine-derived *Penicillium* fungi. *Mini Rev. Med. Chem.* **2021**, *21*, 2000–2019. [CrossRef]
11. Yang, J.Y.; Tanga, M.M.; Chen, L.; Lai, X.Y.; Zhuo, X.; Zhou, X.M.; Chen, G.Y. Study on the secondary metabolites of endophytic *penicillium sclerotiorum* HLL113. *Chin. J. Org. Chem.* **2022**, *42*, 896–900. [CrossRef]
12. Wang, H.C.; Ke, T.Y.; Ko, Y.C.; Lin, J.J.; Chang, J.S.; Cheng, Y.B. Anti-inflammatory azaphilones from the edible alga-derived fungus *Penicillium sclerotiorum*. *Mar. Drugs* **2021**, *19*, 529. [CrossRef] [PubMed]
13. Gao, J.M.; Yang, S.X.; Qin, J.C. Azaphilonoids: Chemistry and biology. *Chem. Rev.* **2013**, *113*, 4755–4811. [CrossRef] [PubMed]
14. Wang, S.; Zeng, Y.B.; Yin, J.J.; Chang, W.J.; Zhao, X.L.; Mao, Y. Two new azaphilones from the marine-derived fungus *Penicillium sclerotiorum* E23Y-1A. *Phytochem. Lett.* **2022**, *47*, 76–80. [CrossRef]
15. Wang, Z.; Zeng, Y.B.; Zhao, W.B.; Dai, F.H.; Chang, W.J.; Lv, F. Structures and biological activities of brominated azaphilones produced by *Penicillium sclerotiorum* E23Y-1A. *Phytochem. Lett.* **2022**, *52*, 138–142. [CrossRef]
16. Gu, B.B.; Wu, J.; Tang, J.; Jiao, W.H.; Li, L.; Sun, F.; Wang, S.P.; Yang, F.; Lin, H.W. Azaphilone and isocoumarin derivatives from the sponge-derived fungus *Eupenicillium* sp. 6A-9. *Tetrahedron Lett.* **2018**, *59*, 3345–3348. [CrossRef]
17. Son, S.; Ko, S.K.; Kim, J.W.; Lee, J.K.; Jang, M.; Ryoo, I.J.; Hwang, G.J.; Kwon, M.C.; Shin, K.S.; Futamura, Y.; et al. Structures and biological activities of azaphilones produced by *Penicillium* sp. KCB11A109 from a ginseng field. *Phytochemistry* **2016**, *122*, 154–164. [CrossRef]
18. Pavesi, C.; Flon, V.; Mann, S.; Leleu, S.; Prado, S.; Franck, X. Biosynthesis of azaphilones: A review. *Nat. Prod. Rep.* **2021**, *38*, 1058–1071. [CrossRef]
19. Luo, X.; Lin, X.; Tao, H.; Wang, J.; Li, J.; Yang, B.; Zhou, X.; Liu, Y. Isochromophilones A–F, cytotoxic chloroazaphilones from the marine mangrove endophytic fungus *Diaporthe* sp. SCSIO 41011. *J. Nat. Prod.* **2018**, *81*, 934–941. [CrossRef]
20. Wang, C.-Y.; Hao, J.-D.; Ning, X.-Y.; Wu, J.-S.; Zhao, D.-L.; Kong, C.-J.; Shao, C.-L.; Wang, C.-Y. Penicilazaphilones D and E: Two new azaphilones from a sponge-derived strain of the fungus *Penicillium sclerotiorum*. *RSC Adv.* **2018**, *8*, 4348–4353. [CrossRef]

21. Yoshida, E.; Fujimoto, H.; Baba, M.; Yamazaki, M. Four new chlorinated azaphilones, helicusins A–D, closely related to 7-*epi*-sclerotiorin, from an ascomycetous fungus, *Talaromyces helices*. *Chem. Pharm. Bull.* **1995**, *43*, 1307–1310. [CrossRef]
22. Wolinski, K.; Hinton, J.F.; Pulay, P. Efficient implementation of the gauge-independent atomic orbital method for NMR chemical shift calculations. *J. Am. Chem. Soc.* **1990**, *112*, 8251–8260. [CrossRef]
23. Grimblat, N.; Zanardi, M.M.; Sarotti, A.M. Beyond DP4: An improved probability for the stereochemical assignment of isomeric compounds using quantum chemical calculations of NMR shifts. *J. Org. Chem.* **2015**, *80*, 12526–12534. [CrossRef] [PubMed]
24. Yamada, T.; Jinno, M.; Kikuchi, T.; Kajimoto, T.; Numata, A.; Tanaka, R. Three new azaphilones produced by a marine fish-derived *Chaetomium globosum*. *J. Antibiot.* **2012**, *65*, 413–417. [CrossRef] [PubMed]
25. Wu, Q.; Wu, C.; Long, H.; Chen, R.; Liu, D.; Proksch, P.; Guo, P.; Lin, W. Varioxiranols A−G and 19-O-methyl-22-methoxypre-shamixanthone, PKS and hybrid PKS-derived metabolites from a sponge-associated *Emericella variecolor* fungus. *J. Nat. Prod.* **2015**, *78*, 2461–2470. [CrossRef]
26. Iwai, T.; Kubota, T.; Kobayashi, J. Absolute configuration of amphidinin A. *J. Nat. Prod.* **2014**, *77*, 1541–1544. [CrossRef]
27. Matsumori, N.; Kaneno, D.; Murata, M.; Nakamura, H.; Tachibana, K. Stereochemical determination of acyclic structures based on carbon-proton spin-coupling constants. A method of configuration analysis for natural products. *J. Org. Chem.* **1999**, *64*, 866–876. [CrossRef]
28. Kim, S.M.; Son, S.; Kim, J.W.; Jeon, E.S.; Ko, S.K.; Ryoo, I.J.; Shin, K.S.; Hirota, H.; Takahashi, S.; Osada, H.; et al. Penidioxolanes A and B, 1, 3-dioxolane containing azaphilone derivatives from marine-derived *Penicillium* sp. KCB12C078. *Nat. Prod. Sci.* **2015**, *21*, 231–236. [CrossRef]
29. O'Boyle, N.M.; Vandermeersch, T.; Flynn, C.J.; Maguire, A.R.; Hutchison, G.R. Confab-systematic generation of diverse low-energy conformers. *J. Cheminformatics* **2011**, *3*, 8. [CrossRef]
30. Bannwarth, C.; Ehlert, S.; Grimme, S.J. GFN2-xTB—An accurate and broadly parametrized self-consistent tight-binding quantum chemical method with multipole electrostatics and density-dependent dispersion contributions. *Chem. Theory Comput.* **2019**, *15*, 1652–1671. [CrossRef]
31. Frisch, M.J.; Trucks, G.W.; Schlegel, H.B.; Scuseria, G.E.; Robb, M.A.; Cheeseman, J.R.; Scalmani, G.; Barone, V.; Petersson, G.A.; Nakatsuji, H.; et al. *Gaussian 16*, Revision C.01; Gaussian, Inc.: Wallingford, CT, USA, 2019.
32. Bruhn, T.; Schaumloffel, A.; Hemberger, Y.; Bringmann, G. SpecDis: Quantifying the comparison of calculated and experimental electronic circular dichroism spectra. *Chirality* **2013**, *25*, 243–249. [CrossRef] [PubMed]
33. Liu, F.Z.; Wang, H.; Li, W.; Yang, L.; Yang, J.L.; Yuan, J.Z.; Wei, Y.M.; Jiang, B.; Mei, W.L.; Dai, H.F. Filarones A and B, new anti-inflammatory dimeric 2-(2-phenethyl) chromones from agarwood of *Aquilaria filaria*. *Phytochem. Lett.* **2021**, *46*, 11–14. [CrossRef]
34. Mosmann, T.J. Rapid colorimetic assay for cellular growth and survival: Application to proliferation and cytotoxicity assays. *J. Immunol. Methods* **1983**, *65*, 53–63. [CrossRef] [PubMed]

marine drugs

MDPI

Article

Structurally Diverse Diterpenes from the South China Sea Soft Coral *Sarcophyton trocheliophorum*

Yu-Ting Song [1,2], Dan-Dan Yu [4], Ming-Zhi Su [4], Hui Luo [5], Jian-Guo Cao [2], Lin-Fu Liang [3,*], Fan Yang [2,*] and Yue-Wei Guo [1,*]

1 State Key Laboratory of Drug Research, Shanghai Institute of Materia Medica, Chinese Academy of Sciences, 555 Zu Chong Zhi Road, Zhangjiang Hi-Tech Park, Shanghai 201203, China
2 College of Life Sciences, Shanghai Normal University, 100 Guilin Road, Shanghai 200234, China
3 College of Materials Science and Engineering, Central South University of Forestry and Technology, 498 South Shaoshan Road, Changsha 410004, China
4 Shandong Laboratory of Yantai Drug Discovery, Bohai rim Advanced Research Institute for Drug Discovery, Yantai 264117, China
5 Key Laboratory of Zhanjiang for Research and Development Marine Microbial Resources in the Beibu Gulf Rim, Marine Biomedical Research Institute, Guangdong Medical University, Zhanjiang 524023, China
* Correspondence: lianglinfu@csuft.edu.cn (L.-F.L.); bayer@shnu.edu.cn (F.Y.); ywguo@simm.ac.cn (Y.-W.G.)

Abstract: The present investigation of the South China Sea soft coral *Sarcophyton trocheliophorum* resulted in the discovery of six new polyoxygenated diterpenes, namely sartrocheliols A–E (**1**, **3**, **5–8**) along with four known ones, **2**, **4**, **9**, and **10**. Based on extensive spectroscopic data analysis, sartrocheliol A (**1**) was identified as an uncommon capnosane diterpene, while sartrocheliols B–E (**3**, **5–8**) were established as cembrane diterpenes. They displayed diverse structural features not only at the distinctly different carbon frameworks but also at the various types of heterocycles, including the epoxide, γ-lactone, furan, and pyran rings. Moreover, their absolute configurations were determined by a combination of quantum mechanical-nuclear magnetic resonance (QM-NMR) approach, modified Mosher's method, and X-ray diffraction analysis. In the anti-tumor bioassay, compound **4** exhibited moderate cytotoxic activities against A549, H1975, MDA-MB-231, and H1299 cells with the IC_{50} values ranging from 26.3 to 47.9 μM.

Keywords: soft coral; *Sarcophyton trocheliophorum*; capnosane; cembrane; absolute configuration; anti-tumor activity

Citation: Song, Y.-T.; Yu, D.-D.; Su, M.-Z.; Luo, H.; Cao, J.-G.; Liang, L.-F.; Yang, F.; Guo, Y.-W. Structurally Diverse Diterpenes from the South China Sea Soft Coral *Sarcophyton trocheliophorum*. *Mar. Drugs* **2023**, *21*, 69. https://doi.org/10.3390/md21020069

Academic Editors: Vassilios Roussis, Bin-Gui Wang and Haofu Dai

Received: 15 December 2022
Revised: 17 January 2023
Accepted: 17 January 2023
Published: 20 January 2023

1. Introduction

A fairly large variety of bioactive secondary metabolites have been found in the soft corals of the genus *Sarcophyton* [1–4]. Over the past few decades, about 20 *Sarcophyton* species have been chemically investigated and more than 500 secondary metabolites have been identified. Based on their structural types, these metabolites could be classified into terpenes [2], steroids [3,5–7], quinones [8,9], prostaglandins [10,11], ceramides [12,13], and other miscellaneous compounds [8,14,15]. As reported in the literature, terpenes have dominated the chemical profile of the genus *Sarcophyton* [2,16]. Moreover, terpenes display rich structural diversities, which could be further categorized as sesquiterpenes [17–19], diterpenes [20–24], and biscembranoids [21,25–28]. Moreover, these metabolites exhibit a wide spectrum of biological activities, including anti-angiogenic [17], antimicrobial [27,29], cytotoxic [29,30], anti-inflammatory [27,31], immunomodulatory [25], antifouling [8,32], and neuroprotective [33,34] effects. The intriguing chemical and biological properties of terpenes have led to extensive attention from global researchers [35].

In the last decades, we systematically carried out chemical and biological studies on South China Sea marine fauna and flora [36]. As a result, numerous terpenes with complex structures, some of which possessed unprecedented carbon frameworks, were found from the genus *Sarcophyton* [20,37,38]. During our continuous research, we frequently encountered the soft coral *Sarcophyton trocheliophorum*, one productive species belonging to the above-mentioned genus. Previously, our group disclosed a vast array of diterpenes with four skeletons, and these metabolites showed a broad spectrum of pharmacological activities such as protein tyrosine phosphatase 1B (PTP1B) inhibitory, antitumor, antibacterial and neuroprotective activities [39]. It was interesting to notice that the chemical profiles of diterpenes from the title soft corals varied upon temporal variations and geographical distributions. The samples collected in Yalong Bay, Hainan Island, South China Sea in February 2006 yielded diterpenes with an unprecedented carbon framework [37] together with sarsolenane, capnosane, and cembrane skeletons [39]. However, the specimen collected in the same water but in May 2006 only yielded cembrane diterpenes [40]. Of more interest, the soft corals collected from another different region, Ximao Island, South China Sea, merely afforded the cembranoids [41]. These findings probably reflected the existence of different metabolic processes in different seasons and inhabiting environments, which needs to be further investigated.

In order to obtain more evidence to disclose that the diterpenoid profile of the title animals was influenced by temporal variations, we made a new collection of *S. trocheliophorum* from Ximao Island. In the current study, six new polyoxygenated diterpenes, namely sartrocheliols A–E (**1**, **3**, **5**–**8**), and four known related analogs **2**, **4**, **9**, and **10** were obtained (Figure 1). These diterpenes displayed two distinctly different carbon frameworks: capnosane and cembrane, the former of which has rarely been found in soft corals. Meanwhile, various types of heterocycles, including the epoxide, γ-lactone, furan, and pyran rings were incorporated in their macrocyclic skeletons. Herein, we report the detailed structural elucidation of these isolates from the title soft corals, especially the challenging stereochemistry determination of the new compounds, which was dissolved by a combination of quantum mechanical-nuclear magnetic resonance (QM-NMR) approach, modified Mosher's method, and X-ray diffraction analysis. In addition, their biological evaluations including cytotoxicity against a panel of cancel cells and antibacterial against an array of bacteria are described.

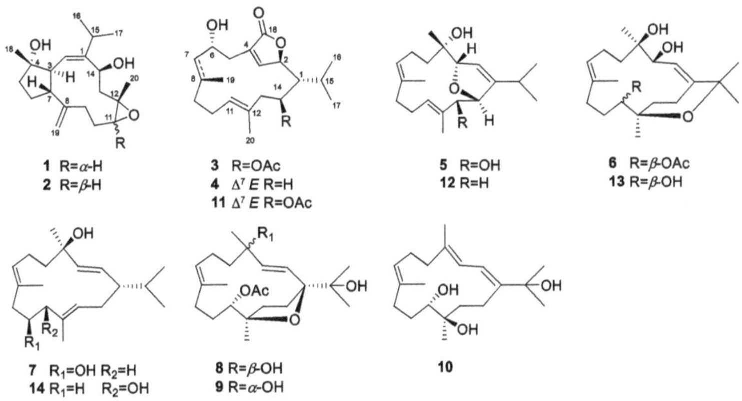

Figure 1. The chemical structures of compounds **1**–**14**.

2. Results and Discussion

The frozen animals were cut into pieces and extracted exhaustively with acetone. Then an Et_2O-soluble portion of the acetone extract was repeatedly chromatographed over silica gel, Sephadex LH-20, and RP-HPLC to yield six new compounds **1**, **3**, **5**–**8**, and four known analogs **2**, **4**, **9**, and **10** (Figure 1). The known compounds were readily identified as sarcophyolide D (**2**) [42], sarcophytonolide H (**4**) [43], sarcophytrol O (**9**) [44], sinulaflexiolide I (**10**) [45], respectively, based on the comparison of their NMR spectral and specific optical data with those reported in the literature. In our previous work [43], the absolute configuration of sarcophytonolide H (**4**) was established by the modified Mosher's method. In the present study, its crystals were obtained, which were suitable for X-ray diffraction experiment with Cu K_α (λ = 1.54178 Å) radiation. The X-ray diffraction analysis allowed the assignment of the absolute configuration of **4** as 1*R*,2*R*,6*R*,14*S* (Flack parameter: –0.03 (9)) (Figure 2, Table S2, CCDC 2205721), which was consistent with our previous study.

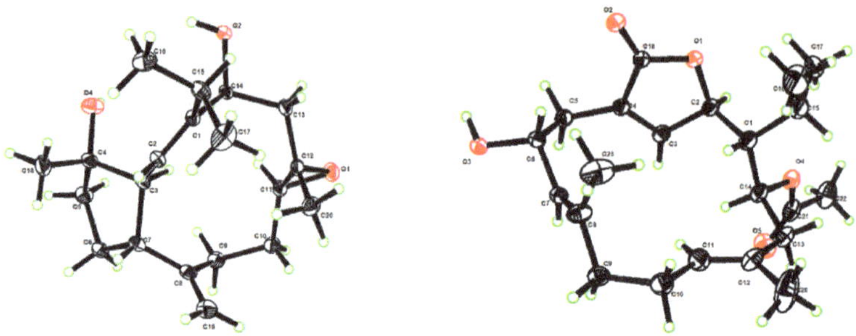

Figure 2. Perspective ORTEP drawing of the X-ray structures of compounds **1** (**left**) and **4** (**right**) (displacement ellipsoids are drawn at the 50% probability level).

Sartrocheliol A (**1**) was obtained as optically active colorless crystals. Its molecular formula $C_{20}H_{32}O_3$ was deduced from the protonated molecule peak at *m/z* 343.2243 ([M + Na]$^+$, calcd. for $C_{20}H_{32}O_3Na$, 343.2244) in the HRESIMS spectrum, implying five degrees of unsaturation. The IR spectrum indicated the presence of hydroxyl (ν_{max} 3441 cm^{-1}) and olefinic (ν_{max} 1660, 910 cm^{-1}) groups. The ^{1}H and ^{13}C NMR spectra displayed the signals of a trisubstituted double bond (δ_H 5.14 (1H, d, *J* = 12.6 Hz, H-2), δ_C 149.90 (qC, C-1), 125.85 (CH, C-2)), a terminal double bond (δ_H 4.88 (1H, s, H-19a), 4.69 (1H, s, H-19b), δ_C 148.54 (qC, C-8), 112.26 (CH_2, C-19)), an epoxide (δ_H 2.89 (1H, dd, *J* = 10.6, 2.9 Hz, H-11), δ_C 62.28 (CH, C-11), 59.06 (qC, C-12)), an oxygenated methine (δ_H 4.88 (1H, dd, *J* = 11.2, 5.1 Hz, H-14), δ_C 69.34 (CH, C-14)), and an oxygenated carbon (δ_C 81.64 (C, C-4)) (Tables 1 and 2). As revealed by the ^{1}H and ^{13}C NMR data, there were two double bonds and one epoxide, accounting for three degrees of unsaturation. The remaining two degrees of unsaturation were due to the presence of two rings in the molecule. Considering the co-isolated secondary metabolite sarcophyolide D (**2**) [42], compound **1** was likely a capnosane-type diterpene. Indeed, the NMR data of **1** was almost identical to those of **2**, except for the chemical shift of C-11 (δ_C 62.28 for **1** vs. δ_C 59.54 for **2**). The interpretation of ^{1}H–^{1}H COSY and HMBC spectra (Figure 3) indicated they shared the same gross structure. Analysis of the NOESY spectrum (Figure S6) revealed that there was lack of correlation between H-11 (δ_H 2.89) and H_3-20 (δ_H 1.12), which revealed the *trans*-orientation of H-11 and H_3-20 in **1**, the orientation of which was different from that of the isolate **2**. In order to check the proposed structure as well as establish the absolute configuration of 1, a suitable single crystal was obtained in MeOH after many attempts. A successful performance of X-ray crystallography study using Cu K_α (λ = 1.54178 Å) radiation firmly confirmed the structure of **1** and determined its absolute configuration ambiguously as 3*S*,4*S*,7*R*,11*S*,12*S*,14*S* (Flack parameter: 0.05 (6)) (Figure 2, Table S1, CCDC 2196207). Hereto, the structure of **1** was established, as shown in Figure 1.

Sartrocheliol B (**3**) was isolated as optically active colorless oil. Its molecular formula $C_{22}H_{34}O_5$ was established by the HRESIMS protonated molecule peak at *m/z* 379.2482 ([M + H]$^+$, calcd. for $C_{22}H_{35}O_5$, 379.2479), which was indicative of six degrees of unsaturation. The IR spectrum of **3** suggested the presence of lactone (ν_{max} 1759 cm^{-1}), ester (ν_{max} 1736 cm^{-1}) and hydroxyl (ν_{max} 3442 cm^{-1}) groups, while a strong UV absorption at λ_{max} 231 nm (logε 3.83) suggested the presence of α,β-unsaturated γ-lactone group [43]. According to the ^{1}H and ^{13}C NMR data of **3** (Tables 1 and 2), five degrees of unsaturation were attributed to one α,β-unsaturated γ-lactone group (δ_H 4.95 (1H, d, *J* = 10.7 Hz, H-2),

7.44 (1H, t, *J* = 1.8 Hz, H-3), δ_C 81.19 (CH, C-2), 151.06 (CH, C-3), 131.39 (qC, C-4), 172.97 (qC, C-18)), one trisubstituted double bond (δ_H 5.05 (1H, t, *J* = 7.7 Hz, H-11), δ_C 129.20 (CH, C-11), 130.80 (qC, C-12)) and one acetyl group (δ_H 2.09 (3H, s), δ_C 21.35 (CH_3), 171.13 (qC)). The remaining one degree of unsaturation implied the monocyclic nature of this molecule. Considering the co-isolated cembranoids **4**, **9** and **10**, metabolite **3** was likely a cembrane-type diterpene. Through detailed literature reviews on diterpenes from the genus *Sarcophyton*, the above-mentioned structural features were reminiscent of previously reported sarcophytonolide H (**11**) [43], a cembrane from South China Sea soft coral *Sarcophyton latum*. Our interpretation of the 2D NMR spectra (Figure 3) suggested they differed by the lack of the double bond Δ^7, which was supported by the significantly up-field shifted chemical shifts of C-7 and C-8 ($\Delta\delta_C$ ca. 80 and 110 ppm, respectively). Similar patterns of NOE correlations in the NOESY spectra of **3** and **11**, especially for the key NOE cross-peaks of H-2 (δ_H 4.95)/H-14 (δ_H 5.02), H-2/H_3-16 (δ_H 1.11), H-3 (δ_H 7.44)/H-5 (δ_H 2.45), and H-11 (δ_H 5.05)/H-13 (δ_H 2.20), indicated they shared the same configurations for the olefinic bonds Δ^3 and Δ^{11} and the chiral centers C-1, C-2, and C-14 (Figure 4). The clear NOE correlation of H-6 (δ_H 4.22)/H_3-19 (δ_H 0.77) suggested that H-6 and H_3-19 possessed the same orientation. Due to the lack of NOE correlations between the two sets: **a**. H-6 and H_3-19 and **b**. H-1, H-,2 and H-14, the absolute configurations of these chiral centers corresponding to the above-mentioned two sets of protons were determined by the modified Mosher's method and CD spectrum, respectively. Treatment of **3** with (*R*)- and (*S*)-α-methoxy-α-trifluoromethylphenyl acetyl chlorides (MTPA-Cl) in dry pyridine successfully afforded the (*S*)- and (*R*)-MTPA ester derivatives **3s** and **3r**, respectively. The distribution pattern of observed $\Delta\delta_{H(S\text{-}R)}$ values (Figure 5) established the absolute configuration *S* for C-6 in **3**. Considering the correlation between H-6 and H_3-19, the absolute configuration of C-8 could be determined as *S*. It might be worth pointing out that the absolute configuration of sarcophytonolide H (**4**) had already been ambiguously confirmed as 1*R*,2*R*,6*R*,14*S* by the single crystal X-ray diffraction experiment in this study (Figure 2). The CD spectrum of **3** displayed the Cotton effect resembling that of co-isolated **4** (Figure 6), suggesting they shared the same absolute configuration *R* for the chrial carbon C-2 of the chromophore α,β-unsaturated γ-lactone. With the relationships of H-1, H-2, and H-14 in hand, the absolute configurations of C-1 and C-14 could be assigned as *R* and *S*, respectively. Consequently, the absolute configuration of **3** could be established as 1*R*,2*R*,6*S*,8*S*,14*S*.

Table 1. ^{1}H NMR spectroscopic data of sartrocheliols A–E (**1**, **3**, **5–8**) in $CDCl_3$.

No.	**1** [a]	**3** [b]	**5** [b]	**6** [b]	**7** [b]	**8** [a]
1		1.55 (d, 10.7)			1.54 (br s)	
2	5.14 (d, 10.6)	4.95 (d, 10.7)	5.51 (d, 1.4)	5.42 (m)	5.51 (dd, 15.6, 9.6)	5.56 (d, 15.6)
3	2.84 (t, 10.6)	7.44 (t, 1.8)	4.66 (dq, 5.5, 1.4)	4.40 (br s)	5.63 (d, 15.6)	6.13 (d, 15.9)
4						
5	1.84 (m)	2.75 (dt, 13.2, 1.2)	1.83 (m)	1.94 (m)	1.89 (ddd, 13.4, 8.6, 1.9)	1.87 (m)
	1.78 (m)	2.45 (dd, 13.3, 10.5)	1.53 (m)	1.76 (m)	1.58 (m)	1.69 (m)
6	1.66 (m)	4.22 (t, 9.8)	2.34 (m)	2.06 (m)	2.24 (m)	2.64 (m)
			1.92 (m)			
7	2.48 (t, 7.1)	1.63 (m)	5.22 (m)	5.24 (t, 6.3)	5.11 (d, 7.8)	5.43 (dd, 10.6, 4.0)
		1.35 (m)				
8		1.29 (m)				
9	2.33 (m)	1.36 (m)	2.10 (m)	1.96 (m)	2.25 (m)	2.00 (m)
	2.00 (t, 13.5)	1.26 (m)		1.67 (m)	2.12 (m)	
10	2.21 (m)	2.08 (m)	2.26 (m)	1.86 (m)	4.07 (d, 10,9)	1.90 (m)
	1.46 (m)		2.11 (m)	1.43 (m)		1.55 (m)
11	2.89 (dd, 10.6, 2.9)	5.05 (t, 7.7)	5.32 (m)	5.04 (d, 10.3)	2.09 (m)	5.08 (d, 9.4)
12						
13	2.25 (m)	2.26 (dd, 13.5, 3.0)	4.29 (t, 4.3)	1.94 (m)	5.24 (t, 7.7)	1.78 (m)
	1.65 (m)	2.20 (dd, 13.5, 10.9)		1.61 (m)		1.59 (m)
14	4.88 (dd, 11.2, 5.1)	5.02 (ddd, 10.8, 4.2, 1.1)	4.84 (td, 5.0, 1.6)	3.04 (m)	2.07 (m)	2.33 (td, 11.8, 7.4)
				2.38 (m)		1.68 (m)
15	2.48 (m)	2.22 (m)	2.65 (m)		1.70 (m)	
16	1.12 (d, 6.5)	1.11 (d, 6.6)	1.16 (d, 6.8)	1.37 (s)	0.96 (d, 6.6)	1.11 (s)
17	1.16 (d, 6.5)	1.12 (d, 6.6)	1.08 (d, 6.9)	1.30 (s)	0.84 (d, 6.7)	1.13 (s)
18	1.16 (s)		1.00 (s)	1.20 (s)	1.30 (s)	1.35 (s)
19	4.88 (s)	0.77 (d, 6.0)	1.56 (s)	1.56 (s)	1.54 (s)	1.71 (s)
	4.69 (s)					
20	1.12 (s)	1.67 (s)	1.56 (s)	1.10 (s)	1.57 (m)	1.17 (s)
OAc		2.09 (s)		2.08 (s)		2.05 (s)

[a] 500 MHz. [b] 600 MHz.

Table 2. ^{13}C NMR spectroscopic data (125 MHz, $CDCl_3$) of sartrocheliols A–E (**1**, **3**, **5–8**).

No.	1	3	5	6	7	8
1	149.90, qC	49.67, CH	150.34, qC	146.95, qC	51.96, CH	91.52, qC
2	125.85, CH	81.19, CH	121.20, CH	121.34, CH	122.10, CH	129.84, CH
3	51.08, CH	151.06, CH	87.63, CH	73.16, CH	141.04, CH	137.41, CH
4	81.64, qC	131.39, qC	74.93, qC	75.86, qC	74.20, qC	72.47, qC
5	41.04, CH_2	37.42, CH_2	39.37, CH_2	39.18, CH_2	44.21, CH_2	41.64, CH_2
6	25.90, CH_2	66.40, CH	22.57, CH_2	21.75, CH_2	23.66, CH_2	22.90, CH_2
7	54.09, CH	47.19, CH_2	128.74, CH	127.40, CH	127.31, CH	130.06, CH
8	148.54, qC	28.31, CH	133.16, qC	134.03, qC	130.05, qC	132.11, qC
9	24.65, CH_2	37.25, CH_2	39.64, CH_2	33.90, CH_2	46.88, CH_2	34.86, CH_2
10	28.07, CH_2	24.31, CH_2	25.00, CH_2	25.03, CH_2	68.82, CH	27.37, CH_2
11	62.28, CH	129.20, CH	130.08, CH	72.70, CH	38.96, CH_2	77.36, CH
12	59.06, qC	130.80, qC	132.19, qC	74.15, qC	132.94, qC	84.58, qC
13	45.46, CH_2	41.43, CH_2	79.67, CH	29.33, CH_2	128.47, CH	35.80, CH_2
14	69.34, CH	73.92, CH	86.05, CH	20.95, CH_2	23.68, CH_2	30.76, CH_2
15	27.09, CH	25.81, CH	27.43, CH	74.60, qC	29.13, CH	72.84, qC
16	21.23, CH_3	24.93, CH_3	21.11, CH_3	30.42, CH_3	21.15, CH_3	26.01, CH_3
17	23.55, CH_3	18.74, CH_3	22.57, CH_3	28.85, CH_3	21.85, CH_3	24.62, CH_3
18	27.70, CH_3	172.97, qC	23.36, CH_3	21.72, CH_3	29.17, CH_3	29.00, CH_3
19	112.26, CH_2	18.25, CH_3	15.84, CH_3	17.41, CH_3	15.52, CH_3	16.33, CH_3
20	23.05, CH_3	18.47, CH_3	12.89, CH_3	23.63, CH_3	15.09, CH_3	20.69, CH_3
OAc		171.13, qC		170.47, qC		170.17, qC
		21.35, CH_3		21.75, CH_3		21.42, CH_3

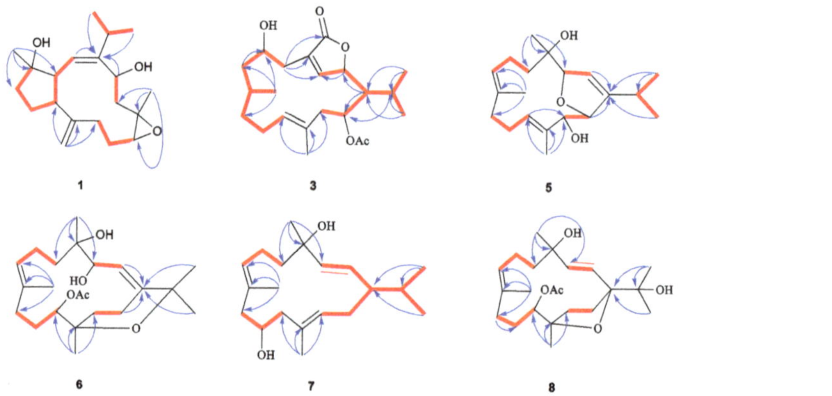

Figure 3. The selected key ^{1}H–^{1}H COSY (red lines) and HMBC (blue arrows, from ^{1}H to ^{13}C) correlations of compounds **1**, **3**, **5–8**.

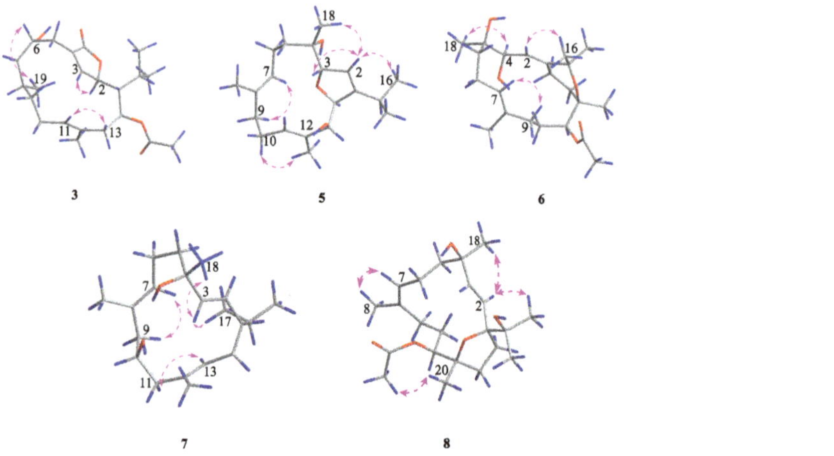

Figure 4. The selected key NOESY (pink dashed lines, from ^{1}H to ^{1}H) correlations of compounds **3**, **5–8**.

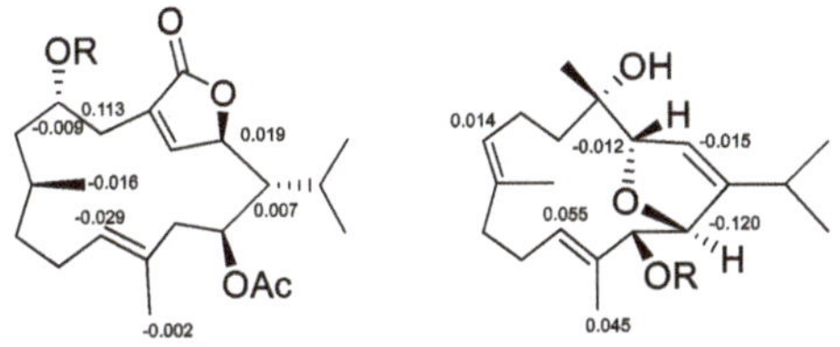

Figure 5. $\Delta\delta$ values (δ_S–δ_R) (ppm) for (*S*)-and (*R*)-MTPA esters of **3** and **5**.

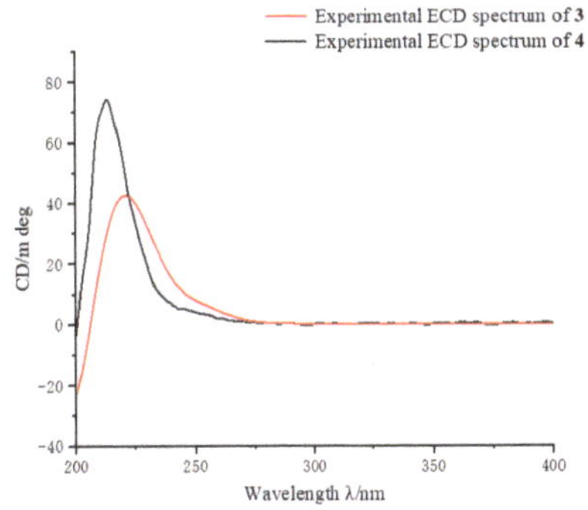

Figure 6. ECD curves of compounds **3** (up) and **4** (down).

The protonated molecule peak at m/z 343.2237 ($[M + Na]^+$, calcd. for $C_{20}H_{32}O_3Na$, 343.2244) displayed in the HRESIMS spectrum of sartrocheliol C (**5**) revealed that compound **5** had the molecular formula $C_{20}H_{32}O_3$, demonstrating the presence of an additional oxygen atom with respect to that of a dihydrofuran cembranoid **12** [46]. The IR spectrum of **5** showed the presence of olefinic (ν_{max} 3726 cm^{-1}) and hydroxyl (ν_{max} 3441 cm^{-1}) groups. A careful analysis of its NMR spectra revealed that the NMR spectroscopic features of **5** (Tables 1 and 2) highly resembled those of **12**. In fact, the main difference between compounds **5** and **12** was that the CH_2-13 in **12** was hydroxylated in **5**. The presence of a hydroxyl group at C-13 was supported by the dramatically down-field shifted carbon chemical shift ($\Delta\delta_C$ 56 ppm), and further confirmed by the diagnostic HMBC correlations of from H-11 (δ_H 5.32) to C-13 (δ_C 79.67) and from H_3-20 (δ_H 1.56) to C-13 (Figure 3). The cross-peak of H_3-18 (δ_H 1.00)/H-2 (δ_H 5.51) observed in the NOESY spectrum of **5** (Figure 4) together with the small coupling constant (1.4 Hz) between H-2 and H-3 (δ_H 4.66) suggested the same orientation of H_3-18 and H-3. Moreover, the coupling constant (4.6 Hz) between H-13 (δ_H 4.29) and H-14 (δ_H 4.84) indicated these two protons were *cis*-orientated. Whereas the lack of the NOE correlation between H-3 and H-14 revealed the *trans*-orientation of these two protons of the 2,5-dihydrofuran ring. Thus, the relative configuration of **5** could be assigned as 3*R**,4*R**,13*S**,14*S** based on the extensive analysis of the NOESY spectrum. Meanwhile, the relative configuration of **5** could also be elucidated via QM-NMR protocol by using the DP4+ method, which has become one of the most popular and reliable methods to find the most likely structure from a set of putative candidates [34,47]. Four possible isomers (2*R**,3*R**,13*R**,14*S**)-**5a**, (2*R**,3*R**,13*S**,14*S**)-**5b**, (2*R**,3*S**,13*R**,14*S**)-**5c** and (2*R**,3*S**,13*S**,14*R**)-**5d** (Table S3) were subjected to QM-NMR calculations. As a result, the experimental NMR data of compound **5** gave the best match for **5b**, with 100% probability (Figure 7, Table S4). Herein, the assigned relative configuration 3*R**,4*R**,13*S**,14*S** was consistent with the observations deduced from the NOESY spectrum. As there was a secondary hydroxyl group at C-13, the modified Mosher's method was applied. The resulting distribution pattern of observed $\Delta\delta_{H(S-R)}$ values (Figure 5) established the absolute configuration *R* for C-13 in **5**. Subsequently, the absolute configuration of **5** could be assigned as 3*S*,4*S*,13*R*,14*R*.

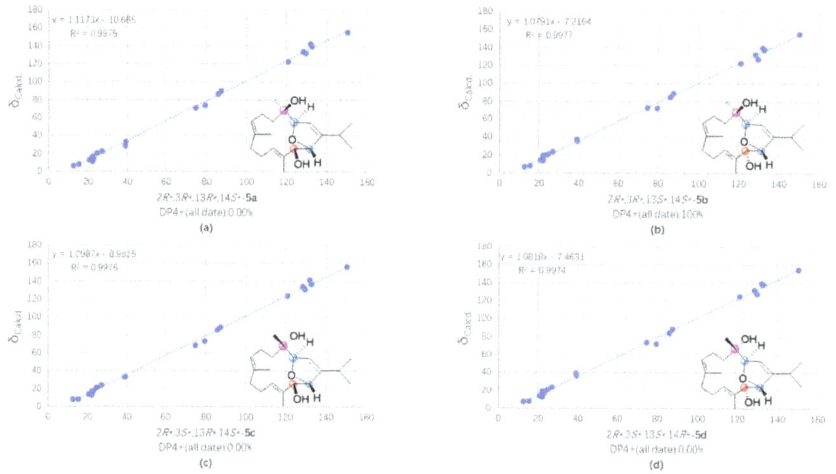

Figure 7. Regression analysis of experimental vs. calculated ^{13}C NMR chemical shifts of (**a**) (2*R**,3*R**,13*R**,14*S**)-**5a**, (**b**) (2*R**,3*R**,13*S**,14*S**)-**5b**, (**c**) (2*R**,3*S**,13*R**,14*S**)-**5c**, and (**d**) (2*R**,3*S**,13*S**,14*R**)-**5d** at the PCM/mPW1PW91/6-31 + G** level, using the DP4+ method.

The HRESIMS spectrum of sartrocheliol D (**6**) displayed a protonated molecule peak at *m/z* 403.2459 ([M + Na]$^+$, calcd. for $C_{22}H_{36}O_5Na$, 403.2455), suggesting that **6** possessed the molecular formula $C_{22}H_{36}O_5$. Thus, five degrees of unsaturation were determined for **6**. The NMR data (Tables 1 and 2) revealed the presence of two trisubstituted double bonds (δ_H 5.42 (1H, m, H-2), δ_C 146.95 (qC, C-1), 121.34 (CH, C-2); δ_H 5.24 (1H, t, J = 6.3 Hz, H-7), δ_C 127.40 (CH, C-7), 134.03 (qC, C-8)), one acetyl group (δ_H 2.08 (3H, s), δ_C 21.75 (CH_3), 170.47 (qC)), two oxygenated methines (δ_H 4.40 (1H, br s, H-3), 5.04 (1H, d, J = 10.3 Hz, H-11), δ_C 72.70 (CH, C-11), 73.16 (CH, C-3)), and three oxygenated quaternary carbons (δ_C 75.86 (qC, C-4), 74.15 (qC, C-12), 74.60 (qC, C-15)), which accounted for three degrees of unsaturation. The remaining two degrees of unsaturation strongly indicated one macrocyclic carbon skeleton fused with an oxacycle. Careful analysis of the 2D NMR spectrum (Figure 3) of **6** revealed this compound had almost the same gross bicyclic framework of sarcophytrol R (**13**) [44] except the hydroxyl group at C-11 in **13** was acetylated in **6**. Due to the acetylation, the chemical shifts of H-11 and C-11 shifted down-field ($\Delta\delta_H$ 1.5 ppm, $\Delta\delta_C$ 3.2 ppm, respectively). It might be worth pointing out that the lists of ^{1}H and ^{13}C NMR data of sarcophytrols R and S in the reference [44] were exchanged inadvertently by the authors. The high similarity of the ^{1}H and ^{13}C NMR data as well as similar patterns of NOE correlations of compounds **6** and **13** suggested they shared the same relative configuration 3*S**,4*R**,11*S**,12*R**. With the relative configuration in hand, we tried to use the Mosher's method to establish the absolute configuration of this compound. However, to our disappointment, the left compound after bioassay was degraded although kept in a fridge.

Sartrocheliol E (**7**) was isolated as colorless oil, and its molecular formula was assigned as $C_{20}H_{34}O_2$ by protonated molecule peak at *m/z* 329.2448 ([M + Na]$^+$, calcd. for $C_{20}H_{34}O_2Na$, 329.2451) in the HRESIMS spectrum. Its ^{13}C NMR data (Table 2), in combination with the DEPT and HSQC spectra, allowed the identification of 20 carbon resonances, involving six olefinic carbons (δ_C 141.04, 132.94, 130.05, 128.47, 127.31, 122.10), two oxygenated carbons (δ_C 74.20, 68.82), and five methyl carbons (δ_C 29.17, 21.85, 21.15, 15.52, 15.09). The presence of six olefinic carbons was attributed to three double bonds, which accounted for three degrees of unsaturation, indicating that **7** had a monocyclic carbon framework. In fact, the NMR spectroscopic characters of **7** were reminiscent of those of cembrendiol (**14**) [48]. Careful comparison of their NMR data disclosed their structures differed at the position of the secondary hydroxyl substituent. The secondary hydroxyl group substituted at C-10 in **7** was supported by the consecutive proton system extending from H_2-9 to H_2-11 through H-10 in the $^1H-^1H$ COSY spectrum (Figure 3). Due to the lack of the evidence regarding the orientations of H-10 and H_3-18, it was hard to figure out the relative configurations of C-4 and C-10. In order to solve this problem, the QM-NMR method was applied (Table S5). By means of this approach, the experimentally observed NMR data for compound **7** gave the best match with the 1*S**,4*R**,10*S** isomer (>90% probability) (Figure 8, Table S6). Thus, the relative configuration of compound **7** was determined as 1*S**,4*R**,10*S**. Unfortunately, the application of Mosher's reaction failed, probably due to the insufficient amounts left after an array of bioassays.

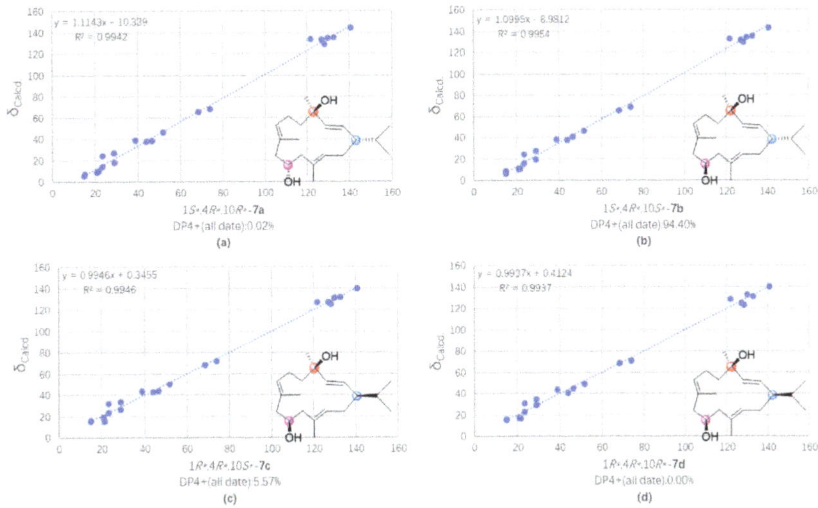

Figure 8. Regression analysis of experimental vs. calculated ^{13}C NMR chemical shifts of (**a**) (1*S**,4*R**,10*R**)-**7a**, (**b**) (1*S**,4*R**,10*S**)-**7b**, (**c**) (1*R**,4*R**,10*S**)-**7c**, and (**d**) (1*R**,4*R**,10*R**)-**7d** at the PCM/mPW1PW91/6-31 + G** level, using the DP4+ method.

The protonated molecule peak at *m/z* 403.2446 ($[M + Na]^+$, calcd. for $C_{22}H_{36}O_5Na$, 403.2455) in the HRESIMS spectrum of sartrocheliol F (**8**) suggested compound **8** and co-isolate **9** [44] had the same molecular formula $C_{22}H_{36}O_5$. Moreover, the NMR data of **8** (Tables 1 and 2) highly resembled those of **9**, implying that they shared the same gross structure. The distinct difference was found as the chemical shift of C-4, which was δ_C 72.47 ppm in **8** whereas δ_C 74.49 ppm in **9**, disclosing the reverse configuration of the hydroxyl group at C-4. This reversion was further deduced from the NOE interactions between H_3-18 and H-2 and between H-2 and H_3-16 (Figure 4). As the absolute configuration of the deacetylation derivative of **9** was determined as 1*S*,4*S*,11*S*,12*R* by the Mosher's method [44], the absolute configuration of **8** was consequently established as 1*S*,4*R*,11*S*,12*R*.

Although chemical investigations of the soft coral *S. trocheliophorum* have been well documented in the literature, the present study of this species collected from Ximao Island provided further intriguing results. In the current study, six new cembranoids diterpenoids, sartrocheliols A–E (**1**, **3**, **5–8**), along with four known related ones (**2**, **4**, **9**, and **10**) were obtained. Among them, compound **1** was a capnosane diterpene, while others were cembrane diterpenes. They displayed diverse structural features not only at the distinctly different carbon frameworks but also at the various types of heterocycles, including the epoxide, γ-lactone, furan, and pyran rings. Compared with the previous research of the Ximao Island specimen [41], the major difference was the discovery of a capnosane diterpene in this study. This observation further indicated the possible impact of temporal variations on the different metabolic processes for the title soft corals.

The anti-tumor effects of all the ten secondary metabolites were evaluated against a list of tumor cells including A549 (human lung cancer cell), H1975 (human lung adenocarcinoma cell), MDA-MB-231 (human breast cancer cell) and H1299 (human non-small cell lung cancer cell). The results showed compound **4** displayed moderate cytotoxic activities against these four cells with the IC_{50} values of 47.9, 26.3, 44.7, 33.1 μM, respectively, while **6** only showed moderate cytotoxicity against H1975 (IC_{50} = 40.4 μM). Further, we also conducted molecular interaction experiments on all the compounds, looking for compounds that have the potential to bind to BRD4 and ROR1 anti-tumor targets, respectively. Unfortunately, we did not obtain satisfactory results. In addition, these compounds were tested for their antibacterial activities against a vast array of bacteria including the human pathogens *Staphylococcus aureus* ATCC27154, *Enterococcus faecium*, *Escherichia coli* ATCC25922, *Enterobacter cloacae* ZR042, *Enterobacter hormaechei* 2R043, *Pseudomonas aeruginosa* ATCC10145, methicillin-resistant *Staphylococcus aureus* (MRSA), and *Candida albicans* ATCC76485 and the marine strains *Streptococcus parauberis* KSP28, *Streptococcus parauberis* SPOF3K, *Lactococcus garvieae* MP5245, *Aeromonas salmonicida* AS42, Phoyobacterium damselae FP2244, *Pseudomonas fulva* ZXM181, *Photobacterium halotolerans* LMG22194T. To our disappointment, all of them were judged as inactive. Other bioassays such as neuroprotective and anti-inflammatory are currently on the way.

3. Materials and Methods

3.1. Subsection

Melting points were measured on an X-4 digital micromelting point apparatus. The X-ray measurements were made on a Bruker D8 Venture X-ray diffractometer with Cu Kα radiation (Bruker Biospin AG, Fällanden, Germany). IR spectra were recorded on a Nicolet iS50 spectrometer (Thermo Fisher Scientific, Madison, WI, USA). Optical rotations were measured on a PerkinElmer 241MC polarimeter (PerkinElmer, Fremont, CA, USA). CD & UV spectra were measured on a JASCO J-810 instrument (JASCO Corporation, Tokyo, Japan). ^{1}H and ^{13}C NMR spectra were acquired on a Bruker AVANCE III 400 and 600 MHz spectrometer (Bruker Biospin AG, Fällanden, Germany). Chemical shifts were reported with the residual $CHCl_3$ (δ_H 7.26; δ_C 77.16) as the internal standard for ^{1}H and ^{13}C NMR spectra. The LREIMS and HREIMS data were recorded on a Finnigan-MAT-95 mass spectrometer (Finnigan-MAT, San Jose, CA, USA). HRESIMS spectra were recorded on Agilent G6250 Q-TOF (Agilent, Santa Clara, CA, USA). Commercial silica gel (Qingdao Haiyang Chemical Co., Ltd., Qingdao, China, 200–300 mesh, 300–400 mesh) was used for column chromatography, and precoated silica gel GF254 plates (Sinopharm Chemical Reagent Co., Shanghai, China) were used for analytical TLC. Sephadex LH-20 (Pharmacia, Piscataway, NJ, USA) was also used for column chromatography. Reversed-phase (RP) HPLC was performed on an Agilent 1260 series liquid chromatography equipped with a DAD G1315D detector at 210 nm (Agilent, Santa Clara, CA, USA). An Agilent semi-preparative XDB-C18 column (5 μm, 250 × 9.4 mm) was employed for the purification. All solvents used for column chromatography and HPLC were of analytical grade (Shanghai Chemical Reagents Co., Ltd., Shanghai, China) and chromatographic grade (Dikma Technologies Inc., Foothill Ranch, CA, USA), respectively.

3.2. Animal Material

The soft coral *Sarcophyton trocheliophorum* was collected by scuba at a depth of 15 m in May 2018 in Ximao Island, Hainan Province, China. The animal material was identified by Prof. Xiu-Bao Li from Hainan University. A voucher specimen (No. 18XD-19) is available for inspection at the Shanghai Institute of Materia Medica, CAS.

3.3. Extraction and Isolation

The frozen animals (551 g, dry weight) were cut into pieces and extracted exhaustively with acetone at room temperature (3 × 3 L, 30 min in ultrasonic bath). The organic extract was evaporated to give a brown residue (80 g), which was partitioned between Et_2O and H_2O. The Et_2O solution was concentrated under reduced pressure to give a dark brown residue (55.3 g), which was fractionated by gradient silica gel (200–300 mesh) column chromatography (0 $\rightarrow$ 100% Et_2O in petroleum ether (PE)), yielding seven fractions (A–G). Fractions E and F were subjected to a column of Sephadex LH-20 eluted with CH_2Cl_2 and PE/CH_2Cl_2/MeOH (2:1:1) to remove the fatty acids and give ten subfractions (EA–EF and FA–FF), respectively. The subfraction EE was purified by semi-preparative HPLC (60% $\rightarrow$ 100% MeCN in 20 min, 2.5 mL/min), yielding compounds **5** (3.6 mg; t_R 10.0 min) and **10** (1.5 mg; t_R 12.0 min). FCA-FCC was got from subfraction FC by silica gel column chromatography (300—400 mesh, PE/Et_2O (100:1 $\rightarrow$ 70:1)). The subfraction FCA afforded compounds **2** (1.0 mg; t_R 7.1 min), **6** (0.9 mg; t_R 17.0 min), **8** (0.9 mg; t_R 18.9 min) and **9** (2.0 mg; t_R 12.0 min) through semi-preparative HPLC (70% MeCN, 2.5 mL/min). While subfraction FCB gave compounds **1** (1.3 mg; t_R 7.0 min) and **7** (1.6 mg; t_R 16.0 min), through semi-preparative HPLC (70% MeCN, 2.5 mL/min) as well. The subfraction FD afforded compounds **3** (3.2 mg; t_R 6.2 min) and **4** (8.0 mg; t_R 5.5 min) through semi-preparative HPLC (70% MeCN, 2.5 mL/min).

3.4. Spectroscopic Data of Compounds

Sartrocheliol A (**1**): colorless crystals; $[\alpha]_D^{20}$ +250.0 (*c* 0.05, MeOH); IR (KBr) ν_{max}: 3441, 2925, 1384 cm^{-1}; ^{1}H and ^{13}C NMR data, see Tables 1 and 2; HRESIMS *m/z* 343.2243 [M + Na]$^+$ (calcd. for $C_{20}H_{32}NaO_3$, 343.2244).

Sartrocheliol B (**3**): colorless oil; $[\alpha]_D^{20}$ +12.7 (*c* 0.05, MeOH); UV (MeOH) λ_{max} (log ε) 240 (3.23) nm; CD (MeOH) λ_{max} ($\Delta\varepsilon$) 240 (+4.06), 285 (−1.92) nm; IR (KBr) ν_{max}: 3442, 2932, 1759, 1736, 1234, 1047, 1023 cm^{-1}; ^{1}H and ^{13}C NMR data, see Tables 1 and 2; HRESIMS *m/z* 379.2479 [M + H]$^+$ (calcd. for $C_{22}H_{34}O_5$, 379.2482).

Sartrocheliol C (**5**): colorless oil; $[\alpha]_D^{20}$ -12.7 (*c* 0.1, MeOH); IR (KBr) ν_{max}: 3726, 3624, 3441, 2960, 1384, 1087, 1032 cm^{-1}; ^{1}H and ^{13}C NMR data, see Tables 1 and 2; HRESIMS *m/z* 343.2237 [M + Na]$^+$ (calcd. for $C_{20}H_{32}NaO_3$, 343.2244).

Sartrocheliol D (**6**): colorless oil; $[\alpha]_D^{20}$ -20.0 (*c* 0.05, MeOH); IR (KBr) ν_{max}: 3443, 2927, 1384, 1038 cm^{-1}; ^{1}H and ^{13}C NMR data, see Tables 1 and 2; HRESIMS *m/z* 403.2459 [M + Na]$^+$ (calcd. for $C_{22}H_{36}NaO_5$, 403.2455).

Sartrocheliol E (**7**): colorless oil; $[\alpha]_D^{20}$ +120.0 (*c* 0.01, MeOH); IR (KBr) ν_{max}: 3446, 2922, 1384, 1142, 1044 cm^{-1}; ^{1}H and ^{13}C NMR data, see Tables 1 and 2; HR-ESIMS *m/z* 329.2448 [M + Na]$^+$ (calcd. for $C_{20}H_{34}NaO$, 329.2451).

Sartrocheliol F (**8**): colorless oil; $[\alpha]_D^{20}$ +40.0 (*c* 0.05, MeOH); IR (KBr) ν_{max}: 3442, 1384 cm^{-1}; ^{1}H and ^{13}C NMR data, see Tables 1 and 2; HR-ESIMS *m/z* 403.2446 [M + Na]$^+$ (calcd. for $C_{22}H_{36}NaO_5$, 403.2455).

3.5. X-ray Crystallographic Analysis for Compounds **1** and **4**

The crystals of **1** and **4** were both recrystallized from methanol at 4 °C. X-ray analysis of **1** and **4** were carried out on a Bruker D8 Venture diffractometer with Cu Kα radiation (λ = 1.54178 Å) at 170 K, respectively. The acquisition parameters for **1** and **4** are provided in the Supplementary Materials, and crystallographic data for compounds **1** and **4** (deposition no. CCDC 2,196,207 and CCDC 2205721) have been deposited at the Cambridge Crystallographic Data Center. Copies of the data can be obtained free of charge via www.ccdc.cam.ac.uk/conts/retrieving.html (accessed on 11 August 2022).

3.6. Esterification of Compounds **3** and **5** with MTPA Chlorides

Compounds **3** and **5** (2.0 mg each) were dissolved in dry pyridine (1.2 mL), divided them into two sets (0.6 mL each), then treated with (*R*)-(–)-2-methoxy-2-(trifluoromethyl) phenylacetyl chloride ((*R*)-(–)-MTPA-Cl) (10 µL) and (*S*)-(+)-2-methoxy-2-(trifluoromethyl) phenylacetyl chloride ((*S*)-(+)-MTPA-Cl) (10 μL), respectively. After stirring overnight at room temperature, the solutions were evaporated in vacuo and the residues were purified by silica gel column chromatography (PE/Et_2O = 90:10) to obtain the *S*-MTPA ester **3r** (0.3 mg), *R*-MTPA ester **3s** (0.3 mg), *S*-MTPA ester **5r** (0.3 mg), and *R*-MTPA ester **5s** (0.3 mg), respectively. The obtained products were then subjected to the ^{1}H NMR experiment.

3.7. QM-NMR Calculation of Compounds **5** *and* **7**

Theoretical calculations of all theoretical stereoisomers were carried out to determine the relative configuration of **5** and **7**, based on the alignment of its ^{1}D NMR chemical shifts (^{13}C NMR chemical shifts herein) and calculation-generated chemical shifts. Confab was used to search the conformational space of **5a–5d** and **7a–7d**. Conformational searches were carried out using the torsional sampling (MCMM) method and OPLS_2005 force field in the Macromodel 9.9.223 software applying an energy window of 21 kJ/mol. Conformers above 1% population were re-optimized with Gaussian 09 at the B3LYP/6-311G(d,p) level with IEFPCM (Polarizable Continuum Model using the Integral Equation Formalism variant) solvent model for acetonitrile. The Boltzmann populations of the conformers were obtained based on the potential energy provided by the OPLS_2005 force field, leading to 11, 9, 13 and 24 conformers for **5a–5d**; 11, 9, 13 and 24 conformers for **7a–7d** above 1% population for further re-optimization, respectively. The obtained conformers were subjected to optimization and frequency calculations on B3LYP/6-311G(d,p) level of theory. GIAO DFT ^{13}C NMR calculations were calculated on mPW1PW91/6-31G* ($CHCl_3$) level of theory, and the calculated shielding tensors were Boltzmann averaged according to Gibbs free energy and then converted into chemical shifts following MSTD protocol. The experimental ^{13}C NMR data of **5** and **7** were compared with the calculated NMR chemical shifts of **5a–5d** and **7a–7d** using the mean absolute error (MAE) values, maximum deviation (MD) values, correlation coefficient (R^2), and DP4+ probability analysis. XYZ data for all conformations and detailed data for DP4+ analysis are provided in the Supplementary Materials.

3.8. Cytotoxic Bioassays

H1975, MDA-MB-231, A549, and H1299 cancer cell lines were purchased from the Procell Life Science & Technology Co., Ltd. Cells were cultured at 37 °C in a 5% CO_2 humidified incubator and maintained in high glucose Dulbecco's Modified Eagle Medium (DMEM, Nissui, Tokyo, Japan) containing 100 mg/mL streptomycin, 2.5 mg/L amphotericin B and 10% heat-inactivated fetal bovine serum (FBS). The cells were inoculated in 96-well culture plates for 12 h and then treated with different concentrations of compounds for 72 h. Water-soluble tezole (WST) reagent was added to each (10 μL) well and cultured at 37 °C for 2 h to assess cell viability. The absorbance was read with a microplate reader at 450 nm. Adriamycin (DOX) was the positive control.

4. Conclusions

In summary, the present investigation of the soft coral *S. trocheliophorum* from Ximao Island provided intriguing results including six new cembranoids diterpenoids, sartrocheliols A–E (**1**, **3**, **5–8**), along with four known related ones (**2**, **4**, **9**, and **10**). This study not only extended the members of capnosane and cembrane diterpenes but also enriched the chemical diversity of the title species. In addition, a whole set of NMR computations, modified Mosher's method, and X-ray diffraction analysis were applied to assign the absolute configurations of compounds **1**, **3**, **4**, and **5**. However, the absolute configurations of other new metabolites remain undefined. To achieve it, a collection of the biological materials and accumulation of these new compounds should be conducted in future, which could supply sufficient amounts of metabolites for either the subsequent chemical transformations or recrystallization experiments. Anti-tumor and antibacterial bioassays were carried out. Among these secondary metabolites, compound **4** displayed moderate cytotoxicity against A549, H1975, MDA-MB-231, and H1299 cells (IC_{50} = 47.9, 26.3, 44.7, 33.1 μM, respectively), while **6** only exhibited moderate cytotoxicity against H1975 (IC_{50} = 40.4 μM). However, none of them were active in the antibacterial bioassay. Moreover, the bioactivities of these compounds such as anti-virus, will be evaluated in future.

Supplementary Materials: The following supporting information can be downloaded at: https://www.mdpi.com/article/10.3390/md21020069/s1, Figures S1–S48: NMR, HRESIMS and IR data of compounds **1**, **3**, **5–8**; Figures S49 and S50: Regression analysis of experimental vs. calculated ^{13}C NMR chemical shifts of different isomers of compounds **5** and **7** at the PCM/mPW1PW91/6-31 + G** level using DP4+ method; Tables S1 and S2: X-ray crystallographic data for compounds **1** and **4**; Tables S3 and S5: Cartesian coordinates of all conformers of isomers **5a–5b** and **7a–7b** used after optimization at the B3LYP/6-311G (d,p) level of theory as required for DP4+ analysis; Tables S4 and S6: DP4+ results obtained using experimental data of compounds **5** *versus* isomers **5a–5b** and 7 *versus* isomers **7a–7b**.

Author Contributions: Conceptualization, L.-F.L., F.Y. and Y.-W.G.; methodology, L.-F.L., F.Y. and Y.-W.G.; validation, Y.-T.S., M.-Z.S., H.L. and J.-G.C.; formal analysis, Y.-T.S. and D.-D.Y.; investigation, Y.-T.S. and D.-D.Y.; data curation, Y.-T.S. and D.-D.Y.; writing—original draft preparation, Y.-T.S.; writing—review and editing, L.-F.L. and Y.-W.G.; supervision, Y.-W.G.; project administration, Y.-W.G.; funding acquisition, L.-F.L. and Y.-W.G. All authors have read and agreed to the published version of the manuscript.

Funding: This research work was financially supported by the National Natural Science Foundation of China (NSFC) (Nos. 81991521 and 41876194), the National Key Research and Development Program of China (No. 2022YFC2804100), and the SKLDR/SIMM Project (No. SIMM2103ZZ-06).

Institutional Review Board Statement: Not applicable.

Informed Consent Statement: Not applicable.

Data Availability Statement: The data presented in this study are available on request from the corresponding author.

Acknowledgments: We thank X.-B. Li from Hainan University for the taxonomic identification of the soft coral material.

Conflicts of Interest: The authors declare no conflict of interest.

References

1. Anjaneyulu, A.S.R.; Rao, G.V. Chemical constituents of the soft coral species of *Sarcophyton* genus: A review. *J. Indian Chem. Soc.* **1997**, *74*, 272–278.
2. Liang, L.-F.; Guo, Y.-W. Terpenes from the soft corals of the genus *Sarcophyton*: Chemistry and biological activities. *Chem. Biodivers.* **2013**, *10*, 2161–2196. [CrossRef]
3. Zubair, M.S.; Al-Footy, K.O.; Ayyad, S.-E.N.; Al-Lihaibi, S.S.; Alarif, W.M. A review of steroids from *Sarcophyton* species. *Nat. Prod. Res.* **2016**, *30*, 869–879. [CrossRef]
4. Elkhawas, Y.A.; Elissawy, A.M.; Elnaggar, M.S.; Mostafa, N.M.; Al-Sayed, E.; Bishr, M.M.; Singab, A.N.B.; Salama, O.M. Chemical diversity in species belonging to soft coral genus *Sacrophyton* and its impact on biological activity: A review. *Mar. Drugs* **2020**, *18*, 41. [CrossRef]
5. Wang, Z.; Tang, H.; Wang, P.; Gong, W.; Xue, M.; Zhang, H.; Liu, T.; Liu, B.; Yi, Y.; Zhang, W. Bioactive polyoxygenated steroids from the South China Sea soft coral, *Sarcophyton* sp. *Mar. Drugs* **2013**, *11*, 775–787. [CrossRef]
6. Chen, W.-T.; Liu, H.-L.; Yao, L.-G.; Guo, Y.-W. 9,11-Secosteroids and polyhydroxylated steroids from two South China Sea soft corals *Sarcophyton trocheliophorum* and *Sinularia flexibilis*. *Steroids* **2014**, *92*, 56–61. [CrossRef]
7. Ngoc, N.T.; Hanh, T.T.H.; Quang, T.H.; Cuong, N.X.; Nam, N.H.; Thao, D.T.; Thung, D.C.; Kiem, P.V.; Minh, C.V. Polyhydroxylated steroids from the Vietnamese soft coral *Sarcophyton ehrenbergi*. *Steroids* **2021**, *176*, 108932. [CrossRef]
8. Wang, C.-Y.; Chen, A.-N.; Shao, C.-L.; Li, L.; Xu, Y.; Qian, P.-Y. Chemical constituents of soft coral *Sarcophyton infundibuliforme* from the South China Sea. *Biochem. Syst. Ecol.* **2011**, *39*, 853–856. [CrossRef]
9. Huang, T.-Y.; Huang, C.-Y.; Chen, S.-R.; Weng, J.-R.; Tu, T.-H.; Cheng, Y.-B.; Wu, S.-H.; Sheu, J.-H. New hydroquinone monoterpenoid and cembranoid-related metabolites from the soft coral *Sarcophyton tenuispiculatum*. *Mar. Drugs* **2021**, *19*, 8. [CrossRef]
10. Anjaneyulu, A.S.R.; Murthy, M.V.R.K.; Gowri, P.M.; Venugopal, M.; Laatsch, H. A rare prostaglandin from the soft coral *Sarcophyton crassocaule* of the Indian Ocean. *J. Nat. Prod.* **2000**, *63*, 1425–1426. [CrossRef]
11. Cheng, Z.-B.; Deng, Y.-L.; Fan, C.-Q.; Han, Q.-H.; Lin, S.-L.; Tang, G.-H.; Luo, H.-B.; Yin, S. Prostaglandin derivatives: Nonaromatic phosphodiesterase-4 inhibitors from the soft coral *Sarcophyton ehrenbergi*. *J. Nat. Prod.* **2014**, *77*, 1928–1936. [CrossRef] [PubMed]

12. Cheng, S.-Y.; Wen, Z.-H.; Chiou, S.-F.; Tsai, C.-W.; Wang, S.-K.; Hsu, C.-H.; Dai, C.-F.; Chiang, M.Y.; Wang, W.-H.; Duh, C.-Y. Ceramide and cerebrosides from the octocoral *Sarcophyton ehrenbergi*. *J. Nat. Prod.* **2009**, *72*, 465–468. [CrossRef] [PubMed]
13. Eltahawy, N.A.; Ibrahim, A.K.; Radwan, M.M.; Zaitone, S.A.; Gomaa, M.; ElSohly, M.A.; Hassanean, H.A.; Ahmed, S.A. Mechanism of action of antiepileptic ceramide from Red Sea soft coral *Sarcophyton auritum*. *Bioorg. Med. Chem. Lett.* **2015**, *25*, 5819–5824. [CrossRef] [PubMed]
14. Řezanka, T.; Dembitsky, V.M. γ-Lactones from the soft corals *Sarcophyton trocheliophorum* and *Lithophyton arboreum*. *Tetrahedron* **2001**, *57*, 8743–8749. [CrossRef]
15. Shaaban, M.; Ghani, M.A.; Issa, M.Y. New naturally occurring compounds from *Sarcophyton trocheliophorum*. *Biointerface Res. App. Chem.* **2022**, *12*, 2285–2331. [CrossRef]
16. Rodrigues, I.G.; Miguel, M.G.; Mnif, W. A brief review on new naturally occurring cembranoid diterpene derivatives from the soft corals of the genera *Sarcophyton*, *Sinularia*, and *Lobophytum* since 2016. *Molecules* **2019**, *24*, 781. [CrossRef]
17. Shaaban, M.; Yassin, F.Y.; Soltan, M.M. Calamusins J-K: New anti-angiogenic sesquiterpenes from *Sarcophyton glaucum*. *Nat. Prod. Res.* **2021**, *35*, 5720–5731. [CrossRef]
18. Sawant, S.S.; Youssef, D.T.A.; Sylvester, P.W.; Wali, V.; Sayed, K.A.E. Antiproliferative sesquiterpenes from the Red Sea soft coral *Sarcophyton glaucum*. *Nat. Prod. Commun.* **2007**, *2*, 117–119. [CrossRef]
19. Anjaneyulu, A.S.R.; Rao, V.L.; Sastry, V.G.; Rao, D.V. Trocheliophorin: A novel rearranged sesquiterpenoid from the Indian Ocean soft coral *Sarcophyton trocheliophorum*. *J. Asian Nat. Prod. Res.* **2008**, *10*, 597–601. [CrossRef]
20. Ye, F.; Li, J.; Wu, Y.; Zhu, Z.-D.; Mollo, E.; Gavagnin, M.; Gu, Y.-C.; Zhu, W.-L.; Li, X.-W.; Guo, Y.-W. Sarinfacetamides A and B, nitrogenous diterpenoids with tricyclo [6.3.1.0^{1,5}] dodecane scaffold from the South China Sea soft coral *Sarcophyton infundibuliforme*. *Org. Lett.* **2018**, *20*, 2637–2640. [CrossRef]
21. Lin, K.-H.; Lin, Y.-C.; Huang, C.-Y.; Tseng, Y.-J.; Chen, S.-R.; Cheng, Y.-B.; Hwang, T.-L.; Wang, S.-Y.; Chen, H.-Y.; Dai, C.-F.; et al. Cembranoid-related diterpenes, novel secoditerpenes, and an unusual bisditerpene from a Formosan soft coral *Sarcophyton tortuosum*. *Bull. Chem. Soc. Jpn.* **2021**, *94*, 2774–2783. [CrossRef]
22. Bu, Q.; Yang, M.; Yan, X.-Y.; Yao, L.-G.; Guo, Y.-W.; Liang, L.-F. New flexible cembrane-type macrocyclic diterpenes as TNF-α inhibitors from the South China Sea soft coral *Sarcophyton mililatensis*. *Int. J. Biol. Macromol.* **2022**, *222*, 880–886. [CrossRef] [PubMed]
23. Wang, C.; Zhang, J.; Shi, X.; Li, K.; Li, F.; Tang, X.; Li, G.; Li, P. Sarcoeleganolides C–G, five new cembranes from the South China Sea soft coral *Sarcophyton elegans*. *Mar. Drugs* **2022**, *20*, 574. [CrossRef] [PubMed]
24. Mohamed, T.A.; Elshamy, A.I.; Abd El-Razek, M.H.; Abdel-Tawab, A.M.; Ali, S.K.; Aboelmagd, M.; Suenaga, M.; Pare, P.W.; Umeyama, A.; Hegazy, M.-E.F. Sarcoconvolutums F and G: Polyoxygenated cembrane-type diterpenoids from *Sarcophyton convolutum*, a Red Sea soft coral. *Molecules* **2022**, *27*, 5835. [CrossRef]
25. Sun, P.; Cai, F.-Y.; Lauro, G.; Tang, H.; Su, L.; Wang, H.-L.; Li, H.H.; Mándi, A.; Kurtán, T.; Riccio, R.; et al. Immunomodulatory biscembranoids and assignment of their relative and absolute configurations: Data set modulation in the density functional theory/nuclear magnetic resonance approach. *J. Nat. Prod.* **2019**, *82*, 1264–1273. [CrossRef] [PubMed]
26. Li, Y.; Li, S.; Cuadrado, C.; Gao, C.; Wu, Q.; Li, X.; Pang, T.; Daranas, A.H.; Guo, Y.; Li, X. Polyoxygenated anti-inflammatory biscembranoids from the soft coral *Sarcophyton tortuosum* and their stereochemistry. *Chin. Chem. Lett.* **2021**, *32*, 271–276. [CrossRef]
27. Yan, X.; Liu, J.; Huang, J.; Wang, Y.; Leng, X.; Li, T.; Ouyang, H.; Yan, X.; He, S. Bistrochelides H−L: Biscembranoids from the South China Sea soft coral *Sarcophyton serenei*. *Phytochemistry* **2022**, *204*, 113438. [CrossRef]
28. Nguyen, N.B.; Chen, L.-Y.; Chen, P.-J.; El-Shazly, M.; Hwang, T.-L.; Su, J.-H.; Su, C.-H.; Yen, P.-T.; Peng, B.-R.; Lai, K.-H. MS/MS molecular networking unveils the chemical diversity of biscembranoid derivatives, neutrophilic inflammatory mediators from the cultured soft coral *Sarcophyton trocheliophorum*. *Int. J. Mol. Sci.* **2022**, *23*, 15464. [CrossRef]
29. Al-Footy, K.O.; Alarif, W.M.; Asiri, F.; Aly, M.M.; Ayyad, S.-E.N. Rare pyrane-based cembranoids from the Red Sea soft coral *Sarcophyton trocheliophorum* as potential antimicrobial–antitumor agents. *Med. Chem. Res.* **2015**, *24*, 505–512. [CrossRef]
30. Mohamed, T.A.; Elshamy, A.I.; Abdel-Tawab, A.M.; AbdelMohsen, M.M.; Ohta, S.; Pare, P.W.; Hegazy, M.-E.F. Oxygenated cembrene diterpenes from *Sarcophyton convolutum*: Cytotoxic sarcoconvolutum A–E. *Mar. Drugs* **2021**, *19*, 519. [CrossRef]
31. Li, J.-F.; Zeng, Y.-B.; Li, W.-S.; Luo, H.; Zhang, H.-Y.; Guo, Y.-W. Xishaglaucumins A—J, new cembranoids with anti-inflammatory activities from the South China Sea soft coral *Sarcophyton glaucum*. *Chin. J. Chem.* **2022**, *40*, 79–90. [CrossRef]
32. Zhang, J.; Tang, X.; Han, X.; Feng, D.; Luo, X.; Ofwegen, L.v.; Li, P.; Li, G. Sarcoglaucins A-I, new antifouling cembrane-type diterpenes from the South China Sea soft coral *Sarcophyton glaucum*. *Org. Chem. Front.* **2019**, *6*, 2004–2013. [CrossRef]
33. Badria, F.A.; Guirguis, A.N.; Perovic, S.; Steffen, R.; Müller, W.E.G.; Schröder, H.C. Sarcophytolide: A new neuroprotective compound from the soft coral *Sarcophyton glaucum*. *Toxicology* **1998**, *131*, 133–143. [CrossRef] [PubMed]
34. Du, Y.-Q.; Chen, J.; Wu, M.-J.; Zhang, H.-Y.; Liang, L.-F.; Guo, Y.-W. Uncommon capnosane diterpenes with neuroprotective potential from South China Sea soft coral *Sarcophyton boettgeri*. *Mar. Drugs* **2022**, *20*, 602. [CrossRef] [PubMed]
35. Carroll, A.R.; Copp, B.R.; Davis, R.A.; Keyzers, R.A.; Prinsep, M.R. Marine natural products. *Nat. Prod. Rep.* **2022**, *39*, 1122–1171. [CrossRef]
36. Liu, J.; Gu, Y.-c.; Su, M.-z.; Guo, Y.-w. Chemistry and bioactivity of secondary metabolites from South China Sea marine fauna and flora: Recent research advances and perspective. *Acta Pharmacol. Sin.* **2022**, *43*, 3062–3079. [CrossRef]
37. Liang, L.-F.; Kurtán, T.; Mándi, A.; Yao, L.-G.; Li, J.; Zhang, W.; Guo, Y.-W. Unprecedented diterpenoids as a PTP1B inhibitor from the Hainan soft coral *Sarcophyton trocheliophorum* Marenzeller. *Org. Lett.* **2013**, *15*, 274–277. [CrossRef]

38. Yang, M.; Li, X.-L.; Wang, J.-R.; Lei, X.; Tang, W.; Li, X.-W.; Sun, H.; Guo, Y.-W. Sarcomililate A, an unusual diterpenoid with tricyclo [11.3.0.0^{2,16}] hexadecane carbon skeleton, and its potential biogenetic precursors from the Hainan soft coral *Sarcophyton mililatensis*. *J. Org. Chem.* **2019**, *84*, 2568–2576. [CrossRef]
39. Yang, Q.-B.; Wu, Q.; Chen, J.-K.; Liang, L.-F. The soft coral *Sarcophyton trocheliophorum*: A warehouse of terpenoids with structural and pharmacological diversity. *Mar. Drugs* **2023**, *21*, 30. [CrossRef]
40. Yao, L.-G.; Zhang, H.-Y.; Liang, L.-F.; Guo, X.-J.; Mao, S.-C.; Guo, Y.-W. Yalongenes A and B, two new cembranoids with cytoprotective effects from the Hainan soft coral *Sarcophyton trocheliophorum* Marenzeller. *Helv. Chim. Acta* **2012**, *95*, 235–239. [CrossRef]
41. Chen, Z.-H.; Gao, T.-R.; Yang, M.; Yao, L.-G.; Guo, Y.-W. Further new cembranoids from the South China Sea soft coral *Sarcophyton trocheliophorum*. *Fitoterapia* **2021**, *151*, 104902. [CrossRef] [PubMed]
42. Xi, Z.; Bie, W.; Chen, W.; Liu, D.; van Ofwegen, L.; Proksch, P.; Lin, W. Sarcophyolides B–E, new cembranoids from the soft coral *Sarcophyton elegans*. *Mar. Drugs* **2013**, *11*, 3186–3196. [CrossRef] [PubMed]
43. Jia, R.; Guo, Y.-W.; Mollo, E.; Gavagnin, M.; Cimino, G. Sarcophytonolides E−H, cembranolides from the Hainan soft coral *Sarcophyton latum*. *J. Nat. Prod.* **2006**, *69*, 819–822. [CrossRef] [PubMed]
44. Liang, L.-F.; Chen, W.-T.; Li, X.-W.; Wang, H.-Y.; Guo, Y.-W. New bicyclic cembranoids from the South China Sea soft coral *Sarcophyton trocheliophorum*. *Sci. Rep.* **2017**, *7*, 46584. [CrossRef]
45. Wen, T.; Ding, Y.; Deng, Z.; van Ofwegen, L.; Proksch, P.; Lin, W. Sinulaflexiolides A-K, cembrane-type diterpenoids from the Chinese soft coral *Sinularia flexibilis*. *J. Nat. Prod.* **2008**, *71*, 1133–1140. [CrossRef]
46. Kobayashi, M.; Kondo, K.; Osabe, K.; Mitsuhashi, H. Marine terpenes and terpenoids. V. Oxidation of sarcophytol A, a potent anti-tumor-promoter from the soft coral *Sarcophyton glaucum*. *Chem. Pharm. Bull.* **1988**, *36*, 2331–2341. [CrossRef]
47. Li, S.-W.; Cuadrado, C.; Yao, L.-G.; Daranas, A.H.; Guo, Y.-W. Quantum mechanical–NMR-aided configuration and conformation of two unreported macrocycles isolated from the soft coral *Lobophytum* sp.: Energy calculations versus coupling constants. *Org. Lett.* **2020**, *22*, 4093–4096. [CrossRef]
48. Raldugin, V.A.; Pleshkov, I.G.; Gatilov, Y.V.; Yaroshenko, N.I.; Salenko, V.L.; Shevtsov, S.A.; Pentegova, V.A. Photosensitized oxidation of isocembrol. VII. Products of reaction at the C_{11} double bond. *Chem. Nat. Compd.* **1984**, *20*, 45–52. [CrossRef]

marine drugs

Article

Eremophilane-Type Sesquiterpenes from a Marine-Derived Fungus Penicillium Copticola with Antitumor and Neuroprotective Activities †

Jianping Zhang [1], Dong Liu [1], Aili Fan [1], Jian Huang [1] and Wenhan Lin [1,2,*]

[1] State Key Laboratory of Natural and Biomimetic Drugs, Peking University, Beijing 100191, China
[2] Institute of Ocean Research, Ningbo Institute of Marine Medicine, Peking University, Ningbo 315010, China
* Correspondence: whlin@bjmu.edu.cn; Tel.: +86-10-82806188
† Dedication to Peter Proksch for his 70th birthday.

Abstract: Chemical examination of a marine sponge-associated *Penicillium copticola* fungus resulted in the isolation of ten undescribed eremophilanes, namely copteremophilanes A–J (**1–10**), along with two new glycosides, 5-glycopenostatin F (**11**) and 5-glucopenostatin I (**12**). Their structures were determined by extensive spectroscopic data, in association with ECD data and chemical conversions for configurational assignments. Analogs **1**, **2**, and **10** represent a group of uncommon skeletons of eremophilanes with an aromatic ring and a methyl migration from C-5 to C-9, and analogs **11** and **12** are characteristic of a PKS scaffold bearing a glucose unit. The incorporation of a chlorinated phenylacetic unit in **3–9** is rarely found in nature. Analog **7** showed neuroprotective effect, whereas **8** exhibited selective inhibition against human non-small cell lung cancer cells (A549). This study enriched the chemical diversity of eremophilanes and extended their bioactivities to neuroprotection.

Keywords: marine fungus; *Penicillium copticola*; copteremophilanes A–J; structure elucidation; antitumor activity; neuroprotection

Citation: Zhang, J.; Liu, D.; Fan, A.; Huang, J.; Lin, W. Eremophilane-Type Sesquiterpenes from a Marine-Derived Fungus Penicillium Copticola with Antitumor and Neuroprotective Activities. *Mar. Drugs* **2022**, *20*, 712. https://doi.org/10.3390/md20110712

Academic Editors: Bin-Gui Wang and Haofu Dai

Received: 1 November 2022
Accepted: 11 November 2022
Published: 13 November 2022

1. Introduction

Farnesyl diphosphate (FPP) biogenetically generates sesquiterpenes with diverse scaffolds via various ring rearrangements. Eremophilanes are one class of sesquiterpenes characterized by the presence of a bicyclic backbone with irregular rule to assemble FPP in association with a methyl migration from C-10 to C-5 [1,2]. Eremophilane analogues extensively distribute in nature. Apart from the eremophilanes derived by plants of which the genus *Ligularia* is dominated to produce eremophilanes [3], terrestrial- or marine-originated fungi emerge as the additional sources to produce relevant analogs. It is noteworthy that most fungal eremophilanes are enantiomeric to the corresponding entities from plants [1], and marine-associated fungi have the potential to generate structurally unique analogs [4–6]. Eremophilanes possess wide range of bioactivities, such as phytotoxins, antimicrobials, protein inhibitors, immunomodulators and cytotoxins. In marine-derived fungi, eremophilanes with chemical diversity are widely distributed in fungal genera of *Acremonium* [6], *Penicillium* [7–9], *Cochliobolus* [10], *Phomopsis* [11], and *Cryptosphaeria* [12]. These findings imply marine-derived fungi as a potential source to generate structurally unique and bioactive eremophilanes. With the aim to continue our discovery of bioactive natural products from marine-associated organisms, a marine sponge (*Xestospongia testudinaria*)-associated fungus strain *Penicillium copticola* WZXY-m122-9 was selected for chemical examination. The LC-MS/MS data of the EtOAc extract of the cultured fungus were processed into a molecular network using MZmine and the GNPS platform (http://gnps.ucsd.edu, accessed on 20 June 2022), which allowed the formation of the spectral nodes into clusters (Figure S103). Annotation of the nodes in a cluster with m/z values of 200 to 300 by GNPS MS/MS spectral library matched eremophilanes, including dehydropetasol (m/z 233 $[M + H]^+$), dihydrosporogen AO-1 (m/z 251 $[M + H]^+$), hydroxyphomenone (m/z 267 $[M + H]^+$), sporogen AO-1 (m/z 248 $[M + H]^+$) and penicilleremophilane X (m/z 279 $[M + H]^+$). In addition, a cluster with the nodes ranging from m/z 430 to 480 presenting chlorine feature ($[M]^+/[M + 2]^+$ = 3:1) did not hit in the database, suggests a group of untapped metabolites. The scaled-up fermentation and extensive chromatographic separation of the EtOAc extract resulted in the isolation of ten new eremophilane-type sesquiterpenes, along with two new PKS glucosides (Figure 1). Herein, we report the structural determination of the new compounds (Figures S1–S102) and the bioactivities of antitumor cell lines and neuroprotection.

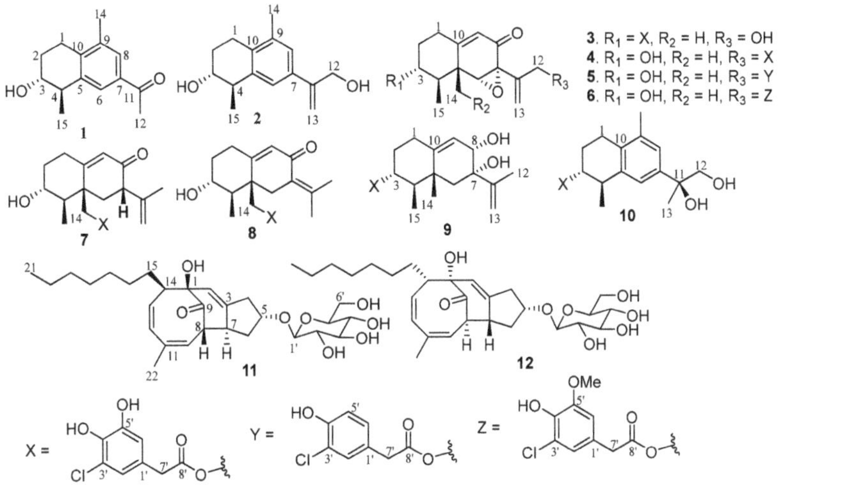

Figure 1. Structures of new compounds.

2. Results

2.1. Structure Elucidation of New Compounds

Copteremophilane A (**1**) has a molecular formula of $C_{14}H_{18}O_2$ as determined by the HRESIMS and NMR data, requiring six degrees of unsaturation. The ^{13}C NMR and DEPT data afforded a total of 14 carbon resonances, including six aromatic carbons for a phenyl group, a ketone, as well as seven alkyl carbons, of which three were classified into methyl groups. A *tetra*-substituted aromatic ring was recognized by the presence of two *meta*-coupled aromatic protons, H-6 (δ_H 7.61, brs) and H-8 (δ_H 7.55, brs). The substitution of a methyl group at C-9 (δ_C 136.4) and an acetyl group at C-7 (δ_C 140.8) was confirmed by the HMBC correlations from H_3-14 (δ_H 2.24, s) to C-8 (δ_C 126.8), C-9, and C-10 (δ_C 134.8), and both H-6 and H-8 to carbonyl carbon C-11 (δ_C 198.3), along with H_3-12 (δ_H 2.53, s) to C-7 (δ_C 140.8) and C-11. The COSY correlations established a segment in ring A from C-1 to C-4, and this unit was fused to C-5 (δ_C 141.1) and C-10 positions of the aromatic ring to form a cyclohexene ring due to the HMBC correlations of C-5 and C-10 to both H_2-1 (δ_H 2.62, 2.69) and H-4 (δ_H 2.78, dq, J = 2.6, 7.2 Hz). In addition, a methyl group at C-4 and a hydroxy group at C-3 were confirmed by the COSY correlations from H-4 to H-3 (δ_H 3.66, dt, J =2.6, 4.9 Hz) and H_3-15 (δ_H 1.23, d, J = 7.2 Hz), along with the HMBC correlations from H_3-15 to C-3 (δ_C 69.8), C-4 (δ_C 41.2), and C-5. The NOE correlation between H_3-15 and H-3 was indicative of a *trans*-orientation of H_3-15 toward OH-3. Based on the modified Mosher method, the (*R*)-MPA and (*S*)-MPA esters of **1** were synthesized. Calculation of the $\Delta\delta$ ($\delta_R - \delta_S$) values resulted in 3*R* configuration (Figure 2). Thus, a 4*R* configuration was suggested with the help of the NOE data. The similar data of the experimental ECD compared to that calculated for (3*R*, 4*R*)-**1** (Figure 3A) further supported the configurational assignment.

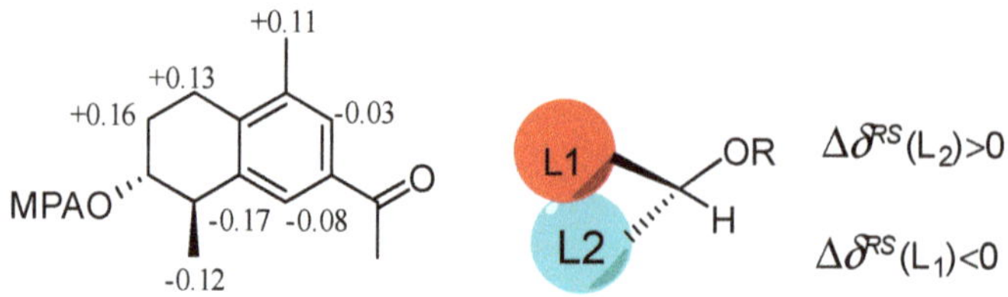

Figure 2. $\Delta\delta$ ($\delta_R - \delta_S$) values of MPA esters of **1**.

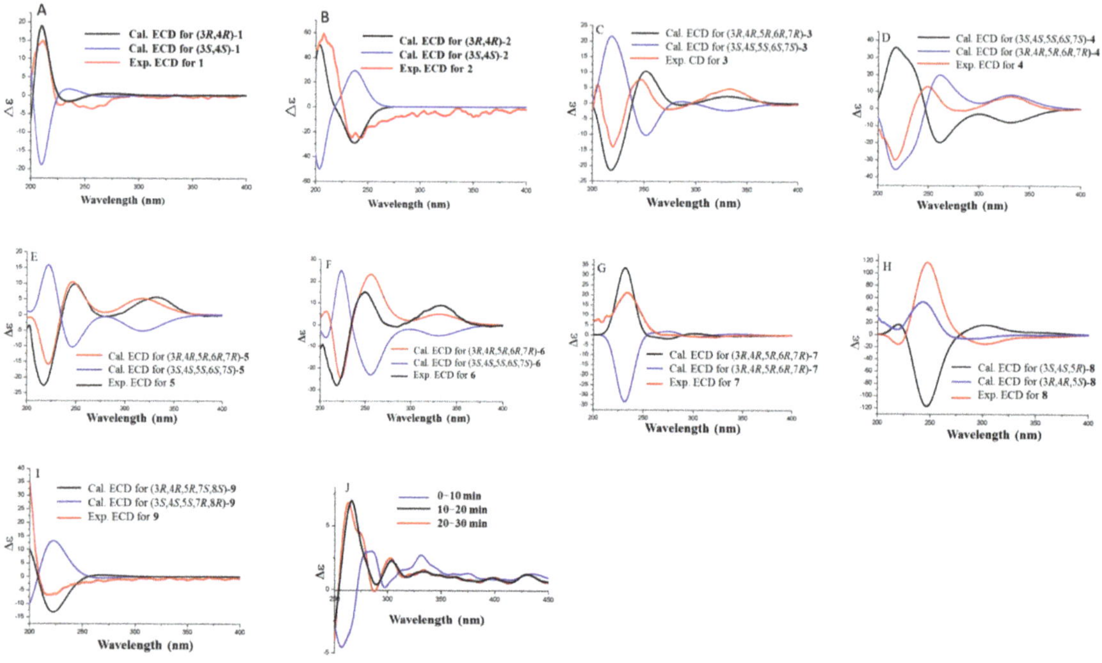

Figure 3. Experimental and calculated ECD spectra of **1–9** (**A–I**) and ICD spectra of **10** (**J**).

The molecular formula ($C_{15}H_{20}O_2$) of copteremophilane B (**2**) was afforded by the HRESIMS and NMR data. The NMR data of both **1** and **2** (Table 1) were comparable, except for the distinction of the substituent at C-7 (δ_C 136.3). A hydroxyisopropene unit was identified by the NMR resonances for two olefinic carbons and a hydroxymethyl group, in association with the HMBC correlations from the hydroxymethyl protons H_2-12 at δ_H 4.28 (s) to C-11 (δ_C 148.3) and C-13 (δ_C 110.1) and between H_2-13 (δ_H 5.22, 5.34) and C-12 (δ_C 63.0). The location of the hydroxyisopropene unit at C-7 was confirmed by the HMBC correlations of H-6 (δ_H 7.06, brs) and H-8 (δ_H 7.03, brs) to C-11 and from H_2-13 to C-7. The same relative configuration of both **1** and **2** was evident from the NOE correlations between H-3 and H_3-15. The similar ECD data suggested the same absolute configuration for both **1** and **2**. This was supported by the calculated ECD data (Figure 3B), in which the experimental ECD data were consistent with those calculated for (3*R*, 4*R*)-**2**.

Table 1. ^{1}H and ^{13}C NMR data of **1**, **2** and **10** in DMSO-d_6.

	1		**2**		**10**	
	δ_H	δ_C	δ_H	δ_C	δ_H	δ_C
1	2.62, dt (4.0, 12.0) 2.69, td (6.0, 12.0)	23.7	2.50, dt (3.5, 12.0) 2.63, td (5.0, 12.0)	23.3	2.55, dt (4.0, 12.0) 2.62, td (4.0, 12.0)	22.4
2	1.80, m; 1.95, m	27.1	1.71, m; 1.90, m	27.7	1.90, m; 1.98, m	23.9
3	3.66, dt (2.6, 4.9)	69.8	3.60, dt (3.0, 5.0)	70.3	4.87, dt (4.4, 6.3)	74.6
4	2.78, dq (2.6, 7.2)	41.2	2.67, dq (3.0, 7.2)	41.2	2.91, dq (4.4, 7.2)	38.1
5		141.1		140.5		138.0
6	7.61, brs	127.1	7.06, brs	124.7	7.12, s	123.8
7		140.8		136.3		145.0
8	7.55, brs	126.8	7.03, brs	124.2	7.06, s	125.1
9		136.4		135.6		134.9
10		134.8		134.1		131.4
11		198.3		148.3		73.9
12	2.53, s	27.1	4.28, s	63.0	3.35, dd (4.0, 12.0) 3.37, dd (4.0, 12.0)	71.0
13			5.22, d (1.7) 5.34, d (1.7)	110.1	1.36, s	26.6
14	2.24, s	19.8	2.16, s	21.3	2.17, s	20.1
15	1.23, d (7.2)	21.4	1.12, d (7.2)	19.9	1.21, d (7.2)	21.7
1′						125.51
2′					6.66, d (1.9)	121.0
3′						120.3
4′						141.1
5′						146.9
6′					6.63, d (1.9)	115.6
7′					3.45, s	40.1
8′						171.4
OH-12					4.59, t (4.0)	

Copteremophilane C (**3**) was determined to have a molecular formula of $C_{23}H_{25}ClO_7$ by the HRESIMS and NMR data, requiring 11 degrees of unsaturation. Diagnostic 2D NMR data revealed **3** to be assembled by two moieties. Based on the 2D NMR data, one of them was identified as a phomenone unit, a co-isolated eremophilane-type sesquiterpene which was structurally characterized by the presence of an α,β-unsaturated ketone with an epoxy group at ring B [13]. The second moiety contained six aromatic carbons for a phenyl unit, a methylene and a carbonyl carbon, along with two *meta*-coupling aromatic protons H-2′ (δ_H 6.70, d, J = 2.1 Hz) and H-6′ (δ_H 6.67, d, J = 2.1 Hz) and the methylene protons H_2-7′ (δ_H 3.50, brs). The HMBC correlations from H_2-7′ to a carbonyl carbon at C-8′ (δ_C 171.2), C-2′ (δ_C 121.0) and C-6′ (δ_C 115.5) and from H-2′ and H-6′ to the aromatic carbons established a tri-substituted phenylacetic segment. Additional HMBC correlations from two phenol protons OH-4′ (δ_H 9.01, brs) and OH-5′ (δ_H 9.76, brs) to the aromatic carbons, respectively, in association with the molecular composition, allowed the assignment of a 3-chloro-4,5-dihydroxyphenylacetic moiety. The linkage of the acyl unit to C-3 for an ester formation was evident from the HMBC correlation between H-3 (δ_H 4.82, dt, J = 4.4, 11.2 Hz) and C-8′. The same relative configuration of **3** as phomenone was identified by the NOE relationships of both H_3-14 (δ_H 1.22, s) and H_3-15 (δ_H 0.90, d, J = 6.8 Hz) with H-3 and H-6 (δ_H 3.40, s) and between H-6 and H_2-13 (δ_H 5.08, 5.20) (Figure 4). Comparison of the experimental and calculated ECD data (Figure 3C) suggested **3** possessing *R* configuration for C-3, C-4, C-5, C-6, and C-7.

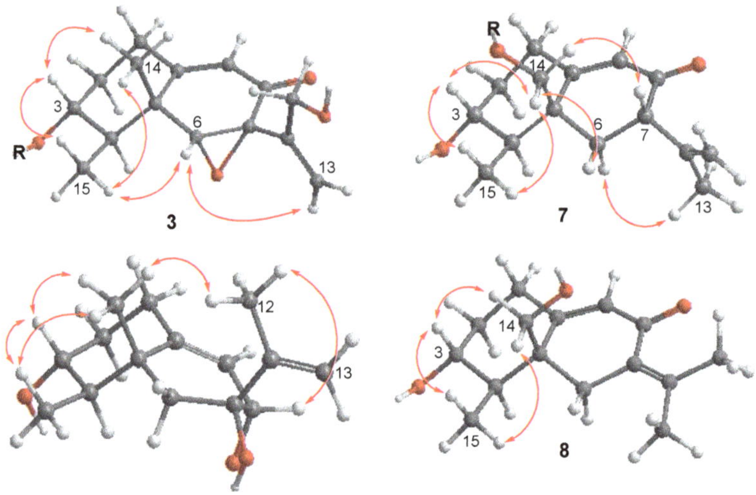

Figure 4. Key NOE correlations of **3** and **7–9**.

The molecular composition of copteremophilane D (**4**) showed the same as that of **3** according to the HRESIMS data. The NMR data of both compounds (Tables 2 and 3) were very similar, and the 2D NMR data established the gross structure of **4** also containing two segments which were identical to those of **3**. The distinction was attributed to the 3-chloro-4,5-dihydroxyphenylacetic moiety location. The chemical shifts of C-3 (δ_C 69.0 vs. δ_C 73.2 of **3**) and C-12 (δ_C 64.9 vs. δ_C 62.1 of **3**) in addition to the HMBC correlation between H_2-12 (δ_H 4.70, 4.71) and a carbonyl carbon C-8′ (δ_C 171.1) allowed the location of the acyl group at C-12. The similar NOE data of both **3** and **4**, along with the comparable ECD data of **4** to those calculated for (3*R*, 4*R*, 5*R*, 6*R*, 7*R*)-**4** (Figure 3D), assigned the same configuration of both **3** and **4**.

The molecular formula of copteremophilane E (**5**) was established as $C_{23}H_{25}ClO_6$ from the HRESIMS (m/z 431.1259 $[M - H]^-$, calcd 431.1261) data, containing 11 degrees of unsaturation. The NMR data of **5** (Tables 2 and 3) were comparable to those of **4**, and the NMR resonances for phomenone moiety of both compounds were identical. The acyl moiety was identified as 3-chloro-4-hydroxyphenylacetic group based on the presence of an ABX spin system for the aromatic protons at H-3′ (δ_H 6.88, d, J = 8.2 Hz), H-2′ (δ_H 6.97, dd, J = 1.5, 8.2 Hz), and H-6′ (δ_H 7.18, d, J = 1.5 Hz), and the HMBC correlations from the methylene protons H_2-7′ (δ_H 3.54, s) to C-1′ (δ_C 126.1), C-2′ (δ_C 131.0), C-6′ (δ_C 129.4) and the carbonyl carbon C-8′ (δ_C 171.2), along with the correlations of H-2′ to the nonprotonated carbons C-3′ (δ_C 119.8) and C-4′ (δ_C 152.5). The similar NOE and ECD data (Figure 3E) suggested the same configuration for both **4** and **5**.

Table 2. ^{1}H NMR data of **3-9** in DMSO-d_6.

	3	4	5	6	7	8	9
1	2.63, dt (2.0, 14.5) 2.36, td (2.0, 14.5)	2.28, td (3.0, 12.0) 2.54, dt (3.0, 12.0)	2.26, td (3.2, 14.1) 2.51, dt (3.2, 14.1)	2.25, td (3.4, 11.2) 2.50, dt (3.4, 11.2)	2.30, m 2.33, m	2.30, td (3.4, 12.0) 2.39 dt (3.4, 11.2)	2.07, m 2.29, m
2	1.36, m 2.04, m	1.25, m 1.99, m	1.19, m 1.98, m	1.20, m 1.98, m	1.31, m 2.00, m	1.29, m 2.01, m	1.27, m 1.97, m
3	4.82, dt (4.4, 11.2)	3.44, dt (5.0, 12.0)	3.42, dt (5.0, 10.5)	3.43, dt (5.0, 10.5)	3.49, m	3.46, td (5.4, 10.6)	4.73, m
4	1.82, dq (6.8,11.2)	1.53, dq (6.8, 12.0)	1.54, dq (6.8, 10.5)	1.54, dq (6.8, 10.5)	1.29, dq (6.9, 10.5)	1.37, dq (6.8, 10.6)	1.44, dq (6.8, 11.0)
6	3.40, s	3.38, s	3.38, s	3.41, s	1.80, t (11.0) 2.20, dd (4.0, 11.0)	2.24, d (15.0) 3.08, d (15.0)	1.68, d (12.6) 1.69, d (12.6)
7					3.19, dd (4.0, 11.0)		
8							3.60, brd (4.5)
9	5.75, d (1.7)	5.71, d (1.7)	5.70, d (1.7)	5.68, d (1.7)	5.78, d (1.3)	5.82, d (1.3)	5.40, d (4.5)
12	4.05, dd (4.0, 12.0) 4.12, dd (4.0, 12.0)	4.70, d (12.0) 4.71, d (12.0)	4.67, d (12.0) 4.74, d (12.0)	4.70, d (12.0) 4.76, d (12.0)	1.60, s	1.80, d (1.5)	1.76, s
13	5.08, d (1.5) 5.20, d (1.5)	5.29, d (0.9) 5.36, d (0.9)	5.28, d (1.1) 5.36, d (1.1)	5.38, d (1.1) 5.29, d (1.1)	4.68, brs 4.83, brs	2.02, d (1.5)	4.76, d (1.0) 4.85, d (1.0)
14	1.22, s	1.12, s	1.12, s	1.11, s	4.19, d (11.5) 4.43, d (11.5)	4.15, s	1.08, s
15	0.90, d (6.8)	1.11, d (6.8)	1.11, d (6.8)	1.11, d (6.8)	0.98, br d (6.9)	1.06, d (6.8)	0.79, d (6.8)
2′	6.70, d (2.1)	6.64, d (2.0)	6.97, dd (1.5, 8,2)	6.79, d (1.9)	6.61, d (1.6)	6.61, d (1.9)	6.68, d (1.9)
3′			6.88, d (8.2)				
6′	6.67, d (2.1)	6.62, d (2.0)	7.18, d (1.5)	6.77, d (1.9)	6.58, d (1.6)	6.57, d (1.9)	6.65, d (1.9)
7′	3.50, s	3.47, s	3.54, s	3.55, s	3.45, s	3.40, s	3.47, s
OH-3		4.70, br					
OH-12	4.83 t (4.0)						
OH-4′	9.01, brs	9.02, s	10.0, brs		9.00, s	8.99, s	
OH-5′	9.76, brs	9.74, s			9.80, s	9.69, s	
MeO				3.78, s			

The structure of copteremophilane F (**6**) was determined to be a 5′-methoxylated analog of **4** due to the comparable NMR data of both analogs, with the exception of an additional methoxy group in **6**. The methoxy protons at δ_H 3.78 (s) showed HMBC correlation to C-5′ (δ_C 148.4) and the NOE relationship with H-6′ (δ_H 6.77, d, J = 1.9 Hz) confirmed a 3-chloro-4-hydroxy-5-methoxyphenylacetic unit, which was linked to C-12 (δ_C 64.5) on the basis of the HMBC correlation between H_2-12 (δ_H 4.70, 4.76) and the carbonyl carbon C-8′ (δ_C 170.6). The similar NOE data of **6** as those of **4** and **5** in association with the comparable ECD data of **6** to those calculated for (3*R*, 4*R*, 5*R*, 6*R*, 7*R*)-**6** identified the same configuration of **6** as **4**.

Copteremophilane G (**7**) has a molecular formula of $C_{23}H_{27}ClO_6$ as determined by the HRESIMS data. The NMR data demonstrated **7** possessing two moieties. Analyses of 2D NMR data (Figure S104) revealed the sesquiterpene unit being identical to JBIR-27 [9], while the second unit was consistent with a 3-chloro-4,5-dihydroxyphenylacetic group. The linkage of the acyl unit to nucleus moiety at C-14 (δ_C 66.6) was deduced by the HMBC correlation between H_2-14 (δ_H 4.19, 4.43) and the carbonyl carbon C-8′ (δ_C 171.2). The NOE relationships from H_2-14 to H_3-15 (δ_H 0.98, d, J = 6.9 Hz) and H-7 (δ_H 3.19, dd, J = 4.0, 11.0 Hz) and between H_3-15 and H-3 (δ_H 3.49, m) (Figure 4) suggested the same relative configuration of the nucleus moiety as JBIR-27 [9]. The comparable ECD data of **7** to those calculated for (3*R*, 4*R*, 5*S*, 7*S*)-**7** (Figure 3G) agreed the absolute configuration of **7** as *R* for C-3 and C-4 and *S* for C-5 and C-7.

Table 3. ^{13}C NMR data of **3-9** in DMSO-d_6.

	3	4	5	6	7	8	9
1	29.9	30.9	30.9	30.3	31.4	31.2	30.1
2	31.7	35.8	35.8	35.4	35.7	35.8	32.9
3	73.2	69.0	69.2	68.4	69.5	69.6	74.7
4	41.9	44.8	44.8	44.3	50.7	49.3	47.9
5	41.2	41.2	41.2	40.7	43.1	44.8	38.6
6	68.3	69.2	68.9	68.6	39.1	38.3	40.6
7	61.7	61.2	61.2	60.7	50.7	127.3	73.7
8	192.2	192.1	192.2	191.6	198.0	190.2	68.2
9	120.7	120.0	120.0	119.4	126.4	128.6	121.2
10	163.7	166.2	166.6	165.6	164.3	163.1	143.0
11	145.3	139.3	139.3	138.8	144.0	142.9	150.9
12	62.1	64.9	64.8	64.5	20.3	22.6	19.4
13	111.9	117.1	117.2	116.8	114.3	23.0	110.4
14	18.1	18.4	18.4	17.9	66.6	66.6	20.2
15	11.3	11.7	11.8	11.2	11.4	11.9	11.0
1′	126.0	125.6	126.1	125.1	125.5	125.5	126.2
2′	121.0	120.3	131.0	122.1	120.9	121.0	121.0
3′	120.3	121.0	119.8	119.5	120.4	120.4	120.3
4′	141.2	141.3	152.5	141.6	141.3	141.2	141.1
5′	146.9	146.9	116.9	148.4	146.9	146.9	146.9
6′	115.5	115.6	129.4	111.8	115.5	115.6	115.4
7′	39.9	39.6	39.2	39.7	39.0	40.0	40.0
8′	171.2	171.1	171.2	170.6	171.2	171.2	171.3
OMe				56.6			

The molecular formula of copteremophilane H (**8**) was the same as **7** due to the same HRESIMS data, requiring ten degrees of unsaturation. The NMR data of **8** were comparable to those of **7**. The distinction was found in ring B, where two methyl groups were recognized for H_3-12 (δ_H 1.80, d, J = 1.5 Hz) and H_3-13 (δ_H 2.02, d, J = 1.5 Hz). The HMBC correlations of H_3-12/H_3-13 to C-7 (δ_C 127.3) and C-11 (δ_C 142.9) indicated **8** to be a stereoisomer of **7** with an olefinic transformation from C-11/C-13 to C-7/C-11. The NOE correlations from H_3-15 (δ_H 1.06, d, J = 6.8 Hz) to H_2-14 (δ_H 4.15, s) and H-3 (δ_H 3.46, td, J = 5.4, 10.6 Hz) clarified the same relative configuration of **8** and **7**. Comparison of the experimental ECD data to those calculated for (3*R*, 4*R*, 5*S*)-**8** (Figure 3H) established 3*R*, 4*R* and 5*S* configurations for **8**.

The molecular formula of copteremophilane I (**9**) was determined as $C_{23}H_{29}ClO_6$ by the HRESIMS data, requiring nine degrees of unsaturation. The NMR data of **9** resembled those of **3** but distinguished by ring B of sesquiterpene unit. Hydroxylations at C-7 (δ_C 73.7) and C-8 (δ_C 68.2) were clarified by the COSY coupling between H-8 (δ_H 3.60, brd, J = 4.5 Hz) and H-9 (δ_H 5.40, brd, J = 4.5 Hz) together with the HMBC correlations from H_2-6 (δ_H 1.68, 1.69) to C-5 (δ_C 38.6), C-7, C-8, C-11 (δ_C 150.9) and C-14 (δ_C 20.2). The NOE correlations from H-3 (δ_H 4.73, m) to H_3-14 (δ_H 1.08, s) and H_3-15 (δ_H 0.79, d, J = 6.8 Hz) and from H_3-12 (δ_H 1.76, s) to H_3-14 and H-8 suggested the same face of both hydroxy groups. The absolute configuration of **9** was assigned by the similar ECD data of **9**, compared to those calculated for (3*R*, 4*R*, 5*R*, 7*S*, 8*S*)-**9**.

Copteremophilane J (**10**) was determined to have a molecular formula of $C_{23}H_{27}ClO_6$ by the HRESIMS data, containing ten degrees of unsaturation. The NMR data featured a homolog of **3**, and the 2D NMR data assigned a 3-chloro-4,5-dihydroxyphenylacetic group at C-3 as the case of **3**. In regard to the sesquiterpene moiety, a phenyl ring was assigned to ring B due to the presence of six aromatic carbons as well as the HMBC correlations of the aromatic protons H-6 (δ_H 7.12, brs) and H-8 (δ_H 7.06, brs) to the aromatic carbons. The HMBC correlations of methyl protons H_3-14 to C-8, C-9, and C-10 in association with the NOE correlation between H_2-1 and H_3-14 substituted a methyl group at C-9. The remaining NMR resonances were attributed to a 1,2-dihydroxypropane unit, which was deduced by the HMBC correlations from H_2-12 (δ_H 3.37, s) to C-11 (δ_C 73.9) and C-13 (δ_C 22.6). The location of this unit at C-7 (δ_C 145.0) was confirmed by additional HMBC correlations from H_3-13 (δ_H 1.36, s) and H_2-12 to C-7. The NOE relationships between H-3 (δ_H 4.87, ddd, J = 2.3, 4.4, 6.3 Hz) and H_3-15 (δ_H 1.21, d, J = 7.2 Hz) suggested a *cis*-orientation of H-3 toward H_3-15. The absolute configuration of C-12 was determined as *S* on the basis of the $Mo(AcO)_4$ induced ECD data (ICD) (Figure 3J), in which a positive Cotton effect at 330 nm was observed.

Compound **11** has a molecular formula of $C_{28}H_{42}O_8$ as established by the HRESIMS data. The DEPT ^{13}C NMR spectrum exhibited a total of 28 carbon resonances, involving 6 olefinic carbons, a ketone carbon, 7 sp^3 oxygenated carbons, and 14 other alkyl carbons. Analyses of the 2D NMR data established a nucleus structure which was identical to penostatin F [13]. This assignment was evident from the COSY correlations along with the HMBC correlations to establish a 4-methylcycloocta-3,5-dien-1-one (Figure 5). Additional COSY relationship between H-7 and H-8 in association with the HMBC correlations from H-7 to C-2 and C-3 and from H-2 to C-1 and C-9 fused a cyclohexene across C-1 and C-8. A spin system from H_2-4 to H-7 via H-5 and H_2-6 along with the HMBC correlation between H-5 and C-3 clarified a cyclopentane ring fused to C-3 and C-7, and C-5 (δ_C 78.7) was oxygenated. The location of an n-heptane unit at C-14 was based on additional COSY and HMBC data. The remaining six oxygenated sp3 carbons were attributed to a sugar unit, which was identified as glucose according to the comparison of the NMR data with those of authentic sample. It was linked to C-5 based on the HMBC correlation between H-5 (δ_H 4.38, t, J = 5.0 Hz) and C-1′ (δ_C 102.1). The NOE correlations between H-5 and H-8 (δ_H 2.86, t, J = 5.6 Hz) and between H-7 (δ_H 3.11, m) and H-14 (Figure 6) suggested the relative configuration of the nucleus moiety to be identical to penostatin F. The $J_{H\text{-}1'/H\text{-}2'}$ value (7.9 Hz) suggested a β-form of Glc unit. Acidic hydrolysis of **11** derived an aglycone and a sugar, the former was identical to penostatin F based on the comparable NMR data and specific rotation and the latter was identical to authentic glucose due to the TLC and HPLC chromatographic comparison.

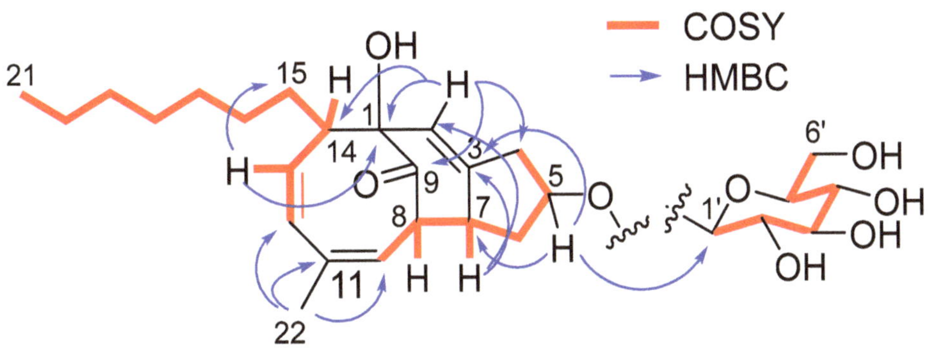

Figure 5. Key ^1H-^1H COSY and HMBC correlations of **11** and **12**.

Compound **12** has the same molecular formula as **11** according to the HRESIMS data. The NMR data of **12** (Table 4) were almost superimposed to those of **11**, suggesting the structural similarity of both compounds. Analysis of 2D NMR data established the same planar structure of **11** and **12**. The similar NOE data with the exception of the correlation between H-5 and H-7 instead of between H-7 and H-1′ of **11** suggested the stereoisomers of both compounds (Figure 6). However, the opposite rotation ($[\alpha]^{20}_D$ −10 for **11** and $[\alpha]^{20}_D$ +12 for **12**) and the opposite Cotton effects (Figure 7) suggested an enantiomeric form with the exception of the chiral center at C-5, implying the nucleus part of **12** to be consistent with penostatin I [13].

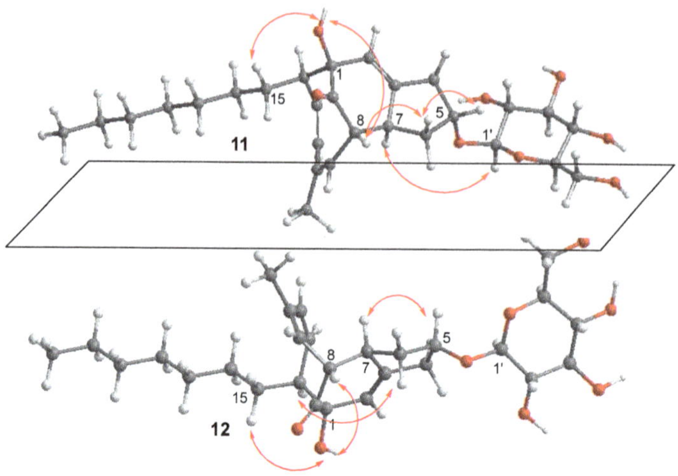

Figure 6. Key NOE correlations of **11** and **12**.

Table 4. ^{1}H and ^{13}C NMR data of 11 and 12 in DMSO-d_6.

No	11		12	
	δ_C	δ_H	δ_C	δ_H
1	82.1		82.3	
2	124.9	5.50, br d (2.6)	125.3	5.50, br d (2.6)
3	146.1		144.9	
4	37.9	2.46, m 2.60, m	37.7	2.27, m 2.68, m
5	78.7	4.38, t (5.0)	78.1	4.34, m
6	39.6	1.44, m 2.43, m	39.5	1.45, m 2.57, m
7	48.6	3.11, m	48.5	2.79, m
8	50.4	2.86, t (5.6)	50.4	2.91, m
9	211.2		210.6	
10	128.3	5.58, dd (0.9, 6.4)	128.2	5.61, dd (0.9, 6.4)
11	129.4		129.5	
12	130.2	5.67, d (11.5)	130.3	5.68, d (11.5)
13	133.9	5.60, dd (9.1, 11.5)	133.9	5.34, dd (9.2, 11.5)
14	43.3	2.57, m	42.7	2.53, m
15	28.4	1.55, m	28.2	1.53, m
16	28.0	1.09, m 1.22, m	27.8	1.08, m 1.21, m
17	29.6	1.22, m	29.2	1.20, m
18	29.1	1.22, m	29.0	1.20, m
19	31.7	1.22, m	31.7	1.21, m
20	22.5	1.25, m	22.5	1.25, m
21	14.4	0.87, t (7.1)	14.4	0.85, t (7.1)
22	25.6	1.75, s	25.6	1.74, s
1′	102.1	4.22, d (7.9)	102.5	4.20, d (7.7)
2′	73.8	2.93, m	73.8	2.89, m
3′	77.2	3.16, m	77.3	3.13, m
4′	70.6	3.05, m	70.5	3.03, m
5′	77.4	3.08, m	77.4	3.03, m 3.08, m
6′	61.6	3.44, dt (5.7, 11.8) 3.67, dd (2.0, 11.8)	61.6	3.43, m 3.67, dd (2.0, 11.8)

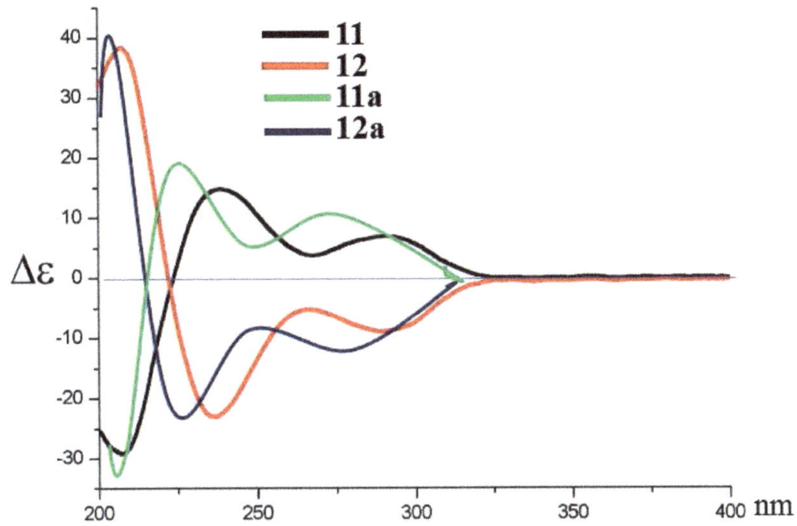

Figure 7. ECD spectra of **11** and **12** and their hydrolyzed products **11a** and **12a**.

In addition, the phenylacetic acids, including sporogen AO-1 [14], phomenone [15], JBIR28 [9], JBIR27 [9], 3-chloro-4,5-dihydroxyphenylacetic acid, 3-chloro-4-hydroxypheny lacetic acid, and 3-chloro-4-hydroxy-5-methoxyphenylacetic acid [16], were isolated and identified on the basis of spectroscopic data.

2.2. Bioassays

Fungus-derived eremophilanes gain potential activities against cancer cell lines for specific aspects in cancer therapies [17,18]. To evaluate the relevant effects of the isolated compounds, an MTT method was performed to detect the cytotoxic activities of compounds toward human non-small cell lung cancer cells (A549), human colon cancer cells (HCT-8) and human breast cancer cells (MCF-7). As shown in Table 5, eremophilane analogs possessing specific selection to inhibit tumor cell lines were observed. Copteremophilane H (**8**) with an acyl unit at C-14 selectively inhibited A549 cells with IC_{50} value of 3.23 μM. However, analog **7** with an olefinic rearrangement from C-7/C-11 of **8** to C-11/C-13 totally attenuated the inhibitory effect. Analogs **4** and **5** with a phenylacetic unit at C-12 of phomenone showed selective inhibition against HCT-8 cell line. It is noteworthy that **4** with a 3-chloro-4,5-dihydroxy-phenylacetic unit exhibited higher activity than **5** which possesses a 3-chloro-4-hydroxyphenylacetic unit, whereas **6** with a 5-methoxy-4-hydroxy-3-chlorophenylacetic unit was inactive toward the three tumor cell lines. These findings suggested that the substituents at the phenylacetic unit are sensitive to affect the activities. In contrast with the bioactive analogs, the eremophilane nuclei and the phenylacetic units alone had weak to no antitumor effects, implying the necessity for the combination of eremophilane moiety with phenylacetic unit to improve antitumor effects. In addition, the irregular eremophilanes **1**, **2** and **10** with an aromatic ring B attenuated the inhibitory activities.

Table 5. Inhibitory effects of **1-10** against tumor cell lines.

	IC_{50} (μM)		
Comps	**A549**	**HCT-8**	**MCF-7**
1	>10	>10	>10
2	>10	>10	>10
3	>10	>10	>10
4	>10	5.4 ± 0.1	>10
5	>10	7.3 ± 0.1	>10
6	>10	>10	>10
7	>10	>10	>10
8	3.2 ± 0.1	>10	>10
9	>10	>10	>10
10	>10	>10	>10
taxol	0.2 ± 0.1	0.7 ± 0.3	0.2 ± 0.1

In previous work, we even reported eremophilanes possessing the inhibitory effects against LPS-induced NO production in macrophages [19]. Oxidative stress has been implicated in the pathology of Alzheimer's disease (AD), and accumulation of β-amyloid (Aβ) causes oxidative stress [20,21]. Aβ is considered as a major pathological factor to induce Alzheimer's disease [22,23]. To evaluate the neuroprotective effects of analogs, we extend the bioassay of analogs to $A\beta_{25\text{-}35}$-stimulated PC12 cell line. PC12 cells are the pheochromocytoma-derived cell line which maintains a differentiated neuroendocrine phenotype [24]. Firstly, the $A\beta_{25\text{-}35}$-induced PC12 cell injury was detected by the CCK8 assay. $A\beta_{25\text{-}35}$ induced cell death in a dose-dependent manner, and around 50% inhibition of PC12 cells was observed after the treatment of $A\beta_{25\text{-}35}$ (40 μM) for 24 hrs. Thus, $A\beta_{25\text{-}35}$ (40 μM) was selected as an optimal dose for subsequent experiments. Alternatively, eremophilanes (50 μM) alone maintained normal PC12 cell morphology and cell numbers, suggesting low cytotoxicity. Most analogs showed inhibitory activities against $A\beta_{25\text{-}35}$-induced cell death, while **7** is the most active to increase the viability of $A\beta_{25\text{-}35}$-induced PC12 cells (Table 6). Lactate dehydrogenase (LDH) as a stable enzyme in the cytosol is quickly released into the medium upon damage of plasma membrane, and malondialdehyde (MDA) as an essential product of lipid peroxidation tends to elevate free radical-mediated myocardial cell injury [25,26]. Thus, low levels of LDH and MDA can protect PC12 cells against oxidative stress. In the $A\beta_{25\text{-}35}$ induced PC12 cell model, LDH release was significantly increased compared to the control group. Analog **7** significantly reduced LDH release dose-dependently. Meanwhile, the intracellular level of MDA, a marker of lipid peroxidation, was also reduced (Figure 8).

Table 6. $A\beta_{25\text{-}35}$-induced PC12 cell viability treated by analogs.

	% PC12 Cell Viability to Control					
μM	**0**	**5.0**	**10.0**	**20.0**	**30.0**	**40.0**
$A\beta_{25\text{-}35}$	100	92.5	81.0	74.6	64.8	49.5
$A\beta_{25\text{-}35}$ + **3**	50.0	53.7	55.8	62.1	63.7	65.5
$A\beta_{25\text{-}35}$ + **4**	50.0	55.2	52.4	67.8	72.2	75.5
$A\beta_{25\text{-}35}$ + **5**	50.0	50.0	50.0	50.0	50.0	50.0
$A\beta_{25\text{-}35}$ + **6**	50.0	51.0	53.0	55.0	57.0	59.2
$A\beta_{25\text{-}35}$ + **7**	50.0	56.2	60.5	74.6	78.6	84.3
$A\beta_{25\text{-}35}$ + **8**	50.0	52.6	55.5	58.4	62.7	65.5
$A\beta_{25\text{-}35}$ + **9**	50.0	50.0	50.0	50.0	50.0	50.0

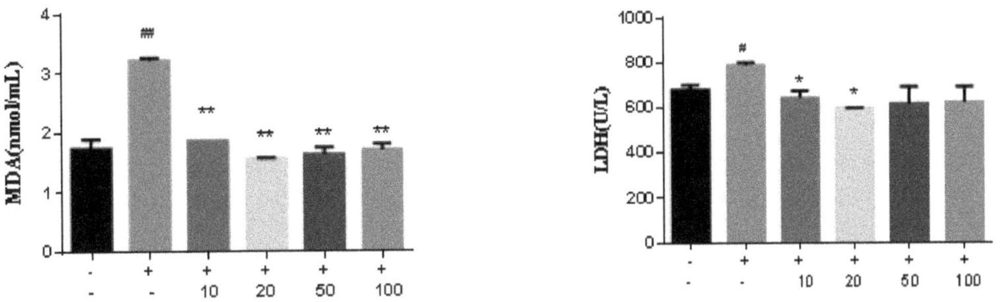

Figure 8. Analog 7 reduced the expression of MDA and LDH in A$\beta_{25\text{-}35}$-induced PC12 cells. Statistical significance values are indicated as *, # $p < 0.05$, **, ## $p < 0.01$.

3. Discussion

Biogenetically, petasol is considered as a precursor to derive relevant analogs. Dehydrogenation of petasol generates 6-dehydropetasol as an intermediate, which follows olefinic rearrangement and methyl migration and subsequent oxidation to afford the aromatic analogs **1**, **2** and **10**. Epoxidation 6-dehydropetasol derives sporogen-OA1, which was isolated from this fungus. 12-Hydroxylation of sporogen-OA1 followed by various acylation generates **3–6**. By the similar manner, 14-hydroxylation of petasol and then acylation derives **7**, which follows olefinic migration affords **8** (Scheme 1).

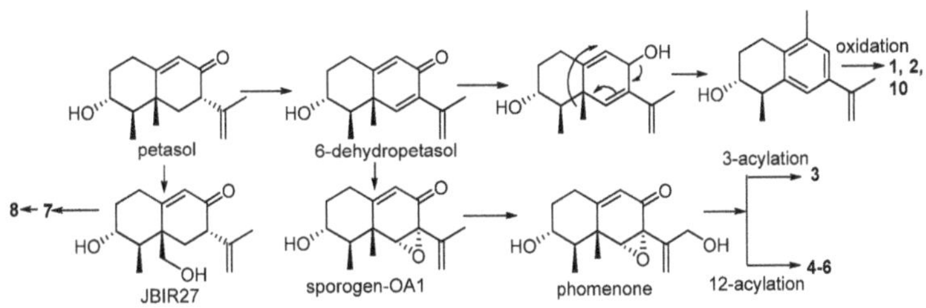

Scheme 1. Biogenetic relationships of the isolated eremophilanes.

4. Materials and Methods

4.1. General Experimental Procedures

Optical rotations were recorded on an Autopol-III automatic polarimeter (Rudolph Research Co., Ltd., Hackettstown, NJ, USA). UV spectra were recorded on a Cary 300 spectrometer. IR spectra were measured on a Thermo Nicolet Nexus 470 FT-IR spectrometer. ECD spectra were measured on a JASCO J-815 spectropolarimeter. ^{1}H and ^{13}C NMR spectra together with 2D NMR spectra were measured on Bruker Advance NMR spectrometers (400 MHz or 600 MHz for ^{1}H and 100 MHz or 125 MHz for ^{13}C, respectively). Chemical shifts are expressed in δ referenced to the solvent peaks of DMSO-d_6 (^{1}H at δ_H 2.50 and for ^{13}C at δ_C 39.5). HRESIMS spectra were obtained on a Bruker APEX IV 70 eV FT-MS spectrometer and on a Waters Xevo G2 Q-TOF spectrometer fitted with an ESI source (acquisition range:100–1000, acquisition: start time 0 and end time 4, voltages: ESI$^+$ 2 kV and ESI$^-$ 1.5–2 kV, external standard: HCOONa). DAD HPLC was performed on Waters e2695 separations module with Waters 2998 photodiode array detector and Thermo BDS column (250 × 4.6 mm, 5 μm). Column chromatography (CC) was performed using ODS (50 μm, Daiso), silica gel (200–300 mesh, Qingdao Marine Chemistry Co., Ltd., Qingdao, China), and Sephadex LH-20 (Amersham Pharmacia Biotech AB, Staffanstorp, Sweden). Precoated silica gel plates (Qingdao Marine Chemistry Co., Ltd.) were used for TLC analyses. Solvents used for isolation are all analytical grade. Semipreparative HPLC was performed on an Alltech 426 pump using a Uvis-201 detector, and the Prevail C_{18} column (semipreparative, 5 μm) was used for separation.

4.2. Fungal Strain and Identification

Fungal strain WZXY-m122-9 was isolated from the sponge of *Xestospongia testudinaria*, which was collected from Weizhou island in May 2016. The fungal identification was based on the ITS gene sequence which showed 100% similarity to a *Penicillium copticola* clone using Blast. The ITS gene sequence was deposited in GenBank. The strain is preserved at the State Key Laboratory of Natural and Biomimetic Drugs, Peking University.

4.3. Fermentation and Extraction

The fungal strain WZXY-m122-9 was cultured on flasks (120 × 500 mL) in rice medium with 45 g rice and 40 mL of 3.3% sea-salt in each flask at 25 °C for 4 days to obtain fresh mycelia and spores. They were then inoculated in 250 mL Erlenmeyer flasks (×10) containing 50 mL PDB medium to obtain seed medium after culture in a rotary shaker set to 120 rpm at 25 °C for 3 days. The seed culture was inoculated in 600 × 250 mL Erlenmeyer flasks, each containing 30 g rice and 30 mL of distilled artificial seawater. After 16 days under static conditions at 25 °C, the fermented culture was extracted with EtOAc (2 × 250 mL) twice. The organic solvent was evaporated to obtain an extract (20.0 g), which was suspended in 10% H_2O of MeOH, and then extracted with cyclohexane to remove lipids. The MeOH solution was evaporated under reduced pressure to obtain an extract (10.0 g). This extract was dissolved in H_2O and then extracted by EtOAc to yield EtOAc extract (4.8 g) after concentrated under vacuum. The EtOAc extract was fractionated upon silica gel column (3 × 25 cm) eluting with Petroether-EtOAc (3:1) to afford three fractions: FA, FB (0.5 g) and FC (1.2 g). Fraction FA (2.4 g) was fractionated upon C_{18} (ODS) column, eluting with MeOH-H_2O (1:3) to yield four subfractions (FA1 to FA4). FA2 (670 mg) was repeatedly separated by semi-preparative HPLC (YMC-packed C_{18}, 5 μm, 250 × 10 mm, 2 mL/min, UV detection at 210 nm) with MeCN-H_2O (1:4, *v*/*v*) as a mobile phase to yield sporogen AO-1 (125 mg, R_t = 27.5 min), phomenone (18.0 mg, R_t = 25.5 min), JBIR28 (4.0 mg, R_t = 23.6 min), and JBIR27 (13.5 mg, R_t = 29.4 min). Subfraction FA2 (195 mg) was separated by semi-preparative HPLC (YMC-packed C_{18}, 5 μm, 250 × 10 mm, 2 mL/min, UV detection at 210 nm) eluting with MeCN-H_2O (1:3, *v*/*v*) to obtain **1** (10.5 mg, R_t = 24.8 min), **2** (5.5 mg, R_t = 29.1 min), **11** (6.9 mg, R_t = 28.5 min), **12** (5.5 mg, R_t = 28.7 min), **3** (9.2 mg, R_t = 30.2 min), **4** (5.8 mg, R_t = 31.1 min), and **5** (10.5 mg, R_t = 30.4 min). Subfraction FA3 (55 mg) was purified by a semi-preparative HPLC (YMC-packed C_{18}, 5 μm, 250 × 10 mm, 2 mL/min, UV detection at 210 nm) eluting with MeOH-H_2O (45:55, *v*/*v*) to afford **6** (7.9 mg, R_t = 22.4 min), **7** (9.2 mg, R_t = 23.6 min), **8** (6.2 mg, R_t = 27.0 min), **9** (7.5 mg, R_t = 25.5 min), and **10** (4.1 mg, R_t = 26.0 min). FA1 (140 mg) was separated upon C_{18} ODS column (C_{18}, 10 μm, 2.5 × 30 cm), eluting 3-chloro-4,5-dihydroxyphenylacetic acid (7.6 mg), 3-chloro-4-hydroxyphenylacetic acid (5.5 mg), and 3-chloro-4-hydroxy-5- methoxyphenylacetic acid (3.1 mg).

Copteremophilane A (**1**): yellow amorphous; $[\alpha]^{20}_{D}$ +62 (*c* = 0.2, MeOH); IR (KBr) ν_{max} 3384, 2933, 2880, 1677, 1601, 1453, 1383, 1209 cm^{-1}; ^1H and ^{13}C NMR data, see Table 1; HRESIMS *m*/*z* 219.1380 $[M + H]^+$ (calcd for $C_{14}H_{19}O_2$, 219.1385).

(*R*)-MPA ester of **1**: white powder; ^1H NMR (400 MHz, DMSO-d_6) δ 2.61 (2H, m, H-1), 1.94–2.08 (2H, m, H-2), 4.93 (1H, m, H-3), 2.85 (1H, m, H-4), 7.50 (1H, s, H-6), 7.58 (1H, s, H-8), 3.28 (3H, s, H-12), 2.25 (3H, s, H-13), 1.11 (1H, d, *J* = 6.5 Hz, H-14).

(*S*)-MPA ester of **1**: white powder; ^1H NMR (400 MHz, DMSO-d_6) δ 2.48 (2H, m, H-1), 1.75–2.92 (2H, m, H-2), 4.99 (1H, m, H-3), 3.02 (1H, m, H-4), 7.58 (1H, s, H-6), 7.61 (1H, s, H-8), 3.28 (3H, s, H-12), 2.14 (3H, s, H-13), 1.23 (1H, d, *J* = 6.5 Hz, H-14).

Copteremophilane B (**2**): yellow amorphous; $[\alpha]^{20}_{D}$ +26 (*c* = 0.2, MeOH); IR (KBr) ν_{max} 3396, 2931, 1685, 1604, 1453, 1384, 1208 cm^{-1}; ^1H and ^{13}C NMR data, see Table 1; HRESIMS *m*/*z* 233.1543 $[M + H]^+$ (calcd for $C_{15}H_{21}O_2$, 233.1542).

Copteremophilane C (**3**): yellow amorphous; $[\alpha]^{20}_{D}$ +90 (*c* = 0.3, MeOH); IR (KBr) ν_{max} 3283, 2929, 1731, 1673, 1596, 1499, 1436, 1208, 1204 cm^{-1}; ^1H and ^{13}C NMR data, see Tables 2 and 3; HRESIMS *m*/*z* 449.1369 $[M + H]^+$ (calcd for $C_{23}H_{26}ClO_7$, 449.1367).

Copteremophilane D (**4**): yellow amorphous; $[\alpha]^{20}_{D}$ +28 (c = 0.2, MeOH); IR (KBr) v_{max} 3260, 2941, 1732, 1670, 1596, 1500, 1436, 1205 cm^{-1}; ^{1}H and ^{13}C NMR data, see Tables 2 and 3; HRESIMS m/z 447.1208 $[M - H]^-$ (calcd for $C_{23}H_{24}ClO_7$, 447.1211).

Copteremophilane E (**5**): yellow amorphous; $[\alpha]^{20}_{D}$ +28 (c = 0.2, MeOH); IR (KBr) v_{max} 3291, 2927, 1735, 1669, 1596, 1510, 1425, 1384, 1153 cm^{-1}; ^{1}H and ^{13}C NMR data, Tables 2 and 3; HRESIMS m/z 431.1259 $[M - H]^-$ (calcd for $C_{23}H_{24}ClO_6$, 431.1261).

Copteremophilane F (**6**): yellow amorphous; $[\alpha]^{20}_{D}$ +46 (c = 0.2, MeOH); IR (KBr) v_{max} 3315, 2926, 1735, 1669, 1505, 1384, 1285, 1145 cm^{-1}; ^{1}H and ^{13}C NMR data, see Tables 2 and 3; HRESIMS m/z 461.1363 $[M - H]^-$ (calcd for $C_{24}H_{26}ClO_7$, 461.1367).

Copteremophilane G (**7**): yellow amorphous; $[\alpha]^{20}_{D}$ +60 (c = 0.2, MeOH); IR (KBr) v_{max} 3268, 2939, 1725, 1674, 1499, 1436, 1291, 1206 cm^{-1}; ^{1}H and ^{13}C NMR data, see Tables 2 and 3; HRESIMS m/z 433.1415 $[M - H]^-$ (calcd for $C_{23}H_{26}ClO_6$, 433.1418).

Copteremophilane H (**8**): yellow amorphous; $[\alpha]^{20}_{D}$ +102 (c = 0.2, MeOH); IR (KBr) v_{max} 3256, 2928, 1685, 1436, 1383, 1297, 1209 cm^{-1}; ^{1}H and ^{13}C NMR data, see Tables 2 and 3; HRESIMS m/z 433.1421 $[M - H]^-$ (calcd for $C_{23}H_{28}ClO_6$, 433.1418).

Copteremophilane I (**9**): yellow amorphous; $[\alpha]^{20}_{D}$ +54 (c = 0.2, MeOH); IR v_{max} (KBr) cm^{-1}: 3386, 2928, 1720, 1436, 1384, 1292, 1206, 1150, 1024; ^{1}H and ^{13}C NMR data, see Tables 2 and 3; HRESIMS m/z 435.1578 $[M - H]^-$ (calcd for $C_{23}H_{28}ClO_6$, 435.1574).

Copteremophilane J (**10**): yellow amorphous; $[\alpha]^{20}_{D}$ +54 (c = 0.2, MeOH); IR (KBr) v_{max} 3432, 2996, 2913, 1661, 1436, 1406, 1312 cm^{-1}; ^{1}H and ^{13}C NMR data, see Table 1; HRESIMS m/z 457.1395 $[M + Na]^+$ (calcd for $C_{23}H_{27}ClO_6Na$, 457.1390).

5-Glycopenostatin F (**11**): yellow amorphous; $[\alpha]^{20}_{D}$ +12 (c 0.2, MeOH); IR v_{max} (KBr) cm^{-1}: 3402, 2926, 2855, 1716, 1674, 1384, 1205; ^{1}H and ^{13}C NMR data, see Table 4; HRESIMS m/z 507.2958 $[M + H]^+$ (calcd for $C_{28}H_{43}O_8$, 507.2958).

5-Glycopenostatin I (**12**): yellow amorphous; $[\alpha]^{20}_{D}$ -10 (c 0.2, MeOH); IR v_{max} (KBr) cm^{-1}: 3372, 2926, 1716, 1456, 1383, 1341, 1218; ^{1}H and ^{13}C NMR data, see Table 4; HRESIMS m/z 507.2958 $[M + H]^+$ (calcd for $C_{28}H_{43}O_8$, 507.2958).

4.4. Hydrolysis of 11 and 12

Compound **11** (1.5 mg) was dissolved in HCl (2 M, 10 mL) to stir at 100 °C for 4 h monitored by TLC. After reaction time, the acidic solution was extracted by 5 mL EtOAc. EtOAc was concentrated under vacuum to yield 0.7 mg hydrolyzed product (**11a**), and the acidic solution was freeze-dried and solubilized in MeOH for further analysis. Compound **12** (1.3 mg) was hydrolyzed by the same protocol as for **11**.

4.5. Mosher Reaction

Compound was dissolved in anhydrous $CHCl_3$, and then *R*-MPA (equal mol), DMAP and DCC were added to react at room temperature for 12 hrs. The product was purified by silica gel column eluting with petroleum ether-EtOAc (2:1) to yield *R*-MPA ester. *S*-MPA ester of compound was synthesized by the same protocol as for *R*-MPA ester. Calculation of the chemical shift difference ($\Delta\delta = \delta_R - \delta_S$) allowed the assignment of configuration of the stereogenic center of secondary alcohol [27].

4.6. Snatzke Method

Compound was weighted to two portions (0.2 mg for each). One portion was dissolved in 0.5 mL DMSO for the detection of ECD curve at wavelength 250–500 nm, and the second portion was dissolved in 0.5 mL DMSO with 0.5 mg/mL $Mo(OAc)_4$ for the detection of the ECD curve within 30 min [28].

4.7. Cytotoxic Detection

Cell viability was evaluated using the MTT assay according to the manufacturer's instruction. In brief, cells were plated at a density of 1.0×10^4 cells per well in 96-well plates and allowed to attach overnight. Then, the cells were treated with or without compound at the indicated concentration. After 48 h, MTT solution containing 1 mg/mL and MTT 100 µL was added into each well and incubated for 2 h at 37 °C. Then, the medium was changed with the same volume of DMSO. After incubation, absorbance is read at 570 nm for MTT by a spectrophotometer and the quantity of formazan product is directly proportional to the number of living cells in culture.

4.8. Cell Viability Assay

Rat pheochromocytoma (PC12) cells were cultured in Dulbecco's modified Eagle medium (DMEM), and supplemented with 10% FBS, 100 U/mL penicillin, and 100 μg/mL streptomycin in a humidified atmosphere at 37 °C with 5% CO_2. The cells were passaged every 3 days. PC12 cells (1×10^4) were cultured for 24 h, and then the medium was replaced with serum-free DMEM medium. The PC12 cells were divided into three groups. In the model group, PC12 cells were treated using $A\beta_{25-35}$ with different concentrations. Then, the cells were incubated for 12 h, 24 h and 48 h until there was a cell viability of up to 50%. PC12 cells without pretreatment were set up as control. The third group of PC12 was treated with different concentrations of compounds to detect the cell viability by CCK8 method. The model groups ($A\beta_{25-35}$ induced PC12 cells) were treated with compounds in different concentrations, and the viability of PC12 cells was measured using cell counting kit-8 (Dojindo, Japan), according to the manufacturer's instructions. Then, the medium was aspirated and cells in each group were incubated with 10 μL of CCK8 at 37 °C for 2 h. Afterwards, the absorbance was determined at a wavelength of 450 nm with a microplate reader (Thermo, MuLTiSKAN MK3, Waltham, MA USA).

4.9. Measurement of Malondialdehyde (MDA)

Cultured PC12 cells were initially seeded in 6-well plates for 24 h. The cells were then pre-incubated with or without compound, followed by incubation with $A\beta_{25-35}$ (40 μM) for 24 h. The cultures were washed with ice-cold PBS and homogenized. The homogenate was centrifuged at 4 °C. The protein concentration in each sample was determined by the MDA Protein Assay Kit as a reference standard. The levels of MDA were determined according to the manufacturer's instructions. Concentrations were normalized to the protein concentration expressed as a percentage of control samples.

4.10. Lactate Dehydrogenase (LDH) Release Assay

Cell injury was assessed through measuring the LDH activity in the supernatant of PC12 cells using an LDH kit according to the manufacturer's protocol. In brief, double-distilled H_2O, 0.2 μM pyruvic acid, matrix buffer and coenzyme I buffer were added in sequence at 48 h after the treatment by compound. The supernatant was collected after incubation for 30 min at room temperature. The absorbance at 450 nm was then measured with a microplate reader.

4.11. Statistical Analysis

Data are expressed as mean ± standard deviation (SD). Biostatistical analyses were conducted with SPSS 16.0 software. Statistical differences among groups were assessed by one-way analysis of variance (ANOVA). Differences were considered to be statistically significant at $p < 0.05$.

5. Conclusions

In summary, this work reports a chemical examination of the marine sponge-derived *P. copticola* fungus to afford 12 undescribed natural products, of which 10 are identified as eremophilanes and 2 are determined as glucosides. Eremophilanes **1**, **2** and **10** are characteristic of the eremophilanes with a phenyl moiety in ring B accompanying a methyl migration to C-9, which are uncommonly found in nature. Eremophilanes **3–9** feature the esterification of a chlorinated phenylacetic unit in backbone, and this is the second isolation of relevant analogs from fungi. Analogs **11** and **12** are characteristic of an unprecedented PKS scaffold bearing a glucose unit. The bioassay results revealed the inhibitory effects of the isolated eremophilanes against tumor cell lines related to the structure variation. Analog **8** only inhibited human non-small cell lung cancer cells (A549), and analogs **4** and **5** selectively inhibited HCT-8. The eremophilane nuclei or acyl moieties showed inactivity towards tumor cell lines. These findings suggest the incorporation of chlorinated phenylacetic units significantly enhances the antitumor effects of eremophilanes, and the substituted positions and the acyl compositions are sensitive for the selective inhibition. In addition, the noncytotoxic analogs, such as **7**, showed a neuroprotective effect. This study implies that eremophilanes have potential for development as antitumor or neuroprotective agents after structure modification.

Supplementary Materials: The following supporting information can be downloaded at: https://www.mdpi.com/article/10.3390/md20110712/s1, Figures S1–S102: spectra of IR, ^{1}H and ^{13}C NMR, COSY, HSQC, HSBC, and NOESY. Figure S103: Part node clusters of EtOAc extract by molecular networking. Figure S104: Key HMBC and COSY correlations of **1–10**.

Author Contributions: Fungal fermentation and chromatography, J.Z.; compound purification, D.L.; bioactive detection, J.H.; biogenetic hypothesis, A.F.; structure elucidation and manuscript editing, W.L. All authors have read and agreed to the published version of the manuscript.

Funding: This research was funded by COMRA DY135-B-05 and NSFC (81991525, 21861142006, 81872793, 81630089), 2022Z144 and 2021Z046.

Data Availability Statement: The authors confirm that the data supporting the findings of this study are available within the article and its supplementary materials.

Conflicts of Interest: The authors declare no conflict of interest.

References

1. Yuyama, K.T.; Fortkamp, D.; Abraham, W.R. Eremophilane-type sesquiterpenes from fungi and their medicinal potential. *Biol. Chem.* **2018**, *399*, 13–28. [CrossRef] [PubMed]
2. Fraga, B.M. Natural sesquiterpenoids. *Nat. Prod. Rep.* **2008**, *25*, 1180–1209. [CrossRef] [PubMed]
3. Wu, L.; Liao, Z.; Liu, C.; Jia, H.; Sun, J. Eremophilane sesquiterpenes from the genus *Ligularia*. *Chem. Biodivers.* **2016**, *13*, 645–671. [CrossRef] [PubMed]
4. Fang, W.; Wang, J.; Wang, J.; Shi, L.; Li, K.; Lin, X.; Min, Y.; Yang, B.; Tang, L.; Liu, Y.; et al. Cytotoxic and antibacterial eremophilane sesquiterpenes from the marine-derived fungus *Cochliobolus lunatus* SCSIO41401. *J. Nat. Prod.* **2018**, *81*, 1405–1410. [CrossRef]
5. Wang, L.; Li, M.; Tang, J.; Li, X. Eremophilane sesquiterpenes from a deep marine-derived fungus, *Aspergillus* sp. SCSIOW2, cultivated in the presence of epigenetic modifying agents. *Molecules* **2016**, *21*, 473. [CrossRef]
6. Cheng, Z.; Zhao, J.; Liu, D.; Proksch, P.; Zhao, Z.; Lin, W. Eremophilane-type sesquiterpenoids from an *Acremonium* sp. fungus isolated from deep-sea sediments. *J. Nat. Prod.* **2016**, *79*, 1035–1047. [CrossRef]
7. Huang, Y.; Qiao, L.; Lv, A.; Pei, Y.; Tian, L. Eremophilane sesquiterenes from the marine fungus *Penicillium* sp. BL27-2. *Chin. Chem. Lett.* **2008**, *19*, 562–564. [CrossRef]
8. Wu, G.; Lin, A.; Gu, Q.; Zhu, T.; Li, D. Four new chloro-eremophilane sesquiterpenes from an Antarctic deep-sea derived fungus, *Penicillium* sp. PR19N-1. *Mar. Drugs* **2013**, *11*, 1399–1408.
9. Motohashi, K.; Hashimoto, J.; Inaba, S.; Khan, S.T.; Komaki, H.; Nagai, A.; Takagi, M.; Shin-ya, K. New sesquiterpenes, JBIR-27 and -28, isolated from a tunicate-derived fungus, *Penicillium* sp. SS080624SCf1. *J. Antibiot.* **2009**, *62*, 247–250. [CrossRef]
10. Xu, J.; Liu, H.; Chen, Y.; Tan, H.; Guo, H.; Xu, L.; Li, S.; Huang, Z.; Li, H.; Gao, X.; et al. Highly substituted benzophenone aldehydes and eremophilane derivatives from the deep-sea derived fungus *Phomopsis lithocarpus* FS508. *Mar. Drugs* **2018**, *16*, 329. [CrossRef]
11. Oh, H.; Jensen, P.R.; Murphy, B.T.; Fiorilla, C.; Sullivan, J.F.; Ramsey, T.; Fenical, W. Cryptosphaerolide, a cytotoxic Mcl-1 inhibitor from a marine-derived ascomycete related to the genus *Cryptosphaeria*. *J. Nat. Prod.* **2010**, *73*, 998–1001. [CrossRef] [PubMed]
12. Riche, C.; Pascard-Billy, C.; Devys, M.; Gaudemer, A.; Barbier, M.; Bousquet, J.F. Crystal and molecular structure of phomenone, phytotoxin from the mushroom *Phoma exigua*. *Tetrahedron Lett.* **1974**, *32*, 2765–2766. [CrossRef]
13. Iwamoto, C.; Minoura, K.; Hagishita, S.; Nomoto, K.; Numata, A. Penostatins F-I, novel cytotoxic metabolites from a *Penicillium* species separated from an *Enteromorpha marine* alga. *J. Chem. Soc. Perkin Trans.* **1998**, *3*, 449–456. [CrossRef]
14. Isaka, M.; Srisanoh, U.; Veeranondha, S.; Choowong, W.; Lumyong, S. Cytotoxic eremophilane sesquiterpenoids from the saprobic fungus *Berkleasmium nigroapicale* BCC 8220. *Tetrahedron* **2009**, *65*, 8808–8815. [CrossRef]
15. Isaka, M.; Jaturapat, A.; Kladwang, W.; Punya, J.; Lertwerawat, Y.; Tanticharoen, M.; Thebtaranonth, Y. Antiplasmodial compounds from the wood-decayed fungus *Xylaria* sp. BCC 1067. *Planta Med.* **2000**, *66*, 473–475. [CrossRef]

16. McDonald, L.A.; Barbieri, L.R.; Bernan, V.S.; Janso, J.; Lassota, P.; Carter, G.T. 07H239-A, a new cytotoxic eremophilane sesquiterpene from the marine-derived xylariaceous fungus LL-07H239. *J. Nat. Prod.* **2004**, *67*, 1565–1567. [CrossRef]
17. Yuan, W.; Goto, M.; Hsieh, K.; Yuan, B.; Zhao, Y.; Morris-Natschke, S.L.; Lee, K. Selective cytotoxic eremophilane-type sesquiterpenes from *Penicillium citreonigrum*. *J. Asian Nat. Prod. Res.* **2015**, *17*, 1239–1244.
18. Chen, Z.; Zhong, C. Oxidative stress in Alzheimer's disease. *Neurosci. Bull.* **2014**, *30*, 271–281. [CrossRef]
19. Bonda, D.J.; Wang, X.; Perry, G.; Nunomura, A.; Tabaton, M.; Zhu, X.; Smith, M.A. Oxidative stress in Alzheimer disease: A possibility for prevention. *Neuropharmacology* **2010**, *59*, 290–294. [CrossRef]
20. Sisodia, S.S.; Price, D.L. Role of the β-amyloid protein in Alzheimer's disease. *FASEB J.* **1995**, *9*, 366–370. [CrossRef]
21. LaFerla, F.M.; Green, K.N.; Oddo, S. Intracellular amyloid-β in Alzheimer's disease. *Nature Rev. Neurosci.* **2007**, *8*, 499–509. [CrossRef] [PubMed]
22. Westerink, R.H.S.; Ewing, A.G. The PC12 cell as model for neurosecretion. *Acta Physiol.* **2008**, *192*, 273–285. [CrossRef] [PubMed]
23. Zhou, Y.; Zhu, L.; Li, H.; Xie, W.; Liu, J.; Zhang, Y.; Li, Y.; Wang, C. In vivo and in vitro neuroprotective effects of maca polysaccharide. *Front. Biosci.* **2022**, *27*, e8. [CrossRef]
24. Zhang, L.; Kong, X.; Wang, Z.; Xu, F.; Zhu, Y. A study on neuroprotective effects of curcumin on the diabetic rat brain. *J. Nutr. Health Aging* **2016**, *20*, 835–840. [CrossRef] [PubMed]
25. Rani, R.; Kumar, V. When will small molecule lactate dehydrogenase inhibitors realize their potential in the cancer clinic? *Future Med. Chem.* **2017**, *11*, 1113–1115. [CrossRef] [PubMed]
26. Shi, Y.; Pinto, B.M. Human lactate dehydrogenase a inhibitors: A molecular dynamics investigation. *PLoS ONE* **2014**, *9*, e86365. [CrossRef] [PubMed]
27. Kusumi, T.; Hamada, T.; Ishitsuka, M.O.; Ohtani, I.; Kakisawa, H. Elucidation of the relative and absolute stereochemistry of lobatriene, a marine diterpene, by a modified Mosher method. *J. Org. Chem.* **1992**, *57*, 1033–1035. [CrossRef]
28. Di Bari, L.; Pescitelli, G.; Pratelli, C.; Pini, D.; Salvadori, P. Determination of absolute configuration of acyclic 1,2-diols with Mo2(OAc)4. 1. Snatzke's method revisited. *J. Org. Chem.* **2001**, *66*, 4819–4825. [CrossRef]

MDPI AG
Grosspeteranlage 5
4052 Basel
Switzerland
Tel.: +41 61 683 77 34

Marine Drugs Editorial Office
E-mail: marinedrugs@mdpi.com
www.mdpi.com/journal/marinedrugs

www.ingramcontent.com/pod-product-compliance
Lightning Source LLC
Chambersburg PA
CBHW061439210926
51850CB00008B/1556
* 9 7 8 3 7 2 5 8 1 7 7 7 1 *